Stochastic Modeling of Scientific Data

STOCHASTIC MODELING SERIES

G. Samorodnitsky and M.S. Taqqu
Stable Non-Gaussian Processes: Stochastic Models with Infinite Variance

K. Sigman
Stationary Marked Point Processes: An Intuitive Approach

P. Guttorp
Stochastic Modeling of Scientific Data

M. Neuts
Algorithmic Probability

A. Shwartz and A. Weiss
Large Deviations for Perfomance Analysis: Queues, Communications and Computing

Stochastic Modeling of Scientific Data

PETER GUTTORP
Professor of Statistics
University of Washington
Seattle, USA

CRC Press
Taylor & Francis Group
Boca Raton London New York

CRC Press is an imprint of the
Taylor & Francis Group, an **informa** business
A CHAPMAN & HALL BOOK

First published 1995 by Chapman & Hall

Published 2019 by CRC Press
Taylor & Francis Group
6000 Broken Sound Parkway NW, Suite 300
Boca Raton, FL 33487-2742

First issued in paperback 2019

ISBN-13: 978-0-367-44900-1 (pbk)
ISBN-13: 978-0-412-99281-0 (hbk)

Visit the Taylor & Francis Web site at
http://www.taylorandfrancis.com

and the CRC Press Web site at
http://www.crcpress.com

Library of Congress Cataloging-in-Publication Data

Catalog record is available from the Library of Congress

To Kenji

Contents

Preface

Traditionally, applied probability texts contain a fair amount of probability theory, varying amounts of applications, and no data. Occasionally an author may touch upon how one would go about fitting a model to data, or use data to develop a model, but rarely is this topic given much weight. On the other hand, the few texts on inference for stochastic processes mostly dwell at length upon the interesting twists that occur in statistical theory when data no longer can be assumed iid. But, again, they rarely contain any data. The intent of this text is to present some probability models, some statistics relevant to these models, and some data that illustrate some of the points made.

My experience as a practicing statistician, specializing in models for dependent data, has been that no real progress can be made without spending substantial time trying to master enough of the subject matter to be able to talk to area scientists in their language, rather than in mathematical language. Consequently, I have tried to give scientific background to many of the applications I present in more detail than is usual in statistics texts. A general scientific background, but no special training, should suffice to enable you to follow the gist of the explanations. From a mathematical point of view you need elementary probability, including conditional distributions and expectations, calculus, ordinary differential equations, and linear algebra. Having encountered statistics would be very useful, although it is not strictly necessary. My students tell me that a first course in stochastic processes is a very useful background, but I have tried to include sufficient material to make this unnecessary. I avoid measure-theoretical arguments, but have to use some L_2-theory to introduce (albeit in an extremely low-key manner) stochastic integration.

As the chapters progress, there are fewer formal proofs and more references to the literature for verification of results. This is because while it is possible to do much discrete time Markov chain theory using elementary probability, it becomes progressively harder to do the proofs as the processes become more complicated. I therefore often resort to special cases and intuitive explanations in many instances.

Picking topics is necessarily a very subjective matter. I have omitted some topics that others may find essential. For example, while renewal theory is a beautiful piece of probability, I have not found many interesting scientific applications. Martingales are not included, in spite of their usefulness in statistical theory. I ignore stationary time series, since in this area there are plenty of books containing both theory and data analysis. On the other hand I try to emphasize the importance of Markov chain Monte Carlo methods, which are having as profound an impact on statistics in the nineties as did the bootstrap in the eighties. I have found the state space modeling approach useful in a variety of situations, as I am sure you can tell.

At first I was intending to write a short introduction, indicating why I think stochastic models are useful. It grew into a chapter of its own. Three appendices contain some material that not everyone may have encountered before. I hope there is enough so that, for example, someone without much mathematical statistics will get an idea of the main tools (that will be used in the text).

The material in this book is suitable for a two-semester course. I have tried to cover most of it in a two-quarter sequence, but that gets a bit rushed. When I teach this course, one of the main teaching tools is laboratory assignments, where the students work on analyzing data sets, simulating processes and statistical procedures, and work through some theory as well. I have included versions of these laboratory assignments in the exercises that follow each chapter. These exercises contain extensions and details of the probabilistic exposition, computational exercises, and data analysis problems. In conjunction with the book, several data sets will be available via anonymous ftp (details are given after the indexes).

There is a variety of examples and applications. The former are attempts to illustrate some concepts while the latter apply the concepts to data. In order to facilitate reference to the book in lectures I have numbered all displayed equations. Theorems, propositions, lemmata, figures, and tables are separately numbered. To simplify life a little for the readers there are, in addition to the regular indexes of terms and notation, indexes of examples/applications and of numbered theorems, propositions, and lemmata. Definitions are indicated with bold typeface (and referred to in the Index of terms). As is common in applied probability (but not in statistics), all vectors are row vectors.

The applications in the text draw on many different sources. Some originate in work by me or my students, others come from colleagues and friends who I have talked into giving me their data, and yet others are quoted from the literature. At the end of each chapter I try to acknowledge my main sources, as well as give occasional hints as to where one can learn more (preferably discussion papers). I apologize for any inadvertent failures to make such acknowledgements.

A large number of students and colleagues have had comments and suggestions regarding the text. Julian Besag, Michael Phelan, and Elizabeth Thompson have been particularly helpful, as well as a host of students over the years. Four reviewers made numerous helpful comments: Simon Tavaré, Michael P. Bailey, David D. Yao and Laurence Baxter. John Kimmel was a patient and helpful editor throughout the project, while Achi Dosanjh and Emma Broomby skillfully guided me through the final stages of manuscript preparation. Several colleagues have generously provided data, figures or other help. I would particularly like to thank João Batista, David Brillinger, Anna Guttorp, David Higdon, Jim Hughes, Stephen Kaluzny, Niels Keiding, Brian Leroux, Hongzhe Li, Iain MacDonald, Roland Madden, Thomas Murray, Michael Newton, Haiganoush Preisler, John Rice, Brian Ripley, Nuala Sheehan, Jim Smith, and Elizabeth Thompson.

The text was produced using Eroff from Elan Computer Group, Inc. The statistical analysis was generally done in Splus (StatSci division of Mathsoft, Inc.) and most figures benefited greatly from the psfig utility written by Antonio Possolo and the setps wrapper written by Frank Harrell. The spatial point process analysis employed the Splancs library produced by B. S. Rowlingson and Peter Diggle at Lancaster University. Software for some specialized analyses was provided by Charlie Geyer and John Rice.

This work had partial support from the National Science Foundation (DMS-9115756), the National Institute of Health (HL 31823), the Environmental Protection Agency, and the Electrical Power Research Institute. This text cannot in any way be construed as official policy of any of these organizations. The final version was produced during a visiting scholar appointment at the Institute for Environmental Studies at the University of Washington, and I am grateful for their generous support during a most difficult time.

My family has patiently suffered through my long preoccupation with this manuscript. June, Eric, and Kenji deserve much more attention (and will, hopefully, get it) from me. They have been very supportive of the effort.

I was fortunate to have Jerzy Neyman and David Brillinger as advisers when I was a graduate student. From them I learned a way of thinking about science, about data, and about modeling. And I learned, in Mr. Neyman's words, that "Life is complicated, but not uninteresting."

Peter Guttorp
University of Washington
Seattle 1995

CHAPTER 1
Introduction

Random behavior of simple or complex phenomena can sometimes be explained in physical terms, with an additional touch of probability theory. We exemplify this with a description of coin tossing. We then define a stochastic process, give some examples, and an overview of the book.

1.1. Randomness

The world is full of unpredictable events. Science strives to understand natural phenomena, in the sense of reducing this unpredictability. There are many ways of doing this. Models, which are abstractions of particular aspects of phenomena, constitute one such way. Experimentation and observation are needed to verify and improve models. These models can be of a variety of types: conceptual, mathematical, or stochastic, to mention a few.

In this book we investigate some simple stochastic models. We shall see how there is an interplay between the model and the science underlying the problem. Sometimes the science enters only conceptually, while in other cases it dictates the precise structure of the model. Common to all models we shall consider is a random element, described very precisely in the language of probability theory.

We can qualitatively distinguish different sources of random behavior.

• *Uncertainty about initial conditions.* In many situations it is very difficult to determine exactly the initial conditions. In some situations one can only determine relative frequencies of different initial conditions. The consequence is that the system exhibits random behavior, in accordance with the random initial conditions.

• *Sensitivity to initial conditions.* Many systems exhibit large changes in output corresponding to very small changes in initial conditions. Such systems are said to display **chaotic** behavior. It is often quite difficult to estimate parameters and study the fit of a deterministic description of a chaotic system, especially when initial conditions are not exactly known.

• *Incomplete description.* Sometimes the theoretical basis for a deterministic description of a system corresponds to only some of the factors that are important in determining outcomes. The lack of complete description renders the behavior unpredictable. This is quite common in, e.g., economic modeling.

• *Fundamental description.* When a single photon hits a glass surface, it may go through it or reflect off it. There is no way to predict what it will do, but the quantum electrodynamic theory can make precise predictions as to the behavior of a large number of single photons, when viewed as an aggregate. Modern quantum theory could not exist without probabilistic descriptions.

To illustrate the first two classes of randomness, let us discuss the simple phenomenon of coin tossing. A circular coin of radius a is tossed upwards, spins around its own axis several times, and falls down. For simplicity we will ignore air resistance, and assume that the coin falls on a sandy surface, so that it does not bounce once it hits the ground. The laws of motion for coin tossing are then elementary (Keller, 1986). Let $y(t)$ be the height of the center of gravity of the coin above the surface at time t, and $\theta(t)$ the angle between the normal to the surface and the normal to the heads side of the coin. Figure 1.1 shows the situation.

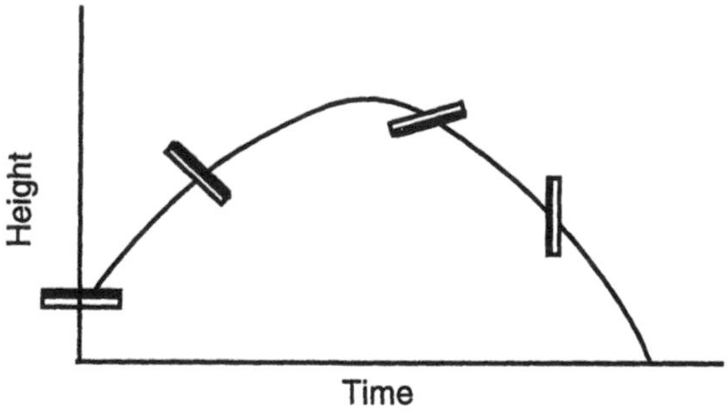

Figure 1.1. The motion of a coin flip. Adapted from Keller (1986); reproduced with permission from the Mathematical Association of America.

Assuming only gravitational force vertically and no acceleration rotationally, the equations of motion are

$$y''(t) = -g \qquad \theta''(t)=0, \tag{1.1}$$

where g is the gravitational acceleration. The coin is assumed to start at a height a above the surface (any other initial position can be assumed without problems) with the heads side facing upwards. It is tossed upwards with vertical velocity u and angular velocity ω. This yields the initial conditions

$$y(0)=a \quad y'(0)=u \quad \theta(0)=0 \quad \theta'(0)=\omega. \tag{1.2}$$

The solution is

$$y(t) = a + ut - gt^2/2 \qquad \theta(t) = \omega t. \tag{1.3}$$

The center of the coin describes a parabola as a function of time, with a maximum value of $a+u^2/(2g)$ at time u/g. The coin lands whenever either end touches the ground, i.e., at the first time t_0 such that

$$y(t_0) = a \,|\sin\theta(t_0)\,|. \tag{1.4}$$

It lands with heads up if it has rotated any number of full rotations (to within 90°), i.e., if for some n

$$(2n-\tfrac{1}{2})\pi < \theta(t_0) < (2n+\tfrac{1}{2})\pi. \tag{1.5}$$

At the extreme points of these intervals $\omega t_0=(2n\pm\tfrac{1}{2})\pi$, so $y(t_0)=a$ or $t_0=2u/g$. Hence, using (1.3), $\omega=(2n\pm\tfrac{1}{2})\pi g/(2u)$. The initial values that ensure heads are contained in successive hyperbolic bands:

$$H = \{(u,\omega): (2n-\tfrac{1}{2})\frac{\pi g}{2u} \le \omega \le (2n+\tfrac{1}{2})\frac{\pi g}{2u}\}. \tag{1.6}$$

Figure 1.2 shows the successive bands as a function of u/g. The band nearest the origin corresponds to heads, the next one to tails, etc. We see that as the velocities increase, the amount of change needed to move from tails to heads decreases. Typical values of the parameters were determined in an experiment by Persi Diaconis (Keller, 1986). They are

$u = 2.4$ m/s (determined from observing the maximum height of typical coin tosses);
ω = 38 rev/s = 238.6 radians/s (determined by wrapping dental floss around the coin and counting the number of unwrappings);
n = 19 rev/toss (computed from $n=\omega t_0$).

On Figure 1.2 this corresponds to u/g of 0.25, and ω of 238.6, far above the top of the picture. It is clear that a very small change in u/g can change the outcome of the toss. Thus the coin toss is predictable precisely when the initial conditions are very well determined. For example, if u/g and ω both are very small, the coin will essentially just fall flat down, and hence will come up heads.

Suppose now that the initial conditions are not under perfect control, but that they can be described by a probability density $f(u,w)$. Then

$$\mathbf{P}(H) = \int_H f(u,\omega)\,du\,d\omega, \tag{1.7}$$

where H is the set depicted in Figure 1.2. We can select f so that $\mathbf{P}(H)$ takes on any value in [0,1]. So how come coins tend to come up heads and tails about equally often? Here is an explanation:

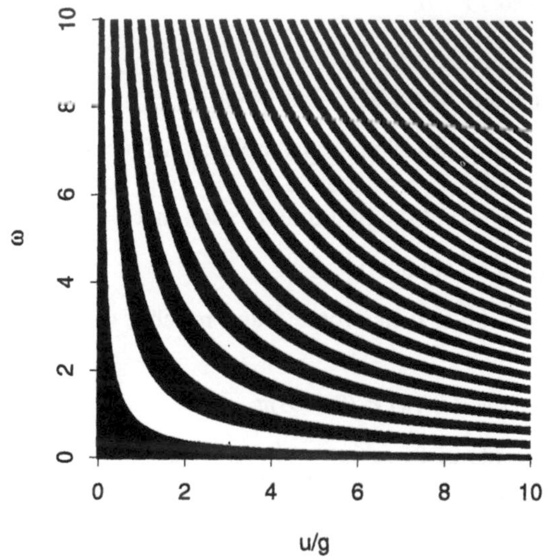

Figure 1.2. The outcome of a coin toss as a function of vertical (u/g) and angular (ω) velocities. Every other band is heads, starting with the lower left-hand corner. Reproduced with permission from Keller (1986), published by the Mathematical Association of America.

Theorem 1.1 For all continuous f such that $f(u,\omega)>0$ for all $u,\omega>0$, and for any $0<\beta\le\pi/2$

$$\lim_{U\to\infty}\iint_H f\left(u-U\cos\beta,\omega-\frac{U}{a}\sin\beta\right)d\omega\,du = \tfrac{1}{2}. \tag{1.8}$$

Proof First change variables to $\omega'=\omega-U\sin\beta/a$, so that $d\omega'=d\omega$ and

$$\mathbf{P}(H) = \int_{U\cos\beta}^{\infty}\sum_{n=0}^{\infty}\int_{(2n-\frac{1}{2})\pi g/2u-U\sin\beta/a}^{(2n+\frac{1}{2})\pi g/2u-U\sin\beta/a} f(u-U\cos\beta,\omega')\,d\omega'\,du. \tag{1.9}$$

For $\beta<\pi/2$ ($\beta=\pi/2$ is argued similarly), $U\to\infty$ implies that $u\to\infty$. The range of each of the inner integrals over ω' is of length $O(1/u)$[1], so we can approximate each of them by the length of the interval times the integrand at the midpoint, with an error of only $o(u^{-1})$. Thus

[1] We say that $g(s)=O(h(s))$ as $s\to s_0$ if $|g(s)/h(s)|$ stays bounded as $s\to s_0$. Here s_0 can be finite or infinite, and the limit can be through all reals or through integers. Also, we write $g(s)=o(h(s))$ if $g(s)/h(s)\to 0$. This notation is due to Landau (1930).

$$\mathbf{P}(H) = \int_{U\cos\beta}^{\infty} \sum_{n=0}^{\infty} f(u-U\cos\beta, \frac{2n\pi g}{2u}) \frac{\pi g}{2u}(1+o(1))du. \tag{1.10}$$

For large U the sum is a Riemann sum for $\frac{1}{2}\int f(u-U\cos\beta, w)dw$, so with $v=u-U\cos\beta$ we get

$$\mathbf{P}(H) = \tfrac{1}{2}\int_0^{\infty}\int_0^{\infty} f(v,w)dwdv(1+o(1)) = \tfrac{1}{2}+o(1), \tag{1.11}$$

proving the result. □

The interpretation of this theorem is that as soon as we make the initial velocity (vertical, angular, or both) large enough, the probability of heads is ½. One would be tempted to argue this result from symmetry, but the initial conditions are not symmetric: we always start with heads upwards.

More precise results of this type are given in section 3.2.2 of Engel (1992). For example, if the vertical velocity $u \sim \mathrm{U}(2.1,2.7)$ m/s and the angular velocity $\omega \sim \mathrm{U}(36,40)$ rev/s, with the two velocities independent, then $|\mathbf{P}(H)-\frac{1}{2}| \leq 0.056$. This is, in essence, a worst-case scenario for the approximation. More realistically, assume that the conditional distribution of u given ω is a mixture of normal distributions, with standard deviations at least 0.15 m/s, while ω is at least 36 rev/s. Then Engel's bound is $|\mathbf{P}(H)-\frac{1}{2}| \leq 1.5\times10^{-11}$.

In practice, it is difficult to study processes of the coin-tossing type. One of the most extensive experiments involved 315,672 throws of dice. These were rolled, twelve at a time, down a fixed slope of cardboard. The roller, W. F. R. Weldon (see Pearson, 1900), found 106,602 instances of the outcome 5 or 6. If the dice were true and the initial conditions ''sufficiently random'', as discussed above, the probability of 5 or 6 should be 1/3. Thus, Weldon found an excess of 0.004366, which is statistically highly significant. The explanation is that the faces with 5 or 6 spots are lighter, because of the larger number of pits for the marking material, thus displacing the center of gravity towards the opposing sides 2 or 1 (this explanation is due to Jeffreys, 1939). In order to avoid this type of a problem, dice would have to have painted faces rather than the pits that are standard. This will, of course, be of little importance to the occasional board game player, but can play an important role in the finances of a major casino.

1.2. Stochastic processes

Classical statistical theory is often concerned with the case of iid random variables $X_0, \ldots, X_n$, i.e.,

$$\mathbf{P}(X_0 \in A_0, \ldots, X_n \in A_n) = \prod_{i=0}^{n} \mathbf{P}(X \in A_i), \tag{1.12}$$

where X is a generic random variable with the same distribution as the X_i. It is straightforward to relax the assumption of identical distributions. But in many cases the independence assumption is violated. For example, if it is raining today, it is more likely to be raining tomorrow than if it is sunny today. There are many different ways of specifying dependent random variables, and we shall see some of them in this book. A **stochastic process** is a collection of random variables $(X_\alpha;\alpha\in T)$ where T is some index set. A formula similar to (1.12) that holds regardless of the dependence between the variables is the following:

$$\mathbf{P}(X_0\in A_0,\ldots,X_n\in A_n)$$
$$=\mathbf{P}\{X_0\in A\}\prod_{i=1}^{n}\mathbf{P}(X_i\in A_i\mid X_0\in A_0,\ldots,X_{i-1}\in A_{i-1}). \qquad (1.13)$$

The theory of stochastic processes provides various specifications of the conditional probabilities on the right-hand side of (1.13). We will often relate such specifications to natural processes, the outcome of which we may be uncertain about. We will be concerned with random variables taking values in a **state space**[1] S, and governed by a probability measure that we call $\mathbf{P}$. Not all specifications of the type (1.13) are necessarily valid. Various kinds of regularity conditions are needed. We give two different examples of such regularity next.

The positivity condition. Consider a discrete multivariate random variable $\mathbf{X}=(X_1,\ldots,X_m)$ with probability distribution $q(\mathbf{x})=\mathbf{P}(\mathbf{X}=\mathbf{x})$, $\mathbf{x}\in S\equiv\{\mathbf{x}:q(\mathbf{x})>0\}$. This choice of S is called the **minimal state space**. Now consider the minimal state spaces for each of the components X_i, i.e., $S_i=\{x:\mathbf{P}(X_i=x)>0\}$. We say that q satisfies the **positivity condition** if

$$S=S_1\times S_2\times\cdots\times S_m, \qquad (1.14)$$

so that if $x_i\in S_i$, $i=1,\ldots,m$, we have $\mathbf{x}=(x_1,\ldots,x_m)\in S$.

Example (The positivity condition) Let $m=2$ and consider binary X_i. If the X_i are independently distributed as Bin(1,½) we have

$$q(0,0)=q(1,1)=q(0,1)=q(1,0)=1/4 \qquad (1.15)$$

which does satisfy the positivity condition. On the other hand, if

$$q(0,0)=q(1,1)=½ \qquad (1.16)$$

we have $S_i=\{0,1\}$, so

$$S_1\times S_2=((0,0),(0,1),(1,0),(1,1)) \qquad (1.17)$$

while

$$S=((0,0),(1,1)), \qquad (1.18)$$

[1] The state space is often called the sample space in introductory probability. We use the term state space which is more common in stochastic process theory.

violating the condition. □

The conditional distribution of X_i, given the values of all the other variables which we denote $\mathbf{X}_{-i}$, is

$$q_i(x_i \mid \mathbf{x}_{-i}) = \mathbf{P}(X_i{=}x_i \mid \mathbf{X}_{-i}{=}\mathbf{x}_{-i}) = \frac{\mathbf{P}(\mathbf{X}{=}\mathbf{x})}{\mathbf{P}(\mathbf{X}_{-i}{=}\mathbf{x}_{-i})} = \frac{q(\mathbf{x})}{\sum_{x_i \in S_i} q(\mathbf{x})}. \qquad (1.19)$$

Clearly, if q satisfies the positivity condition these q_i are well defined and positive for any $\mathbf{x} \in S$. Note that the q_i are univariate probability distributions. It turns out that under the positivity condition knowledge of the q_i suffices to determine q. We shall see uses of this later on. Let $\mathbf{x}_s^t$ be shorthand for $x_s, \ldots, x_t$, interpreted as an empty set if $s>t$. The following expansion is due to Brook (1964).

Proposition 1.1 Suppose that q satisfies the positivity condition. Then

$$\frac{q(\mathbf{x})}{q(\mathbf{y})} = \prod_{i=1}^{m} \frac{q_i(x_i \mid \mathbf{x}_1^{i-1}, \mathbf{y}_{i+1}^m)}{q_i(y_i \mid \mathbf{x}_1^{i-1}, \mathbf{y}_{i+1}^m)}. \qquad (1.20)$$

Proof

$$\begin{aligned}
q(\mathbf{x}) &= q_m(x_m \mid \mathbf{x}_1^{m-1}) \mathbf{P}(\mathbf{X}_1^{m-1}{=}\mathbf{x}_1^{m-1}) \\
&= q_m(x_m \mid \mathbf{x}_1^{m-1}) \mathbf{P}(\mathbf{X}_1^{m-1}{=}\mathbf{x}_1^{m-1}) \frac{q_m(y_m \mid \mathbf{x}_1^{m-1})}{q_m(y_m \mid \mathbf{x}_1^{m-1})} \\
&= \frac{q_m(x_m \mid \mathbf{x}_1^{m-1})}{q_m(y_m \mid \mathbf{x}_1^{m-1})} q(\mathbf{x}_1^m, y_m) \qquad (1.21) \\
&= \frac{q_m(x_m \mid \mathbf{x}_1^{m-1})}{q_m(y_m \mid \mathbf{x}_1^{m-1})} q_{m-1}(x_{m-1} \mid \mathbf{x}_1^{m-2}, y_m) \mathbf{P}(\mathbf{X}_1^{m-2}{=}\mathbf{x}_1^{m-2}, X_m{=}y_m) \\
&= \frac{q_m(x_m \mid \mathbf{x}_1^{m-1})}{q_m(y_m \mid \mathbf{x}_1^{m-1})} \frac{q_{m-1}(x_{m-1} \mid \mathbf{x}_1^{m-2}, y_m)}{q_{m-1}(y_{m-1} \mid \mathbf{x}_1^{m-2}, y_m)} q(\mathbf{x}_1^{m-2}, \mathbf{y}_{m-1}^m).
\end{aligned}$$

Continue this process for each i, "replacing" x_i with y_i by multiplying and dividing by $q_i(y_i \mid \mathbf{x}_1^{i-1}, \mathbf{y}_{i+1}^m)$ and regrouping terms. Positivity assures that if $q(\mathbf{x})>0$ and $q(\mathbf{y})>0$ then $q(\mathbf{x}_1^{i-1}, \mathbf{y}_i^m)>0$ for $i=1, \ldots, m$ (Exercise 1). □

The consequence of this expansion is that q is uniquely determined by the conditional distributions, since $\sum_{\mathbf{x} \in S} q(\mathbf{x})=1$.

The Kolmogorov consistency condition. Consider a stochastic process $(X_i;i=0,1,...)$, having a distribution such that

$$\mathbf{P}(X_{i_1}\in A_1,\ldots,X_{i_n}\in A_n)=\mathbf{P}(X_{i_1}\in A_1,\ldots,X_{i_n}\in A_n,X_{i_{n+1}}\in S) \tag{1.22}$$

for all $i_1,\ldots,i_{n+1}\in\{0,1,...rb$, $n\in\mathbf{Z}_+$, and $A_1,\ldots,A_n$ events (measurable subsets of S). This condition guarantees the existence of a probability measure corresponding to this stochastic process, and is due to Kolmogorov[1] (1933).

Remark It is nontrivial that any stochastic processes exist, in the sense of allowing the description of statments about it in terms of probabilities. In fact, one must impose some structure on S. This is the subject of extension theorems in measure-theoretic probability theory. □

Example (An iid process) The standard iid model so common in probability and statistics satisfies the Kolmogorov condition. In the case of random variables with density $f(x)$, taking values on the real line, so $S=\mathbf{R}$, we have

$$\begin{aligned}
&\mathbf{P}(X_{\alpha_1}\in A_1,\ldots,X_{\alpha_n}\in A_n,X_{\alpha_{n+1}}\in\mathbf{R})\\
&=\mathbf{P}(X_{\alpha_1}\in A_1)\cdots\mathbf{P}(X_{\alpha_n}\in A_n)\mathbf{P}(X_{\alpha_{n+1}}\in\mathbf{R})\\
&=\mathbf{P}(X_{\alpha_1}\in A_1)\cdots\mathbf{P}(X_{\alpha_n}\in A_n)\int_{-\infty}^{\infty}f(x)dx \qquad (1.23)\\
&=\mathbf{P}(X_{\alpha_1}\in A_1)\cdots\mathbf{P}(X_{\alpha_n}\in A_n)=\mathbf{P}(X_{\alpha_1}\in A_1,\ldots,X_{\alpha_n}\in A_n).
\end{aligned}$$

□

Depending on the structure of the state space S and the index set T one has different classifications of stochastic processes. T is often called "time" in a generic sense.

Example (Some interpretations of time) 1. X_t is the number of earthquakes of magnitude above 5 near Mount St. Helens in the time period $(0,t]$ where 0 is the beginning of recording. This is called a **counting process**. We have $T=\mathbf{R}_+$ and $S=\mathbb{N}$. Time is **continuous** and the state space is **discrete**.

2. $X_k=(B_k,D_k)$ are the number of births and deaths on day k in a population of insects. Here $T=\mathbb{N}$ and $S=\mathbb{N}^2$. Time and state space are both **discrete**.

---

[1] Kolmogorov, Andrei Nikolaievich (1903–1987). Russian probabilist of the Moscow School, student of Luzin. Developed the axiomatization of probability theory. Made immense contributions to the theory of stochastic processes, turbulence, and complexity theory.

3. $X_{y,t}$ is the amount of SO_4^{2-} in precipitation at location y at time t. Here $T=\mathbf{R}^2\times\mathbf{R}_+$ (or some appropriate subset thereof), $\alpha=(y,t)$, and $S=[0,\infty)$. This is called a **random field** (because T is more than one-dimensional). The state space is **continuous**. Here clock time is only one component of "time".

4. X_t is the thickness of an optical fiber at a distance t from the origin. Here both state space and time are **continuous** with $T=S=\mathbf{R}_+$. "Time" really is distance.

□

Much of the history of the subject of stochastic processes is rooted in particular physical, biological, social, or medical phenomena. The first occurrence of what is now called a Markov chain may have been as an alternative to the simple iid model in explaining rainfall patterns in Brussels (Quetelet, 1852). The simple branching process was invented by Bienaymé (1845) to compute the probability of extinction of a family surname among nobility. Rutherford and Geiger (1910) enrolled the help of the mathematician H. Bateman to describe the disintegration of radioactive substances using what we now call a Poisson process. Einstein (1905) presented a stochastic process that described well the Brownian motion of gold particles in solution, and Bachelier (1900) had used the same process to describe bond prices. The birth-and-death process was introduced by McKendrick (1924; in a special case in 1914) to describe epidemics, and Gibbs (1902) used nearest-neighbor interaction models to describe the behavior of large systems of molecules. The literature lacks a systematic account of the history of stochastic processes, but it should be clear from this short list of examples that an important aspect of such a history is the scientific problems that generated these different stochastic processes. The historical development of the probability theory associated with these processes rapidly moves beyond the scope of this book.

1.3. Purposes of stochastic models

The use of statistical methods to draw scientific conclusions will be illustrated repeatedly in this text. As a first, and quite simple, example, consider the question of mutation in bacteria. It was well known before 1943 that a pure culture of bacteria could give rise to a small number of cells exhibiting different, and inheritable, behavior. For example, when a plate of *E. coli* becomes infected by the bacteriophage T1, most of the cells are destroyed. A few may, however, survive. These, as well as all their offspring, are now resistant to T1 infection. In order to explain this resistance, two hypotheses were suggested. The **adaptation hypothesis** was that some bacteria became immune to T1 upon exposure to the bacteriophage. The key idea in this hypothesis is that the exposure is necessary for the resistance to develop. Alternatively, the bacteria may become immune through **spontaneous mutation**.

In order to distinguish between these hypothesis, Luria and Delbrück (1943) grew small individual cultures of T1-sensitive *E. coli*. From each culture equal quantities were added to several Petri dishes containing T1. Each dish was scored for the number of surviving (hence phage-resistant) colonies.

If the adaptation hypothesis is correct, each bacterium would have a small probability of developing resistance after being exposed to T1. Hence, using the Poisson limit to the binomial distribution, we would expect all the dishes to show Poisson variability (variance equal to the mean), regardless of which culture they came from. If, on the other hand, spontaneous mutation occurs at a constant rate, we would expect large numbers of resistant colonies originating from cultures in which mutation took place early in the experiment, and would therefore expect super-Poisson variability (variance larger than the mean) between dishes from different cultures.

In order to develop a control group, Luria and Delbrück also computed numbers of mutations in dishes that were all originating in one large culture. Since a large culture is constantly mixed, we would expect Poisson variability in these dishes, regardless of which hypothesis is true. Table 1.1 contains some results.

Table 1.1 Mutation numbers

Dish	Same culture	Different cultures
1	14	6
2	15	5
3	13	10
4	21	8
5	15	24
6	14	13
7	26	165
8	16	15
9	20	6
10	13	10

The sample mean for the control group was 16.7, with a sample variance of 15.0, while the experimental group had a sample mean of 26.2 with a sample variance of 2178. The difference between the means is not statistically significant, as assessed by a two-sample t-test. From comparing the variances it is, however, very clear that the mutation hypothesis is much better supported than the adaptation hypothesis, as has been verified by later experiments.

Many natural scientists are of the opinion that most problems on a macroscopic scale can be solved using deterministic models, at least in principle. Modern chaos theory suggests that some of these models are nonlinear. Techniques of applied probability are often thought of as what may be called **statistical** models, in the sense of being reasonably approximate descriptions of the phenomenon under study, while lacking much physical basis. Our point of view will be somewhat different. There are frequently reasons for using stochastic models (as opposed to deterministic ones) to *model*, rather than just describe, various natural phenomena. A distinction is sometimes made between **forecast** models and **structural** models. The former do not aim at describing the phenomenon under study, but rather try to predict outcomes, while the latter are more focused on describing the processes that produce the phenomenon. Our interest lies mainly in the latter. Some of the potential advantages in using stochastic models are listed below.

Aids understanding of the phenomenon studied. We will see (section 3.8) an example from neurophysiology where a stochastic model can rule out several proposed mechanisms for communication between nerve cells. While it is not possible to observe directly the underlying kinetics, it is possible to ascertain whether or not a certain neurological channel is open or closed. The time spent open can be used to estimate, based on simple assumptions, the actual kinetics of the channel.

More versatile than deterministic models. An example from entomology illustrates this. In a long series of experiments, the Australian entomologist A. J. Nicholson observed the population dynamics of blowflies, an insect that lays its eggs in the skin of sheep. The larvae feed off the flesh of the sheep, and can kill it if the infestation is severe enough. Nicholson's experiments used varying regimes of food, population density, and initial population. The data consist of observations only on the number of flies that were born and died, respectively, on a given day (Exercise 2.D6 in the next chapter gives one such data set). Guckenheimer et al. (1976) developed a model of the form

$$\frac{\partial n(t,a)}{\partial t} + \frac{\partial n(t,a)}{\partial a} = -d(a,n_t)n(t,a), \tag{1.24}$$

where $n(t,a)$ is the number of flies aged a at time t, while n_t is the total population size at time t. This model, even for fairly simple choices of death rate d, can exhibit chaotic behavior. However, since the age distribution of the population was not observed, this model could not directly be fitted to the data, and arbitrary assumptions about the age distribution could only be investigated by the qualitative behavior of numerical solutions of the partial differential equation (1.24). On the other hand, using a stochastic model, Brillinger et al. (1980) were able to reconstruct (with a specified amount of uncertainty) the age distribution of the population, and deduce that the population dynamics was both age and density dependent. In addition, one could infer from the stochastic model that the physiology of the population was changing by natural selection over the course of the experiment (Guttorp, 1980).

Allows assessment of variability. When modeling the long range transport of pollutants in the atmosphere, one important problem is to identify the source from observations at the point of deposition. There are both simple and complicated deterministic models in use that attempt to perform this task (Guttorp, 1986). Such models must take into account differing emission sources, transportation paths, chemical transformations in the atmosphere, and different types of deposition. The output of these models can be used to allocate sources as a fraction of total deposition. While such numbers give an impression of objective scientific analysis, it is extremely difficult to establish their uncertainty. Such uncertainty results from uncertainty in emissions data, model errors in describing the complicated processes involved, measurement errors in the field, laboratory errors, etc. (Guttorp and Walden, 1987). If, on the other hand, a carefully built stochastic model were used (Grandell, 1985, is a general introduction to such models) it would be possible to set confidence bands for source apportionment.

Extension of deterministic models. We saw in section 1.1 how a simple deterministic model of the motion of a coin can be combined with a stochastic model of the initial conditions to yield a probabilistic description of the outcome of coin tosses. It is a lot easier to test the combined model than to test the purely deterministic model, in which the initial conditions must be exactly realized.

Data compression. In the early days of mathematical statistics there was a great deal of emphasis on obtaining the maximum amount of information out of a given, often small, set of data. Many current data-gathering techniques yield vast data sets, which need to be described compactly. The emphasis is on compression while losing the minimum amount of information. For example, such techniques are needed to analyze satellite images (Baddeley et al., 1991). This area, compression of data sets using parametric models describing images, is likely to become one of the most important areas of statistical research over the next few decades in view of the rapid increase in remote sensing applications.

1.4. Overview

In Chapter 2 we introduce Markov chains with discrete time and discrete state space. The concept of reversibility of Markov chains has important consequences, and can sometimes be related to physical considerations of detailed balance in closed systems. Results on classification of states are needed to see under what circumstances statistical inference is possible. We work through the asymptotics of transition probabilities in some detail, partly to establish some necessary foundations for Markov chain Monte Carlo methods that will be very important in Chapter 4. We do both nonparametric and parametric statistical theory, present a linear model for higher order Markov chains, and finally look at hidden Markov processes and their reconstruction from noisy data. This is a first (and very simple) stab at some Bayesian techniques that will be very useful in image reconstruction later.

Chapter 3 switches to continuous time, while still retaining the Markov property. Many of the results from the previous chapter can be used on appropriate discrete time subchains to establish properties of the continuous time models. As in the case of discrete time, we do both parametric and non-parametric inference, and study partially observed as well as completely hidden Markov models.

In the fourth chapter we move to random fields. We look at nearest neighbor interaction potentials and relate them to the Markov property. We encounter the phenomenon of phase transition in a simple example. In this chapter we use parametric statistical inference, making heavy use of Markov chain Monte Carlo techniques. Hidden processes are reconstructed using the Bayesian approach first encountered at the end of Chapter 2, and a general formulation of Markov random fields on graphs is found to be useful.

In the previous chapters, non-Markovian processes were introduced as noisy functions of an underlying (but perhaps not observable) Markovian structure. In Chapter 5 we turn to a class of mostly non-Markovian processes which do not have any underlying Markovian structure, but are developing dynamically depending on their entire previous history. These processes model events that occur separately over time. A variety of ways of thinking about such point processes are discussed, and we encounter both parametric and nonparametric statistical inference.

The Brownian motion process is the foundation of continuous time Markov processes with continuous paths. Such processes are discussed in Chapter 6. We illustrate how one can build diffusion processes using stochastic differential equations. The statistical inference becomes quite complicated here, since it is no longer straightforward to define a likelihood.

1.5. Bibliographic remarks

Much of the material in section 1.1 follows Keller (1986) and Engel (1992). An elementary introduction to the topic of chaotic dynamics is Ruelle (1991). More general discussion of chaos, indeterminism, and randomness is in a report from the Royal Statistical Society Meeting on Chaos (1992).

The importance of the Brook expansion was made clear by Besag (1974). Any measure-theoretic introduction to probability (such as Billingsley, 1979, section 36) explains the use of the Kolmogorov consistency condition.

The *E. coli* example follows Klug and Cummings (1991), pp. 409–412.

1.6. Exercises

Theoretical exercises

1. Suppose that q satisfies the positivity condition (1.14). Show that if $\mathbf{x}$ and $\mathbf{y}$ are outcomes such that $q(\mathbf{x})>0$ and $q(\mathbf{y})>0$, then $q(\mathbf{x}_1^{i-1},\mathbf{y}_i^m)>0$.

2. For $0\le t_1<\cdots<t_n$ and $0\le r_1\le\cdots\le r_n$, define a stochastic process $X(t)$ by assuming that

$$\mathbf{P}(X(t_1)=r_1,\ldots,X(t_n)=r_n)$$
$$=\frac{\lambda^{r_n}e^{-\lambda t_n}t_1^{r_1}(t_2-t_1)^{r_2-r_1}\cdots(t_n-t_{n-1})^{r_n-r_{n-1}}}{r_1!(r_2-r_1)!\cdots(r_n-r_{n-1})!}$$

Show that this process satisfies the Kolmogorov consistency condition.

3. For the process in Exercise 2, write $X_i=X(t_i)$. Show that $q(\mathbf{x})=\mathbf{P}(\mathbf{X}=\mathbf{x})$ satisfies the positivity condition.

4. Consider a (theoretical) roulette wheel divided into n sections that are alternating red and black. Assume that the arcs are all of equal length. The wheel is spun using a large initial impulse, and the angle X in radians (with respect to a fixed point outside the wheel) of a given spot on the wheel (at one end of a black section) is assumed to have a density $f(x)$.

(a) Show that the probability of black, i.e., that the fixed point is nearest a black section, can be computed as

$$\mathbf{P}(\text{black}) = \mathbf{P}(\frac{nX}{4\pi}\bmod(1)\le ½).$$

(b) Poincaré (1896) showed that under mild regularity conditions on the density of X one has that tX mod(1) converges in distribution to a uniform random variable on (0,1) as $t\to\infty$. Apply this result to deduce the limiting probability of black as $n\to\infty$.

Remark: The method used here uses no physics, only the existence of a sufficiently regular density. It is called the method of **arbitrary functions** (see Engel, 1992, for some historical background).

Computing exercises

C1. Simulate the process described in Exercise 4(a), and assess the convergence of $\mathbf{P}$(black) to ½ as a function of the number n of sectors. Modify the process so that the red sections have arc length R and the black have arc length B. How does that affect the probability? The convergence?

Data exercises

D1. Suppose we are observing a roulette wheel of the type described in Exercise 4 (so it has no zero, or, equivalently, we ignore the zeroes) with $n = 32$.

(a) Derive the distribution of runs, i.e., successive strings of the same color, assuming that the probability of black is ½.

(b) Pearson (1900) reported data on runs from the casino in Monte Carlo in July of 1892. The data are given in Table 1.2.

Table 1.2 Runs of the same color in Monte Carlo

Run length	1	2	3	4	5	6
Count	2462	945	333	220	135	81

Run length	7	8	9	10	11	12
Count	43	30	12	7	5	1

Do they agree with the distribution in (a)? If not, can you explain the discrepancy?

CHAPTER 2

Discrete time Markov chains

Starting from a very simple model of daily precipitation, we build some theory for discrete time Markov chains. We develop some estimation and testing theory. and look at the goodness of fit of this simple model in different ways. A parsimonious model for higher order of dependence is applied to some meteorological data. Harmonic functions are introduced as a tool towards computing how long it takes a random walk on a graph to hit a subset of the boundary states. We analyze some problems in epidemiology and genetics using branching processes. A hidden Markov model categorizing atmospheric variables yields an improved fit to the precipitation data. A similar model is used to describe whether a chemical transmission channel in a nerve cell is open or closed.

2.1. Precipitation at Snoqualmie Falls

The US Weather Service maintains a large number of precipitation monitors throughout the United States. One station is located at the Snoqualmie Falls in the foothills of the Cascade Mountains in western Washington. A day is defined as wet if at least 0.01 inches of precipitation falls during a precipitation day: 8 a.m. through 8 a.m. the following calendar day. To start with, we shall ignore the amounts of rainfall, and just look at the pattern of wet and dry days. Using data from 1948 through 1983, and looking at January rainfall only, there were 325 dry and 791 wet days. Let $X_{ij}=1(\text{day } i \text{ of year } j \text{ wet})$, where $1(A)$ is 1 if the event A occurs, and 0 otherwise. A very simple model, which we can call the Bernoulli model, is that $X_{ij}\sim\text{Bin}(1,p)$, with the X_{ij} independent, i.e., an iid model, and with p being the probability of rain at Snoqualmie Falls on a January day. The **likelihood** (probability of the observed data as a function of p) is

$$L(p) \propto p^{791}(1-p)^{325}. \tag{2.1}$$

Appendix A contains a brief review of likelihood theory for multinomial data to illustrate some of the central ideas. Edwards (1985) is a good reference for more general likelihood theory. The maximum point of $L(p)$ is the maximum likelihood estimate (mle) $\hat{p}$ of p. Letting n be the number of days observed, it is easy to see that $\hat{p}=\sum x_{ij}/n=0.709$. A standard error for this estimate is $(p(1-p)/n)^{½}$ which we estimate (using $\hat{p}$ in place of p) to be 0.014.

In order to assess the fit of the binomial model for rainfall, we first try to see if the independence assumption seems reasonable. We may suspect a certain amount of persistence, i.e., stretches of like weather, in the data. This would be induced by the relatively slow movement of large weather systems through an area. In the winter, a typical front may take up to three days to pass through from the Pacific Ocean. In order to study this hypothesis, let us look at consecutive pairs of days. Figure 2.1 shows the pattern of rainfall.

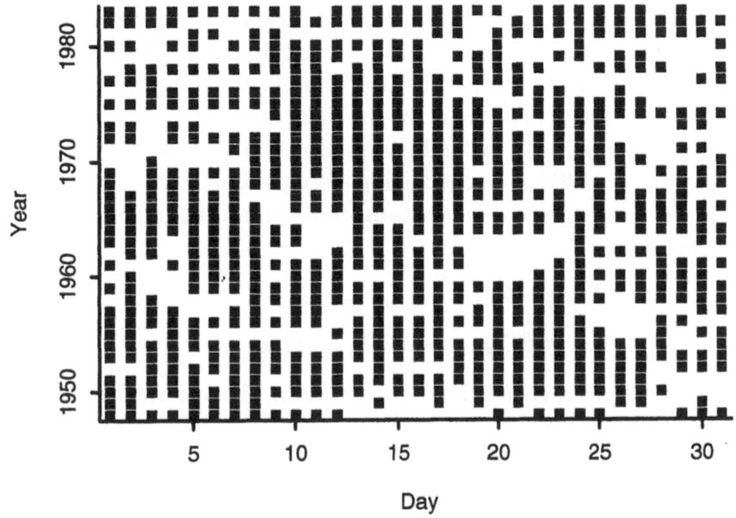

Figure 2.1. The pattern of January precipitation at Snoqualmie Falls. Each square is a day with measurable precipitation. Rows correspond to years, columns to days.

If the independence model is correct, we would expect to see $36\times30\times\hat{p}(1-\hat{p})=223$ dry days following wet days, since we have 36 years of data, and 30 consecutive pairs of days for each January. Table 2.1 contains the total counts, with expected counts under the independence assumption shown in parenthesis. There seems to be a lot more dry days followed by dry days, and wet days followed by wet days, than what the simple iid model predicts. To build a better model of this phenomenon, let us introduce two parameters:

$$p_w = \mathbf{P}(\text{wet today} \mid \text{wet yesterday}) \tag{2.2}$$

$$p_d = \mathbf{P}(\text{wet today} \mid \text{dry yesterday}). \tag{2.3}$$

Table 2.1 Observed precipitation

	Today dry		Today wet		Total
Yesterday dry	186	(91)	123	(223)	309
Yesterday wet	128	(223)	643	(543)	771
Total	314		766		1080

If the X_i are not independent, we must specify the conditional probabilities

$$\mathbf{P}(X_{i+1}=l \mid X_0=k_0, \ldots, X_i=k_i) \tag{2.4}$$

for all i, l, and $k_1, \ldots, k_i$. Note that we will assume unless otherwise specified that the process is observed from time 0. A simple (and perhaps natural) way to specify the probabilities in (2.4) is to assume that the conditional probability only depends on what happened at the previous time point. This assumption was first studied systematically by the Russian probabilist Markov[1] in a sequence of papers, starting in 1907, on generalizing various limit laws to dependent data. Formally we write the **Markov assumption** for a random process (X_n) with discrete state space

$$\mathbf{P}(X_{n+1}=l \mid X_0=k_0, \ldots, X_n=k_n)$$
$$= \mathbf{P}(X_{n+1}=l \mid X_n=k_n) = p_{k_n l}(n). \tag{2.5}$$

If (X_n) satisfies (2.5) it is called a **Markov chain**. Two seemingly more general forms of (2.5) are outlined in Exercise **1**: in part (a) we show that the conditional distribution of the process at any set of future times, given any set of times up to and possibly including the present, only depends on the last of the times in the condition, and in part (b) we show that an equivalent, and rather colorful, way of stating the Markov property is that the future is independent of the past, given the present.

The functions $p_{ij}(n)$ are called **transition probabilities**. We can write the transition probabilities in matrix form. The matrices $\mathbb{P}(n) = (p_{ij}(n))$ are called **transition matrices**.

In order to prove the existence of a Markov chain with a given set of transition matrices and distribution of X_0 one has to verify the Kolmogorov consistency condition (1.22). This is made precise, e.g., in Freedman (1983, pp. 7–8). Here is a simple fact about transition matrices:

---

[1] Markov, Andrei Andreevich (1856–1922). Russian probabilist in the St Petersburg School. He was a student of Chebyshev, and proved the law of large numbers rigorously in a variety of cases, including dependent sequences.

Proposition 2.1 If $\mathbb{P}$ is a sequence of transition matrices for a Markov chain with state space $S=\{0,\ldots,K\}$, where K may be finite or infinite, then $\sum_{j=0}^{K} p_{ij}(n)=1$ for any n.

Proof We have that $p_{ij}(n)=\mathbf{P}(X_{n+1}=j \mid X_n=i)$, so

$$\begin{aligned}\sum_{j=0}^{K} p_{ij}(n) &= \sum_{j=0}^{K} \mathbf{P}(X_{n+1}=j \mid X_n=i) \\ &= \mathbf{P}(\bigcup_{j=0}^{K} \{X_{n+1}=j\} \mid X_n=i) = \mathbf{P}(X_{n+1} \in S \mid X_n=i) = 1 \end{aligned} \quad (2.6)$$

since the process must go somewhere. □

It is often a reasonable simplifying assumption that the transition probabilities are independent of time; such Markov chains are said to have **stationary transition probabilities**. In that case we just need a single transition matrix $\mathbb{P}=\mathbb{P}(1)$. For our rainfall model, we are only considering January. This makes the assumption of stationary transition probabilities reasonable, if we believe (at least approximately) that this month is meteorologically homogeneous. The state space is $\{\text{dry,wet}\}$, which we can map into $\{0,1\}$. Then, using (2.2) and (2.3), $p_{00}=1-p_d$, $p_{01}=p_d$, $p_{10}=1-p_w$, and $p_{11}=p_w$. In matrix notation,

$$\mathbb{P} = \begin{pmatrix} p_{00} & p_{01} \\ p_{10} & p_{11} \end{pmatrix} = \begin{pmatrix} 1-p_d & p_d \\ 1-p_w & p_w \end{pmatrix}. \quad (2.7)$$

A matrix of non-negative elements with all row sums equal to one is often called a **stochastic** matrix. From now on we will, unless specifically stating otherwise, assume that all transition probabilities are stationary. Here are some elementary properties of stochastic matrices.

Proposition 2.2 (i) A stochastic matrix has at least one eigenvalue equal to one.

(ii) If $\mathbb{P}$ is stochastic, then $\mathbb{P}^k$ is also stochastic for all $k=1,2,3,\ldots$.

Proof (i) is a consequence of the definition of a stochastic matrix, which can be written $\mathbb{P}\mathbf{1}^T = \mathbf{1}^T$, where $\mathbf{1}$ is a row vector of ones (recall that all vectors are assumed to be row vectors). Hence $(\mathbb{I}-\mathbb{P})\mathbf{1}^T = \mathbf{0}$, where $\mathbb{I}$ is the identity matrix, so $\mathbf{1}$ is a right eigenvector of $\mathbb{P}$ corresponding to the eigenvalue 1. Now (ii) follows easily, writing

$$\mathbb{P}^k\mathbf{1}^T = \mathbb{P}^{k-1}\mathbb{P}\mathbf{1}^T = \mathbb{P}^{k-1}\mathbf{1}^T = \cdots = \mathbf{1}^T. \quad (2.8)$$

□

The likelihood for a Markov chain can be written, using (2.5) and (1.13), as

$$L(\mathbb{P}) = \mathbf{P}(X_0=x_0)\prod_{i=0}^{n-1}\mathbf{P}(X_{i+1}=x_{i+1} \mid X_i=x_i)$$
$$= \mathbf{P}(X_0=x_0)\prod_{i=0}^{n-1}p_{x_i x_{i+1}} = \mathbf{P}(X_0=x_0)\prod_{k,l=0}^{K}p_{kl}^{n_{kl}}, \tag{2.9}$$

where n_{kl} is the number of transitions from k to l observed in the chain. In our example we have the additional complication that we are considering 36 years. A simple model is to assume that years are independent. While quasi-periodic large-scale meteorological oscillations such as El Niño may make this hypothesis somewhat suspect (cf. Woolhiser, 1992), it nevertheless allows us to proceed. Furthermore, we shall be able to test it later (Exercise **D1**). Under the assumption of year-to-year independence the likelihood is a product of 36 factors, each of the form (2.9). Clearly, the product collapses, and we can use the data in Table 2.1 to compute

$$L(\mathbb{P}) = L(p_{01},p_{11}) = \left[\prod_{i=1}^{36}\mathbf{P}(X_0^i=x_0^i)\right]p_{00}^{186}p_{01}^{123}p_{10}^{128}p_{11}^{643} \tag{2.10}$$

Assuming that the starting values X_0^i for each year i are fixed (this assumption will be discussed in more detail in section 2.7), so that the beginning term in the right-hand side of (2.10) is 1, we find that L is maximized by

$$\hat{p}_{01} = \frac{123}{309} = 0.398 \quad \hat{p}_{11} = \frac{643}{771} = 0.834 \tag{2.11}$$

so

$$\hat{\mathbb{P}} = \begin{pmatrix} 0.398 & 0.602 \\ 0.166 & 0.834 \end{pmatrix}. \tag{2.12}$$

These estimates are substantially different from the estimate $\hat{p}$=0.709 from the iid model. However, we may question whether such a difference could occur by chance. At a first glance this seems very unlikely, since $\hat{p}_{01}$ is 22 standard errors (of $\hat{p}$) away from $\hat{p}$. For a formal test of significance we use the likelihood ratio test. Recall (or see Appendix A) that under suitable regularity conditions, the **log likelihood ratio** $2(\log L(\hat{\mathbb{P}})-\log L(\hat{p}))$ has a χ^2 distribution with degrees of freedom equal to the difference in the dimension of the parameter spaces; in this case 2-1=1. Although this result was developed for iid processes, it is also true in the Markov chain case. We will return to it in section 2.7. In order to be able to compare the likelihoods we need to exclude the January 1 measurements when computing the iid mle, since those observations cannot be used to compute the Markov chain mle's. This yields $\hat{p}$=771/1080=0.714, slightly higher than the 0.709 we obtained from the full data set. Computing the log likelihood ratio we get

$$2(\log L(\hat{\mathbb{P}})-\log L(\hat{p})) = 2(643\log 0.834+128\log 0.166 \\ +123\log 0.398+186\log 0.602 \\ -771\log 0.714-309\log 0.286) = 184.5. \tag{2.13}$$

which under the null hypothesis of the Bernoulli model is distributed $\chi^2(1)$, corresponding to a P-value of 0. We therefore reject the iid model at all reasonable levels.

2.2. The marginal distribution

Although the Markov assumption tells us how to compute conditional probabilities, one often wants marginal probabilities. It is relatively straightforward to compute these. For example, in a 0-1 chain we have that

$$\begin{aligned} \mathbf{P}(X_{n+1}=1) &= \mathbf{P}(X_{n+1}=1, X_n=0) + \mathbf{P}(X_{n+1}=1, X_n=1) \\ &= \mathbf{P}(X_n=0)\, p_{01} + \mathbf{P}(X_n=1)\, p_{11} \\ &= \mathbf{P}(X_n=1)\,(p_{11}-p_{01}) + p_{01}. \end{aligned} \tag{2.14}$$

Define the **initial distribution** $\mathbf{p}_0 = (p_0(0), \ldots, p_0(K))$ where $p_0(i) = \mathbf{P}(X_0=i)$. In the 0-1 case we write $p_0(1)\equiv p_1$. Then (2.14) can be written

$$\begin{aligned} \mathbf{P}(X_1=1) &= p_1(p_{11}-p_{01}) + p_{01}, \\ \mathbf{P}(X_2=1) &= (p_{11}-p_{01})\mathbf{P}(X_1=1) + p_{01} \\ &= p_1(p_{11}-p_{01})^2 + p_{01}(1+(p_{11}-p_{01})) \\ &\cdots \\ \mathbf{P}(X_n=1) &= (p_{11}-p_{01})^n p_1 + p_{01}\sum_{j=0}^{n-1}(p_{11}-p_{01})^j. \end{aligned} \tag{2.15}$$

If $p_{00}=p_{11}=1$ we have $\mathbf{P}(X_n=1)=p_1$. If $p_{01}\neq p_{11}$ we can write

$$\mathbf{P}(X_n=1) = \frac{p_{01}}{1-(p_{11}-p_{01})} + \left[p_1-\frac{p_{01}}{1-(p_{11}-p_{01})}\right](p_{11}-p_{01})^n. \tag{2.16}$$

Notice that the effect of the initial distribution p_1 is dampened exponentially, and disappears completely when $p_1=p_{01}/(1-(p_{11}-p_{01}))$. In that situation $\mathbf{P}(X_n=1)$ is the same for each n. This choice of p_1 is called the **stationary initial distribution**. We will return to this in section 2.4.

More generally, let the state space S be an arbitrary countable set, which we identify with the integers $\mathbf{Z}$, and define $p_{jk}^{(n)} = \mathbf{P}(X_n=k \mid X_0=j)$. Here is an important computation, called the **Chapman[1]–Kolmogorov equation**,

[1] Chapman, Sydney (1888–1970). Leading British astro- and geophysicist. Major contributions to the understanding of the aurora; space physics; and convection in the atmosphere.

although it was discovered independently by many workers, including Bachelier (1900) and Einstein (1905).

Lemma 2.1

$$p_{jk}^{(n)} = \sum_{l\in S} p_{jl}^{(m)} p_{lk}^{(n-m)}, \quad 1\le m\le n-1. \tag{2.17}$$

Proof Since the process must be somewhere in S at time m, we have

$$\begin{aligned} p_{jk}^{(n)} &= \mathbf{P}(\{X_n=k\}\cap\bigcup_{l\in S}\{X_m=l\}\mid X_0=j) \\ &= \sum_{l\in S}\mathbf{P}(X_n=k,X_m=l\mid X_0=j) \\ &= \sum_{l\in S}\mathbf{P}(X_n=k\mid X_m=l,X_0=j)\mathbf{P}(X_m=l\mid X_0=j) \qquad (2.18)\\ &= \sum_{l\in S}\mathbf{P}(X_n=k\mid X_m=l)\mathbf{P}(X_m=l\mid X_0=j). \end{aligned}$$

□

In matrix notation we rewrite (2.17) as

$$\mathbb{P}_n \equiv (p_{jk}^{(n)}) = \mathbb{P}_{n-m}\mathbb{P}_m. \tag{2.19}$$

But $\mathbb{P}_1=\mathbb{P}$ so $\mathbb{P}_n=\mathbb{P}^n$. Let $\mathbf{p}_n=(...,\mathbf{P}(X_n=0),\ldots,\mathbf{P}(X_n=k),...)$ denote the probability distribution of X_n. Since

$$\mathbf{p}_n = \mathbf{p}_{n-1}\mathbb{P} \tag{2.20}$$

(recall the computation of $\mathbf{P}(X_n=1)$ earlier) we see that

$$\mathbf{p}_n = \mathbf{p}_0\mathbb{P}^n. \tag{2.21}$$

Application (Snoqualmie Falls precipitation) Suppose that we accept the Markov chain model developed in section 2.1 for the Snoqualmie Falls precipitation data, and that it happened to rain on January 1 this year. What would be the probability of rain on January 6, i.e., five days hence? To compute this probability, we need to determine $\hat{\mathbb{P}}^5$, where

$$\hat{\mathbb{P}} = \begin{bmatrix} 0.602 & 0.398 \\ 0.166 & 0.834 \end{bmatrix} \tag{2.22}$$

so that

$$\hat{\mathbb{P}}^5 = \begin{bmatrix} 0.305 & 0.695 \\ 0.290 & 0.710 \end{bmatrix}. \tag{2.23}$$

Notice that the two rows of $\hat{\mathbb{P}}^5$ are much more similar than those of $\hat{\mathbb{P}}$. This will be explained in section 2.5. The desired probability is obtained by setting

$\mathbf{p}_0=(0,1)$ in (2.21), so that

$$\hat{\mathbf{p}}_5=(0,1)\hat{\mathbb{P}}^5=(0.29,0.71). \tag{2.24}$$

In other words, the probability of rain at Snoqualmie Falls on January 6, given rain on January 1, is 0.71. □

A different type of question is how long a Markov chain would be expected to stay in a given state. Clearly, if $p_{ii}=0$ it is certain not to stay. If $p_{ii}>0$, the time spent in the state has a geometric distribution with mean $1/(1-p_{ii})$(Exercise **2**). For the Snoqualmie Falls application this translates to a mean of 2.5 consecutive dry days and 6.0 consecutive wet days in January. We return to this in section 2.9.

2.3. Classification of states

Let $A\subset S$. The **hitting time** T_A of A is

$$T_A = \begin{cases} \min\{n>0: X_n\in A\} \text{ if } X_n \text{ ever hits } A \\ \infty \qquad\qquad\qquad\qquad \text{otherwise} \end{cases} \tag{2.25}$$

If $A=\{a\}$ we write T_a. Denote the distribution of the chain, starting from the state x (i.e., $p_0(x)=1$ and $p_0(y)=0$ for any $y\neq x$), by $\mathbf{P}^x$. More generally, we write the distribution of the chain starting from the initial distribution $\mathbf{p}_0$ as $\mathbf{P}^{\mathbf{p}_0}$, and compute it using the formula

$$\mathbf{P}^{\mathbf{p}_0}(A) = \sum_{i\in S} p_0(i)\mathbf{P}^i(A). \tag{2.26}$$

This amounts to first choosing the initial state i at random from $\mathbf{p}_0$, and then running the chain starting from state i.

Proposition 2.3 $\quad p_{jk}^{(n)} = \sum_{m=1}^{n} \mathbf{P}^j(T_k=m)p_{kk}^{(n-m)}.$

Proof Write $\{X_n=k\} = \sum_{m=1}^{n}\{T_k=m,X_n=k\}$, where the summation sign stands for a union of disjoint sets. Now

$$\begin{aligned} p_{jk}^{(n)} &= \mathbf{P}^j(X_n=k) = \sum_{m=1}^{n} \mathbf{P}^j(T_k=m,\ X_n=k) \\ &= \sum_{m=1}^{n} \mathbf{P}^j(T_k=m)\mathbf{P}^j(X_n=k \mid T_k=m) \\ &= \sum \mathbf{P}^j(T_k=m)\mathbf{P}(X_n=k \mid X_0=j,X_1\neq k,\ldots,X_{m-1}\neq k,X_m=k) \\ &= \sum \mathbf{P}^j(T_k=m)\mathbf{P}(X_n=k \mid X_m=k) = \sum \mathbf{P}^j(T_k=m)p_{kk}^{(n-m)}. \end{aligned} \tag{2.27}$$

□

Call a state **absorbing** if $p_{kk}=1$. If the chain ever reaches k it stays there forever.

Corollary For an absorbing state k we have that $p_{jk}^{(n)}=\mathbf{P}^j(T_k\leq n)$.

Proof The content of this equation is really trivial: in order to go from j to k in n steps we need to hit k no later than time n. A formal proof follows from the observation that $p_{kk}^{(n-m)}=1$ for all $m<n$ and Proposition 2.3. □

T_k is an example of a particularly interesting class of random times. Call the random time τ a **Markov time** if the event $\{\tau=n\}$ is completely determined by the values of $X_0,\ldots,X_n$. The **strong Markov property** asserts that the Markov property holds also at Markov times. More formally, let $f_i(k)=\mathbf{P}^k(X_1=i)$. Then

$$\mathbf{P}(X_{\tau+1}=i \mid X_0,\ldots,X_\tau) = f_i(X_\tau). \tag{2.28}$$

A proof of this can be found, e.g., in Freedman (1983, Theorem 1:21).

Say that i **reaches** j, written $i\rightarrow j$, if there is an n such that $p_{ij}^{(n)}>0$. If $i\rightarrow j$ and $j\rightarrow i$ we say that i and j **communicate**, denoted $i\leftrightarrow j$.

Theorem 2.1 $\leftrightarrow$ is an equivalence relation.

Proof $i\leftrightarrow i$ since $p_{ii}^{(0)} = \mathbf{P}(X_n=i \mid X_n=i)=1$. Next, $i\leftrightarrow j$ implies that $j\leftrightarrow i$ by definition. Finally, if $i\leftrightarrow j$ and $j\leftrightarrow k$ there are integers m and n such that $p_{ij}^{(n)}>0$ and $p_{jk}^{(m)}>0$. Thus

$$p_{ik}^{(n+m)} = \sum_r p_{ir}^{(n)}p_{rk}^{(m)} \geq p_{ij}^{(n)}p_{jk}^{(m)} > 0 \tag{2.29}$$

and $i\rightarrow k$. To show that $k\rightarrow i$ uses a similar argument. □

We can partition all states into equivalence classes with respect to the relation $\leftrightarrow$. A Markov chain is **irreducible** if there is only one equivalence class, i.e., if all states communicate.

Example (A model for radiation damage) A finite **birth and death chain** is a Markov chain on $\{0,\ldots,K\}$ in which a particle in state i can either stay or move to one of the neighboring states $i+1$ or $i-1$. Reid and Landau (1951) proposed this chain as a model for the transmission of radiation damage following the initial damage due to the absorption of radiation quanta. The mechanism by which this transmission takes place was assumed to be the depolymerization of macromolecules associated with the sensitive volume of the organism. State 0 corresponds to a healthy organism, and state K to one with visible radiation damage. The intermediate states correspond to amplification or healing of the initial damage, which is taken to be state 1. The extreme states are assumed absorbing, so the transition matrix for this process is

$$\mathbb{P} = \begin{bmatrix} 1 & 0 & 0 & 0 & \cdots & 0 & 0 \\ q_1 & r_1 & p_1 & 0 & \cdots & 0 & 0 \\ 0 & q_2 & r_2 & p_2 & \cdots & 0 & 0 \\ \cdots & \cdots & \cdots & \cdots & \cdots & \cdots & \cdots \\ 0 & 0 & 0 & 0 & \cdots & r_{K-1} & p_{K-1} \\ 0 & 0 & 0 & 0 & \cdots & 0 & 1 \end{bmatrix} \tag{2.30}$$

where r_i is the conditional probability of staying in state i, p_i is the conditional probability of moving to state $i+1$ (amplification of damage), and q_i the conditional probability of moving to state $i-1$ (recovery). Reid and Landau suggested to use r_i=0, p_i=i/K, and q_i=1−i/K. This chain has three classes: {0}, {K}, and {1, . . . ,K−1}. Starting from state 1, we may want to compute the recovery probability (λ_0), i.e., the probability of reaching state 0 before state K. By conditioning on the last step, which must be from 1 to 0, we can write

$$\lambda_0 = p_{10} \sum_{n=0}^{\infty} p_{11}^{(n)}. \tag{2.31}$$

For example, if K=3 so

$$\mathbb{P} = \begin{bmatrix} 1 & 0 & 0 & 0 \\ 2/3 & 0 & 1/3 & 0 \\ 0 & 1/3 & 0 & 2/3 \\ 0 & 0 & 0 & 1 \end{bmatrix} \tag{2.32}$$

we find that $p_{11}^{(2n)} = (\frac{1}{3}\times\frac{1}{3})^n$ while $p_{11}^{(2n+1)} = 0$ (note that the only way to achieve a transition from 1 to 1 in $2n$ steps is to go 1-2-1-2-1 $\cdots$). Hence λ_0=3/4. Generally, $\lambda_0=1-2^{-(K-1)}$ (Exercise **3**). □

We say that a state i has **period** d if $p_{ii}^{(n)}$=0 for all n not divisible by d, and d is the greatest such integer. This means that if the chain is in state i at time n it can only return there at times of the form $n+kd$ for some integer k. If $p_{ii}^{(n)}$=0 for all n, we say that state i has infinite period. A state with period 1 is called **aperiodic**.

Theorem 2.2 Periodicity is an equivalence class property, i.e., if $i\leftrightarrow j$ then $d(i)=d(j)$.

Proof Let m, n be such that $p_{ij}^{(m)}>0$, $p_{ji}^{(n)}>0$, and assume that $p_{ii}^{(s)}>0$. Then

$$p_{jj}^{(m+n)} \geq p_{ji}^{(n)} p_{ij}^{(m)} > 0 \tag{2.33}$$

and

$$p_{jj}^{(m+n+s)} \geq p_{ji}^{(n)} p_{ii}^{(s)} p_{ij}^{(m)} > 0 \tag{2.34}$$

so $d(j)$ must divide $m+n$ and $m+n+s$. Hence it must divide their difference s for any s such that $p_{ii}^{(s)}>0$. Therefore $d(j)$ divides $d(i)$. Similarly, $d(i)$ is seen to divide $d(j)$, so the two numbers must be equal. □

Example (A model for radiation damage, continued) In the Reid–Landau radiation damage model described earlier we have

$$\mathbb{P} = \begin{bmatrix} 1 & 0 & 0 & 0 & \cdots & 0 & 0 \\ q_1 & r_1 & p_1 & 0 & \cdots & 0 & 0 \\ 0 & q_2 & r_2 & p_2 & \cdots & 0 & 0 \\ \cdots & \cdots & \cdots & \cdots & \cdots & \cdots & \cdots \\ 0 & 0 & 0 & 0 & \cdots & r_{K-1} & p_{K-1} \\ 0 & 0 & 0 & 0 & \cdots & 0 & 1 \end{bmatrix}. \tag{2.35}$$

Therefore the period of the class $\{1,\ldots,K-1\}$ is 2. □

To prove the next result, we need a number-theoretic lemma:

Lemma 2.2 Given positive integers n_1 and n_2 with greatest common divider (gcd) 1, any integer $n>n_1n_2$ can be written $n=ln_1+kn_2$ for non-negative integers l and k.

Proof Consider the modulo n_2 residue classes of the n_2 distinct positive integers n, $n-n_1$, $n-2n_1,\ldots,n-(n_2-1)n_1$. Either these residue classes are all different, in which case one residue class must be 0, so the corresponding number $n-kn_1$ is divisible by n_2, i.e., $n=kn_1+ln_2$, or at least two residue classes are the same. If the common residue class is 0 the preceding argument applies. Otherwise we can write $n-sn_1=a+bn_2$ and $n-tn_1=a+cn_2$ for $0\le t<s\le n_2-1$, $b<c$, and $0<a<n_2$. Hence

$$n-sn_1-(n-tn_1)=(s-t)n_1=(c-b)n_2. \tag{2.36}$$

Since $\gcd(n_1,n_2)=1$ we must have $s-t$ containing all prime factors of n_2. But then $s-t\ge n_2$ which is a contradiction. □

Proposition 2.4 If i and j are states of an irreducible aperiodic chain, then there is an integer $N=N(i,j)$ such that $p_{ij}^{(n)}>0$ for all $n\ge N$.

Proof Since $d(j)=1$ there are integers n_1, n_2 with gcd 1 such that $p_{jj}(n_k)>0$, $k=1,2$. From Lemma 2.2 we see that any sufficiently large n can be written ln_1+kn_2, whence

$$p_{jj}^{(n)}=p_{jj}^{(ln_1+kn_2)}\ge\left[p_{jj}^{(n_1)}\right]^l\left[p_{jj}^{(n_2)}\right]^k>0. \tag{2.37}$$

Finally, for each pair i,j there is an n_0 such that $p_{ij}^{(n_0)}>0$. Hence

$$p_{ij}^{(n+n_0)}\ge p_{ij}^{(n_0)}p_{jj}^{(n)}>0. \tag{2.38}$$

□

Corollary Let X and Y be iid irreducible aperiodic Markov chains. Then $Z=(X,Y)$ is an irreducible Markov chain.

Proof It is clear that Z is Markov, with transition probabilities

$$\tilde{p}_{ij,kl} = \mathbf{P}(Z_{t+1}=(k,l) \mid Z_t=(i,j)) = \mathbf{P}(X_{t+1}=k, Y_{t+1}=l \mid X_t=i, Y_t=j)$$
$$= \frac{\mathbf{P}(X_{t+1}=k, X_t=i)}{\mathbf{P}(X_t=i)} \frac{\mathbf{P}(Y_{t+1}=l, Y_t=j)}{\mathbf{P}(Y_t=j)} = p_{ik}p_{jl}. \tag{2.39}$$

By the Proposition we can find an $N=N(i,j,k,l)$ such that $p_{ik}^{(n)}>0$ and $p_{jl}^{(n)}>0$ for all $n>N$. Thus $\tilde{p}_{ij,kl}^{(n)}>0$ and Z is therefore irreducible. □

Let $f_{ij}^{(n)}=\mathbf{P}^i(T_j=n)$ be the **first passage distribution** from state i to state j. We have $f_{ij}^{(0)}=0$ and

$$f_{ij}^{(n)} = \mathbf{P}(X_n=j, X_k \neq j, k=1,..,n-1 \mid X_0=i). \tag{2.40}$$

Define $f_{ij}=\sum_{n=0}^{\infty} f_{ij}^{(n)}=\mathbf{P}^i(T_j<\infty)$. The state i is called **persistent** (also called **recurrent** by some authors) if $f_{ii}=1$, **transient** otherwise. Think of a persistent state as one that the process will eventually return to, while a transient state is one with positive probability of no return.

Theorem 2.3 A state i is persistent iff $\sum_{n\geq 1} p_{ii}^{(n)} = \infty$.

Proof If i is transient, let M be the number of returns to i. Then $M=\sum_n 1(X_n=i)$. By the strong Markov property (2.28) $\mathbf{P}^i(M\geq k)=f_{ii}^k$, so $\mathbf{E}^iM=\sum_{k\geq 1}\mathbf{P}^i(M\geq k)=f_{ii}/(1-f_{ii})$, where $\mathbf{E}^i$ is expectation with respect to $\mathbf{P}^i$. Since $f_{ii}<1$, $\mathbf{E}^iM<\infty$. But

$$\infty > \mathbf{E}^iM = \sum_{n=1}^{\infty} \mathbf{E}^i 1(X_n=i) = \sum_{n=1}^{\infty} p_{ii}^{(n)}. \tag{2.41}$$

Conversely, if i is persistent it returns with probability 1. By the strong Markov property it starts over again, and hence returns with probability one. Thus it returns an infinite number of times with probability one, so $\mathbf{P}^i(M=\infty)=1$, i.e., $\mathbf{E}^iM=\infty$. □

Remark The proof of Theorem 2.3 shows that $M-1$ has a geometric distribution with parameter f_{ii}.

Example (A simple random walk) A **simple random walk** is a birth and death chain on the integers with $p_j\equiv p$, $r_j\equiv 0$ and $q_j\equiv q$, so $q=1-p$. This is an irreducible Markov chain with countable state space. One interpretation is the chain corresponding to the number of heads in successive tosses of a coin with probability p of heads. We will look at the persistence or transience of state 0. A binomial computation shows that

$$p_{00}^{(2n)} = \binom{2n}{n} p^n q^n \sim \frac{(4pq)^n}{(\pi n)^{\frac{1}{2}}} \tag{2.42}$$

using Stirling's formula $n! \sim n^{n+\frac{1}{2}} e^{-n} (2\pi)^{\frac{1}{2}}$. Hence $\sum p_{00}^{(2n)} = \infty$ iff $p = q = \frac{1}{2}$. In other words, 0 is a persistent state iff the coin is fair. □

Remark One can define a simple random walk in higher dimensions by requiring that at any point on the k-dimensional integer lattice the process has the same probabilities of going to its nearest neighbors, regardless of which point it is at. The process is **fair** if the probability is the same to go to each of its neighbors. A similar computation (Exercise **4**) to the one in the example above shows that if $k=2$ the origin (and thus any state) is persistent. However, if $k>2$ it is transient. In three dimensions, with probability 1/6 of going up, down, east, west, north, or south, we get

$$\begin{aligned} p_{00}^{(2n)} &= \left[\frac{1}{6}\right]^{2n} \sum_{j+k\leq n} \frac{2n!}{(j!)^2 (k!)^2 (n-j-k)!^2} \\ &= \left[\frac{1}{2}\right]^{2n} \binom{2n}{n} \sum_{j+k\leq n} \left[\frac{1}{3^n} \frac{n!}{j!k!(n-j-k)!}\right]^2 \\ &\leq \left[\frac{1}{2}\right]^{2n} \binom{2n}{n} \max_{j+k\leq n} \frac{1}{3^n} \frac{n!}{j!k!(n-j-k)!} \sum_{j+k\leq n} \frac{1}{3^n} \frac{n!}{j!k!(n-j-k)!}. \end{aligned} \tag{2.43}$$

The sum is one, being the sum of all probabilities in a trinomial distribution with probability ⅓ of each category, and the maximum is obtained when $j=k=(n-j-k)=n/3$ (or as close as possible to this if n is not divisible by 3). Applying Stirling's formula we see that an upper bound to $p_{00}^{(2n)}$ is, to within an order of n^{-2},

$$2^{-2n} \times \frac{2^{2n}}{\sqrt{n\pi}} \times 3^{-n} \frac{3^{n+3/2}}{2\pi n} = \frac{1}{2} \left[\frac{3}{\pi n}\right]^{3/2}, \tag{2.44}$$

so $\sum p_{00}^{(2n)} < \infty$, whence the walk has positive probability not to return. In fact, the probability of return is about 0.35. In other words, the three-dimensional lattice is a huge place, in which it is easy to get lost. We return to more general random walks in section 2.10. □

Corollary For a transient state i, $p_{ii}^{(n)} \to 0$.

Proof Immediate since $\sum p_{ii}^{(n)} < \infty$. □

For a persistent state $f_{ii}^{(n)}$ is a probability distribution with mean $\mu_i = \sum_n n f_{ii}^{(n)}$, the **mean recurrence time**. If $\mu_i = \infty$, state i is called **null**, otherwise it is called **positive**. This somewhat puzzling nomenclature will be explained in section 2.5. An irreducible aperiodic positive chain is called **ergodic**.

Application (Snoqualmie Falls precipitation, continued) For $n \geq 2$

$$\begin{aligned} f_{11}^{(n)} &= \mathbf{P}^1(X_n=1, X_{n-1}=0, \ldots, X_1=0) \\ &= \mathbf{P}^1(X_n=1 \mid X_{n-1}=0, \ldots, X_1=0)\mathbf{P}^1(X_{n-1}=0, \ldots, X_1=0) \quad (2.45) \\ &= p_{01}(1-p_{11})(1-p_{01})^{n-2}. \end{aligned}$$

Also, $f_{11}^{(1)} = p_{11}$, so the mean recurrence time is $\mu_1 = \sum k f_{11}^{(k)} = 1 + (1-p_{11})/p_{01}$, which we estimate to be 1.42, using $\hat{p}_{01} = 0.398$ and $\hat{p}_{11} = 0.834$. Given a wet day, the mean number of dry days to follow is $\mu_1 - 1$, which we estimate to be 0.42 days. This is a weighted average of wet days inside a wet spell (with no dry days following) and starts of dry spells (with mean duration $1/p_{01}$; cf. Exercise **2**). The variance of the recurrence time is $(1-p_{11})(1-p_{01})/p_{01}^2$. Plugging in the estimated transition probabilities and taking the square root we compute a standard deviation of 0.79. Looking at the actual data, eliminating dry periods that overlap Jan. 1 or 31, the mean dry spell length is 2.21, with a standard deviation of 1.64. In order to compare this to the model estimate of $\mu_1 - 1$, we must multiply by the observed proportion of wet–dry transitions, or 0.166, yielding 0.37, only slightly below the model estimate. □

Theorem 2.4 Persistence is an equivalence class property.

Proof Let $i \leftrightarrow j$ and assume that j is persistent. Then there are integers n and m so that $p_{ij}^{(n)} > 0$ and $p_{ij}^{(m)} > 0$. For any $s \geq 0$

$$p_{ii}^{(n+m+s)} \geq p_{ij}^{(n)} p_{jj}^{(s)} p_{ji}^{(m)} \qquad (2.46)$$

so

$$\sum_s p_{ii}^{(n+m+s)} \geq p_{ij}^{(n)} p_{ji}^{(m)} \sum_s p_{jj}^{(s)} = \infty, \qquad (2.47)$$

and the result follows from Theorem 2.3. □

Let $a_0, a_1, \ldots$ be a sequence of real numbers. If $A(s)=\sum_{k=0}^{\infty} a_k s^k$ converges in some interval $|s|<s_0$ we call $A(s)$ the **generating function** of (a_i). It is easy to show that if $\sum a_k<\infty$ then $A(1-)=\lim_{s\uparrow 1} A(s)=\sum_k a_k$, and for $a_k \geq 0$, if $A(1-)=a<\infty$, then $\sum a_k=a$.

Example (Probability generating functions) If $(p_k; k\geq 0)$ is a probability distribution, the generating function $P(s)=\sum_k^{\infty} p_k s^k$ converges for all $|s|\leq 1$. P is called a **probability generating function** (pgf). If X has pgf P, define the kth **factorial moment** $m_{(k)}=\mathbf{E}X(X-1)\cdots(X-k+1)$. By differentiating under the summation sign in the definition of P we see that

$$m_{(k)} = \sum_{i=k}^{\infty} i(i-1)\cdots(i-k+1)p_k = \frac{d^k}{ds^k}P(1-). \tag{2.48}$$

A similar computation shows that we can recover the probabilities from either the pgf or the factorial moments:

$$p_k = \frac{d^k}{ds^k}\frac{P(s)}{k!}\Big|_{s=0} = \sum_{i=k}^{\infty}(-)^{i-k}\frac{m_{(i)}}{k!(i-k)!}. \tag{2.49}$$

□

Given two sequences (a_k) and (b_k) with generating functions A and B, respectively, we define the **convolution** of the sequences as the sequence (c_k) given by

$$c_k = \sum_{i=0}^{k} a_i b_{k-i}. \tag{2.50}$$

It is easy to see that (c_k) has generating function $C(s)=A(s)B(s)$.

Example (Probability generating functions, continued) If $X_1, \ldots, X_n$ are iid positive random variables with pgf P(s), then the sum $S_n=\sum_1^n X_i$ has pgf $P(s)^n$. For example, $\mathbf{P}(S_n=0)=P(0)^n=p_0^n$, and

$$\mathbf{E}S_n = \frac{dP(s)^n}{ds}\Big|_{s=1} = nP'(1)\,P(1)^{n-1} = n\mathbf{E}X \tag{2.51}$$

since $P(1)=1$. □

Recall from Proposition 2.4 that

$$p_{ii}^{(n)} = \sum_{k=0}^{n} f_{ii}^{(k)} p_{ii}^{(n-k)} \tag{2.52}$$

for any $n\geq 1$. Define the generating functions

$$P_{ij}(s) = \sum_{n=0}^{\infty} p_{ij}^{(n)} s^n \tag{2.53}$$

and

$$F_{ij}(s) = \sum_{n=0}^{\infty} f_{ij}^{(n)} s^n. \tag{2.54}$$

Then, noting that (2.52) is a convolution, we see that

$$F_{ii}(s)P_{ii}(s) = P_{ii}(s) - 1 \tag{2.55}$$

since $p_{ii}^{(0)}=1$. Thus

$$P_{ii}(s) = \frac{1}{1-F_{ii}(s)}. \tag{2.56}$$

Likewise

$$P_{ij}(s) = F_{ij}(s)P_{jj}(s). \tag{2.57}$$

It is worth noting that

$$F_{ii}'(1-) = \mu_i. \tag{2.58}$$

Remark We can use this to give an alternative proof of the result in Theorem 2.3 that i is persistent iff $\sum p_{ii}^{(n)}=\infty$. Assume first that $\sum f_{ii}^{(n)}=1$. Then $F_{ii}(1-)=1$ so $P_{ii}(1-)=\infty$, or $\sum p_{ii}^{(n)}=\infty$. Conversely, if $\sum f_{ii}^{(n)}<1$ we have that $F_{ii}(1-)<1$, so $P_{ii}(1-)<\infty$, and $\sum p_{ii}^{(n)}<\infty$. We can interpret $P_{ii}(1-)$ as the expected number of visits to i, starting from i. □

Example (Coin-tossing) The computation above can be modified to show that for the fair coin-tossing random walk, state 0 (and hence any state) is null persistent. Since $F_{00}(s)=1-P_{00}(s)^{-1}$ we see that $F_{00}'(1-)=P_{00}'(1-)/P_{00}^2(1-)$. Now notice that

$$A_N = \sum_{n=0}^{N} np_{00}^{(n)} = O(N^{3/2}), \tag{2.59}$$

$$B_N = \sum_{n=0}^{N} p_{00}^{(n)} = O(N^{1/2}), \tag{2.60}$$

and $A_N \to P_{00}'(1-)$, $B_N \to P_{00}(1-)$, so that

$$F_{00}'(1-) = \lim_{N\to\infty} \frac{A_N}{B_N^2} = \lim_{N\to\infty} O(N^{1/2}) = \infty, \tag{2.61}$$

showing that the mean recurrence time is infinite. □

Lemma 2.3 Suppose that j is persistent. Then it is positive persistent iff $\pi_j = \lim_{s\uparrow 1}(1-s)P_{jj}(s) > 0$, and then $\pi_j = 1/\mu_j$.

Proof From (2.56) we have

$$1 - F_{jj}(s) = P_{jj}(s)^{-1} \tag{2.62}$$

so

$$\frac{1-F_{jj}(s)}{1-s} = \frac{1}{(1-s)P_{jj}(s)}. \tag{2.63}$$

But the left-hand side of (2.63) converges to $F_{jj}'(1) = \mu_j$ as $s\uparrow 1$. Hence the limit of the right-hand side is the same, and the result follows. □

Remark This almost proves the convergence of averages of transition probabilities, namely

$$(1/n)\sum_1^n p_{jj}^{(k)} \to \mu_j^{-1} \text{ as } n \to \infty. \tag{2.64}$$

Consider

$$(1-s)P_{jj}(s) = \frac{\sum_{k=0}^{\infty} s^k p_{jj}^{(k)}}{\sum_{k=0}^{\infty} s^k} \tag{2.65}$$

for any $s \le 1$. If we could take the limit as $s\uparrow 1$ under the summations, the right-hand side would converge to $(1/n)\sum_1^n p_{jj}^{(k)}$, while the left-hand side would converge to $1/\mu_j$ by the Lemma. There is, however, no elementary result allowing us to take the limit under the summation sign. We need a so-called **Tauberian** theorem, such as that given in Feller (1971, Theorem 5 in section XIII.V). We will be able to deduce the result using less difficult mathematics in section 2.5. □

Call a set C of states **closed** if $f_{jk} = F_{jk}(1-) = 0$ for $j \in C$, $k \notin C$. Then $P_{jk}(1-) = 0 = \sum p_{jk}^{(n)}$ and we must have $p_{jk}^{(n)} = 0$ for all n. In fact, in order to verify that a set of states is closed we need only show that $p_{jk} = 0$ for $j \in C$, $k \notin C$, since, e.g.,

$$p_{jk}^{(2)} = \sum_{s\in S} p_{js}p_{sk} = \sum_{s \in C} p_{js}p_{sk} = 0, \tag{2.66}$$

the case for general n following by induction. If C is closed and the process starts in C it will never leave it. An absorbing state is closed. We call a set C of states **irreducible** if $x \leftrightarrow y$ for all $x,y \in C$. The irreducible closed sets are precisely the equivalence classes under $\leftrightarrow$. If A and B are disjoint sets we write $A+B$ for their union.

Theorem 2.5 If S_T={transient states} and S_P={persistent states}, we have that

$$S = S_T + S_P \tag{2.67}$$

and

$$S_P = \sum C_i \text{ of disjoint, irreducible, closed sets.} \tag{2.68}$$

Proof Let $x \in S_P$, and define $C=\{y \in S_P : x \to y\}$. By persistence $f_{xx}=1$, so $x \in C$. We first show that C is closed. Suppose that $y \in C$, $y \to z$. Since y is persistent, z must also be persistent. Since $x \to y \to z$ we have $z \in C$ so that C is closed.

Next we show that C is irreducible. Choose y and z in C. We need to show that $z \leftrightarrow y$. Since $x \to y$, $y \to x$ by persistence. But $x \to z$ by definition of C, so $y \to x \to z$. The same argument, with y and z transposed, shows that $z \to y$.

Now let C and D be irreducible closed subsets of S_P, and let $x \in C \cap D$. Take $y \in C$. Since C is irreducible, $x \to y$. Since D is closed, $x \in D$, and $x \to y$ we have that $y \in D$. Thus $C \subset D$. Similarly $D \subset C$, so they are equal. □

It follows from this theorem that if a chain starts in C_i it will stay there forever (and we may as well let $S=C_i$). On the other hand, if it starts in S_T it either stays there forever, or moves into one of the C_i in which it stays forever.

Theorem 2.6 Within a persistent class either all the μ_i are finite or all are infinite.

Proof As before we can find k, m such that $p_{ij}^{(k)}>0$, $p_{ji}^{(m)}>0$. Since

$$p_{jj}^{(n+k+m)} \geq p_{ji}^{(m)} p_{ii}^{(n)} p_{ij}^{(k)} \tag{2.69}$$

we see by averaging and anticipating the result (2.64) that

$$\pi_j \geq p_{ji}^{(m)} \pi_i p_{ij}^{(k)}, \tag{2.70}$$

so if $\pi_i>0$ then $\pi_j>0$, while if $\pi_j=0$ then $\pi_i=0$. The converse obtains by interchanging i and j in the argument. □

Proposition 2.5 If S is finite, then at least one state is persistent, and all persistent states are positive.

Proof Assume that all states are transient. Then $\sum_j p_{ij}^{(n)}=1$ for all n. In particular,

$$\lim_{n\to\infty}\sum_{j\in S} p_{ij}^{(n)} = 1.$$

But from the corollary to Theorem 2.3, we have that each term in the sum, and therefore the entire sum, goes to zero. Hence at least one state is persistent. Assume that one such is state j. Consider $C_j = \{i:j\to i\}$. According to Theorem 2.5, once the process the process enters C_j it will stay there forever. For every $i\in C_j$ we can find a finite n with $p_{ij}^{(n)}>0$. For $i\neq j$ let ν_i denote the expected number of visits to i between two visits to j, i.e.,

$$\nu_i = \mathbf{E}^j \sum_{n=0}^{T_j-1} 1(X_n=i) = \sum_{n=0}^{\infty} \mathbf{P}^j(X_n=i,T_j>n). \tag{2.71}$$

Define $\nu_j=1$ in accordance with the definition of ν_i. Let $i\neq j$, and note that $\{X_n=i,T_k>n\}$ then is the same as $\{X_n=i,T_j>n-1\}$. Hence compute

$$\begin{aligned}
\nu_i &= \sum_{n=1}^{\infty} \mathbf{P}^j(X_n=i,T_j>n) = \sum_{n=1}^{\infty} \mathbf{P}^j(X_n=i,T_j>n-1)\\
&= \sum_{n=1}^{\infty}\sum_{k\in S} \mathbf{P}^j(X_n=k,T_j>n-1,X_{n-1}=i)\\
&= \sum_{n=1}^{\infty}\sum_{k\in S} \mathbf{P}^j(X_n=k \mid T_j>n-1,X_{n-1}=k)\,\mathbf{P}^j(T_j>n-1,X_{n-1}=k) \qquad (2.72)\\
&= \sum_{k\in S} p_{kj} \sum_{n=1}^{\infty} \mathbf{P}^j(X_n=k,T_j>n-1) = \sum_{k\in S}\sum_{m=0}^{\infty} \mathbf{P}^j(X_m=k,T_j>m)\\
&= \sum_{k\in S} p_{kj}\nu_k.
\end{aligned}$$

Since C_j is closed, the sum over $k\in S$ only has contributions from the states in C_j. For $i=j$ we have, since j is persistent, that

$$\begin{aligned}
\nu_j \equiv 1 &= \sum_{n=1}^{\infty} \mathbf{P}^j(T_j=n) = \sum_{n=1}^{\infty}\sum_{k\in S} \mathbf{P}^j(T_j=n,X_{n-1}=k)\\
&= \sum_{n=1}^{\infty}\sum_{k\in S} \mathbf{P}^j(T_j>n-1,X_n=j,X_{n-1}=k) \qquad (2.73)\\
&= \sum_{n=1}^{\infty}\sum_{k\in S} p_{kj}\mathbf{P}^j(T_j>n-1,X_n=k) = \sum_{k\in S} p_{kj}\nu_k.
\end{aligned}$$

Writing $\mathbf{v}=(\nu_1,\nu_2,...)$ we have shown that $\mathbf{v}\mathbb{P}=\mathbf{v}$. By iterating we see that $\mathbf{v}\mathbb{P}^n=\mathbf{v}$ for all $n=1,2,\cdots$ In particular, for $i\in C_j$,

$$\nu_i p_{ij}^{(n)} \le \nu_j \equiv 1, \tag{2.74}$$

so $\nu_i \le 1/p_{ij}^{(n)} < \infty$ for some n. Finally we compute

$$\mu_j = \sum_{n=0}^{\infty} \mathbf{P}^j(T_j > n) = \sum_{i \in S} \sum_{n=0}^{\infty} \mathbf{P}^j(X_n = i, T_j > n) = \sum_{i \in S} \nu_i. \tag{2.75}$$

Again, the sum over $i \in S$ is really only over $i \in C_j$. Since S is finite, C_j must be finite, and we can pick n so large that $\nu_i < \infty$ for all i in C_j. The final sum in (2.75) therefore is a finite sum of finite elements, so $\mu_j < \infty$. □

Proposition 2.6 If i is a null persistent state, then $p_{ii}^{(n)} \to 0$ as $n \to \infty$.

This result was first proved by Erdös, Feller and Pollard (1949) using generating function techniques. The details are somewhat involved, and not of a probabilistic nature, so we shall omit the proof which can be found, e.g., in Feller (1968, sec. XIII.11). Incidentally, this proposition explains how null persistent states were named. Correspondingly, positive persistent states have $p_{ii}^{(n)} > 0$ for all n large enough.

2.4. Stationary distribution

A large portion of the theory of stochastic processes focuses on processes that have marginal distributions that are not time-dependent. Looking back at equation (2.16) we see that if we choose $p_1 = p_{01}/(1-(p_{11}-p_{01}))$, we obtain a marginal distribution which is independent of n, and simply equal to the initial distribution. We will denote such a distribution (when it exists) by $\boldsymbol{\pi}$. By letting $\mathbf{p}_n = \boldsymbol{\pi}$ in relation (2.20) we must have $\boldsymbol{\pi} = \boldsymbol{\pi}\mathbb{P}$, or equivalently

$$\boldsymbol{\pi}(\mathbb{I} - \mathbb{P}) = 0. \tag{2.76}$$

Thus $\boldsymbol{\pi}$ is a left eigenvector of $\mathbb{P}$, corresponding to the eigenvalue 1 (recall from Proposition 2.2 that such an eigenvalue always exists). The solution to (2.76) is called the **stationary distribution** of the Markov chain. If $S = \{0,1\}$ we saw that $\pi_1 = p_{01}/(1-(p_{11}-p_{01}))$. Thus if $p_{11} = p_{01}$, so that the occurrence of state 1 is independent of the previous state, then $\pi_1 = p_{01}$. Otherwise π_1 is between the smaller and the larger of p_{01} and p_{11}. To see that this choice of $\boldsymbol{\pi}$ indeed satisfies (2.76), note that

$$\left[\frac{1-p_{11}}{1-(p_{11}-p_{01})}, \frac{p_{01}}{1-(p_{11}-p_{01})} \right] \begin{bmatrix} p_{01} & -p_{01} \\ -(1-p_{11}) & 1-p_{11} \end{bmatrix} = (0,0). \tag{2.77}$$

When we use the stationary distribution as initial distribution we see that

$$\begin{aligned} \mathbf{p}_1 &= \boldsymbol{\pi}\mathbb{P} = \boldsymbol{\pi} \\ \mathbf{p}_2 &= \mathbf{p}_1\mathbb{P} = \boldsymbol{\pi}, \end{aligned} \tag{2.78}$$

etc., so that, indeed,

$$\mathbf{p}_n = \boldsymbol{\pi} \quad \text{for all } n. \tag{2.79}$$

We then say that the one-dimensional distributions $\mathbf{p}_n$ are **time invariant** (another name for this is stationary). Therefore $\boldsymbol{\pi}$ is also known as the **stationary initial distribution**. In fact, starting from $\boldsymbol{\pi}$ all finite-dimensional distributions are time invariant, in the sense that

$$(X_{k_1},X_{k_2},\ldots,X_{k_n}) \stackrel{d}{=} (X_{k_1+k},X_{k_2+k},\ldots,X_{k_n+k}) \tag{2.80}$$

for all non-negative integers n, k, $k_1,\ldots,k_n$ (Exercise **5**). Processes satisfying (2.80) are called **strictly stationary**.

The strength of the dependence in a Markov chain can be computed from the transition matrix. By repeated conditioning we see that

$$\mathbf{E}X_nX_{n+k} = \mathbf{E}(\mathbf{E}(X_{n+k}\mid X_n)X_n) = \sum_{j,l\in S} jlp_{lj}^{(k)}p_n(l). \tag{2.81}$$

If the chain is strictly stationary the right-hand side of (2.81) simplifies to $\sum jlp_{lj}^{(k)}\pi_l$, so the covariance between X_n and X_{n+k} is

$$\mathbf{Cov}(X_n,X_{n+k}) = \sum_{j,l\in S} jlp_{lj}^{(k)}\pi_l - (\sum_{j\in S} j\pi_j)^2. \tag{2.82}$$

Anticipating the next section we see that if the chain has a limiting distribution, so $p_{lj}^{(k)}\to\pi_j$, the covariance goes to zero as k goes to infinity.

Application (Snoqualmie Falls precipitation, continued) In the 0-1 case the sums in (2.82) only have one term. The correlation function for a two-state Markov chain thus becomes

$$\mathbf{Corr}(X_n,X_{n+k}) = \frac{p_{11}^{(k)}-\pi_1}{1-\pi_1} = (p_{11}-p_{01})^k \tag{2.83}$$

using the following induction argument. If $k=1$ we have

$$\frac{p_{11}-\pi_1}{1-\pi_1} = \frac{p_{11}(1-(p_{11}-p_{01}))-p_{01}}{1-p_{11}} = p_{11}-p_{01} \tag{2.84}$$

as required. Assuming the formula (2.83) is correct for $k=n$, then we can write it as

$$p_{11}^{(n)} = \pi_1 + (1-\pi_1)(p_{11}-p_{01})^n. \tag{2.85}$$

Since $p_{11}^{(n+1)}=(1-p_{11}^{(n)})p_{01}+p_{11}^{(n)}p_{11}$ we have

$$\begin{aligned}\frac{p_{11}^{(n+1)}-\pi_1}{1-\pi_1} &= \frac{p_{01}+(1-\pi_1)(p_{11}-p_{01})^{n+1}+\pi_1(p_{11}-p_{01})-\pi_1}{1-\pi_1}\\ &= (p_{11}-p_{01})^{n+1} + \frac{p_{01}-\pi_1(1-p_{11}+p_{01})}{1-\pi_1}\\ &= (p_{11}-p_{01})^{n+1}\end{aligned} \tag{2.86}$$

where the last equality is from the definition of π_1. This completes the induction. We see that the correlation function is geometrically decreasing. For the Snoqualmie Falls data, where $\hat{p}_{11}-\hat{p}_{01}$ is 0.436, we get the estimated correlations given in Table 22.

Table 2.2 Estimated correlations for Snoqualmie Falls data

Lag	1	2	3	4	5	6	7
Corr	0.436	0.190	0.083	0.036	0.016	0.007	0.003

□

Equation (2.76) shows that if a chain has a stationary distribution, it must be the eigenvector of $\mathbb{P}$ corresponding to the eigenvalue 1. Sometimes we can be a bit more explicit. Recall that if k is persistent, then ν_j is the expected number of visits to j before returning to k.

Lemma 2.4 An irreducible positive persistent chain has a stationary distribution given by $\pi_i=\nu_i/\mu_k$ for a fixed state k.

Proof We need to show that $\boldsymbol{\pi}\mathbb{P}=\boldsymbol{\pi}$, i.e., that $\sum_{i\in S}\nu_i p_{ij}=\nu_j$ and that $\sum_{i\in S}\nu_i=\mu_k$ (in order for $\boldsymbol{\pi}$ to be a probability distribution). But this was established in the proof of Proposition 2.5 (without using the assumption of a finite state space). □

Which Markov chains have a stationary distribution? The answer is quite simple. We will restrict attention to irreducible chains, since any other chain can be be decomposed into irreducible subclasses. The quantity $\pi_i=1/\mu_i$, which arose in our criterion for positive persistence in Lemma 2.3, now assumes a more important role.

Theorem 2.7 An irreducible chain has a stationary distribution if and only if it is positive persistent. The stationary distribution is unique and given by $\pi_i=\mu_i^{-1}$.

Proof Suppose that $\boldsymbol{\pi}$ is a stationary distribution and the chain is transient or null persistent. Then $p_{ij}^{(n)}\to 0$ as $n\to\infty$ by the corollary to Theorem 2.3 and by Proposition 2.6, respectively. Hence for any i and j, if we are allowed to take limits under the summation sign,

$$\pi_j = \sum_{i\in S}\pi_i p_{ij}^{(n)} \to 0 \text{ as } n\to\infty \tag{2.87}$$

so $\boldsymbol{\pi}$ is not a distribution. To see that this argument is valid, let (S_m) be a sequence of finite subsets of S, such that $S_m\uparrow S$ as $m\to\infty$. Then

$$\pi_j = \sum_{i\in S_m} \pi_i p_{ij}^{(n)} + \sum_{i\notin S_m} \pi_i p_{ij}^{(n)} \leq \sum_{i\in S_m} \pi_i p_{ij}^{(n)} + \sum_{i\notin S_m} \pi_i. \tag{2.88}$$

For each m the first term goes to zero as $n\to\infty$, since we can always take the elementwise limit of a finite sum. The second term can be made arbitrarily small by taking m large, since $\boldsymbol{\pi}$ is summable. The key to this argument is that the $p_{ij}^{(n)}$ are bounded. Thus the existence of a stationary distribution implies that all states are persistent.

Let the initial distribution be $\boldsymbol{\pi}$. Using Exercise **5**, all finite-dimensional probabilities are time invariant. Thus

$$\begin{aligned}\mathbf{P}^{\pi}(X_n=i,T_j\geq n+1) &= \mathbf{P}^{\pi}(X_n=i,X_1,\ldots,X_n\neq j)\\ &= \mathbf{P}^{\pi}(X_{n-1}=i,X_0,\ldots,X_{n-1}\neq j) \qquad (2.89)\\ &= \mathbf{P}^{\pi}(X_{n-1}=i,T_j\geq n) - \mathbf{P}^{\pi}(X_{n-1}=i,X_0=j,T_j\geq n)\\ &= \mathbf{P}^{\pi}(X_{n-1}=i,T_j\geq n) - \pi_j\mathbf{P}^{j}(X_{n-1}=i,T_j\geq n).\end{aligned}$$

Summing over $n\leq N$ and rearranging terms yields

$$\begin{aligned}\sum_{n=1}^{N}\pi_j\mathbf{P}^{j}(X_{n-1}=i,T_j\geq n) &= \sum_{n=1}^{N}(\mathbf{P}^{\pi}(X_{n-1}=i,T_j\geq n) - \mathbf{P}^{\pi}(X_n=i,T_j\geq n+1))\\ &= \pi_i - \mathbf{P}^{\pi}(X_N=i,T_j\geq N+1) \qquad (2.90)\end{aligned}$$

since the sum telescopes. Letting $N\to\infty$ the last term on the right-hand side of (2.90) disappears since j is ergodic, and we get

$$\sum_{n=1}^{\infty}\pi_j\mathbf{P}^{j}(X_{n-1}=i,T_j\geq n) = \pi_i. \tag{2.91}$$

Summing (2.91) over all $i\in S$ we see that

$$\pi_j\sum_{n=1}^{\infty}\mathbf{P}^{j}(T_j\geq n) = 1. \tag{2.92}$$

But $\sum P^j(T_j\geq n)=\mu_j$ and we see that $\pi_j\mu_j=1$. Suppose that $\pi_i=0$. Then

$$0 = \pi_i = \sum_j \pi_j p_{ji}^{(n)} \geq \pi_j p_{ji}^{(n)}, \tag{2.93}$$

so whenever $j\to i$ we have $\pi_j=0$. But then all the π_i are 0 by irreducibility, and $\boldsymbol{\pi}$ is not a distribution. Hence all the π_i are positive, so $\mu_j<\infty$. Therefore, the existence of a stationary distribution for an irreducible chain implies that it is uniquely given by $\pi_i=\mu_i^{-1}$, and that the chain must be positive persistent. Conversely, for a positive persistent chain the distribution in Lemma 2.4 is a stationary distribution. □

We can now evaluate the expected number of visits to i between successive visits to k.

Corollary $$\nu_i = \mathbf{E}^k \sum_{n=0}^{T_k-1} 1(X_n=i) = \mu_k/\mu_i = \pi_i/\pi_k$$

Proof This follows directly from Lemma 2.4 and the uniqueness in Theorem 2.7. □

The equation (2.76) for the stationary distribution can be written

$$\pi_j = \sum_{i\in S} \pi_j p_{ji} = \sum_{i\in S} p_{ij}\pi_i. \tag{2.94}$$

We can interpret $\sum_i \pi_j p_{ji}$ as the probability flux out of state j, and $\sum_i \pi_i p_{ij}$ as the probability flux into state j. Consider a large number of independent particles following the same Markov chain. Then, if the system is in equilibrium, the number of particles moving into and out of state i at any time should be approximately the same. In other words, the proportion of particles moving out (the flux out of the state) should be the same as the proportion of particles moving in (the flux into the state). In this interpretation, it is natural to think of (2.94) as an equation of **full balance**.

Many physical systems, obeying classical mechanics, have a physical description that is symmetric with respect to past and future. In the context of stochastic processes, the corresponding requirement is that the probabilistic structure of the process run forward in time must be the same as the structure of the process run backward in time.

Let $(X_k, k\in \mathbf{Z})$ be an ergodic chain, defined for both positive and negative time. We may consider the chain Y defined by $Y_k = X_{-k}$. Then Y is a Markov chain, although not necessarily with stationary transition probabilities:

$$\begin{aligned}
&\mathbf{P}(Y_{k+1}=j \mid Y_k=i, Y_{k-1}=i_1, \ldots, Y_{k-n}=i_n) \\
&= \mathbf{P}(X_{-(k+1)}=j \mid X_{-k}=i, X_{-(k-1)}=i_1, \ldots, X_{-(k-n)}=i_n) \\
&= \frac{\mathbf{P}(X_{-(k+1)}=j, X_{-k}=i, X_{-(k-1)}=i_1, \ldots, X_{-(k-n)}=i_n)}{\mathbf{P}X_{-k}=i, X_{-(k-1)}=i_1, \ldots, X_{-(k-n)}=i_n)} \\
&= \frac{\mathbf{P}(X_{-(k-n)}=i_n, \ldots, X_{-(k-1)}=i_1 \mid X_{-k}=i)\mathbf{P}(X_{-k}=i \mid X_{-(k+1)}=j)}{\mathbf{P}(X_{-(k-n)}=i_n, \ldots, X_{-(k-1)}=i_1 \mid X_{-k}=i)} \\
&\times \frac{\mathbf{P}(X_{-(k+1)}=j)}{\mathbf{P}(X_{-k}=i)} = p_{ji}\frac{p_j^{(-(k+1))}}{p_i^{(-k)}}.
\end{aligned} \tag{2.95}$$

If X has the stationary marginal distribution $\boldsymbol{\pi}$ for all k we see that Y has stationary transition probabilities q_{ij} given by

$$q_{ij} = \mathbf{P}(Y_{k+1}=j \mid Y_k=i) = p_{ji}\frac{\pi_j}{\pi_i}. \tag{2.96}$$

Note that since X is defined for all $k \in \mathbf{Z}$, it is not enough to set $\mathbf{p}^{(0)}=\boldsymbol{\pi}$ in order to have marginal distribution $\boldsymbol{\pi}$ for all k. This only works for $k \geq 0$; e.g., X_{-1} does not necessarily yield the right distribution. We call X **reversible** if X and Y have the same transition matrix, i.e. if

$$p_{ij} = p_{ji}\frac{\pi_j}{\pi_i} \tag{2.97}$$

or, equivalently,

$$\pi_i p_{ij} = p_{ji}\pi_j. \tag{2.98}$$

This is called the law of **detailed balance**, stating that the probability flux from i to j in equilibrium is the same as that from j to i. Detailed balance is a property of isolated systems in both classical and quantum mechanics. It was first noted in chemical reaction kinetics. A proof of the detailed balance property for closed classical systems is in Van Kampen (1981, section V.6). The conditions of detailed balance can sometimes be used to find the stationary distribution of a chain.

Theorem 2.8 If, for an irreducible Markov chain, a distribution $\boldsymbol{\pi}$ exists, satisfying the law of detailed balance (2.98) for all $i,j \in S$, then the chain is reversible and positive persistent with stationary distribution $\boldsymbol{\pi}$.

Proof Using Theorem 2.7 we need only show that $\boldsymbol{\pi}$ is a stationary distribution. But

$$\sum_i \pi_i p_{ij} = \sum_i \pi_j p_{ji} = \pi_j \sum_i p_{ji} = \pi_j, \tag{2.99}$$

so $\boldsymbol{\pi} = \boldsymbol{\pi}\mathbb{P}$. □

Example (A Birth and death chain) Assume that $p_j>0$, $q_{j+1}>0$ for all $j \geq 0$, while $q_0=0$, so that all states communicate. Then we will show that the detailed balance equation

$$p_j\pi_j = q_{j+1}\pi_{j+1} \tag{2.100}$$

holds, and that the equilibrium distribution is given by

$$\pi_j = \pi_0 \prod_{i=1}^{j} \frac{p_{i-1}}{q_i} \tag{2.101}$$

where

$$\pi_0^{-1} = \sum_{j=0}^{\infty} \prod_{i=1}^{j} \frac{p_{i-1}}{q_i} \tag{2.102}$$

provided that the sum converges. This convergence is a condition for the persistence of the chain as well as for its reversibility.

To see that (2.100) holds, note first that the full balance equation for π_j is

$$\pi_j = p_{j-1}\pi_{j-1} + r_j\pi_j + q_{j+1}\pi_{j+1},\ j>0, \tag{2.103}$$

while for $j=0$

$$\pi_0 = r_0\pi_0 + q_1\pi_1. \tag{2.104}$$

Since $q_0=0$, $r_0=1-p_0$ and (2.104) becomes

$$p_0\pi_0 = q_1\pi_1 \tag{2.105}$$

which is the detailed balance equation for $j=0$. Assume that (2.100) holds for $j=k$. From (2.104) we see that

$$\pi_{k+1} = p_k\pi_k + r_{k+1}\pi_{k+1} + q_{k+2}\pi_{k+2}. \tag{2.106}$$

Since by the induction hypothesis $p_k\pi_k=q_{k+1}\pi_{k+1}$ (2.106) becomes

$$p_{k+1}\pi_{k+1} = q_{k+2}\pi_{k+2} \tag{2.107}$$

whence the detailed balance equation holds. The evaluation of π is now immediate.

The particular case of a random walk reflected at the origin has $p_i=1-q_i=p$, so

$$\pi_0^{-1} = \sum\left(\frac{p}{1-p}\right)^j = \frac{1-p}{2-p}, \tag{2.108}$$

provided that $p<½$. The stationary distribution is then geometric. If $p\geq½$ the process is transient. □

Example (The Ehrenfest model for diffusion) Consider two containers, labeled 0 and 1, in contact with each other. We have N molecules that move between the containers. At each time one molecule is chosen at random, and moved to the other container. We can describe this system using a binary number of N digits, for a total of 2^N possible states. The transition probabilities for this **micro-level** process X are

$$p_{x,x'} = \begin{cases} 1/N & \text{if } x \text{ and } x' \text{ only differ in one location} \\ 0 & \text{otherwise.} \end{cases} \tag{2.109}$$

Consider the case $N=2$, so the states are 00, 01, 10, and 11 (or in decimal notation 0,...,3). Then

$$\mathbb{P} = \begin{pmatrix} 0 & ½ & ½ & 0 \\ ½ & 0 & 0 & ½ \\ ½ & 0 & 0 & ½ \\ 0 & ½ & ½ & 0 \end{pmatrix}. \tag{2.110}$$

Clearly $\mathbb{P}$ is doubly stochastic, whence the stationary distribution is $\pi_i=2^{-N}$, $i=0,\ldots,2^N-1$. The chain is periodic with period 3. It satisfies not only the detailed balance equation, but also the stronger **micro-reversibility** condition that

$$p_{x,x'} = p_{x',x} \quad \text{for all } x,x'. \tag{2.111}$$

If we regard the molecules as indistinguishable, we get a **macro-level** description of the process. Let Y_k be the number of molecules in container 0. Then Y_k is a Markov chain with non-zero transition probabilities

$$p_{Y;i,i-1} = \frac{i}{N} \quad p_{Y;i,i+1} = \frac{N-i}{N}. \tag{2.112}$$

Since this is a birth and death chain we know from the previous example that the process is reversible and the stationary distribution satisfies

$$\pi_j = \pi_0 \prod_{i=1}^{j} \frac{N-i+1}{i} = \pi_0 \binom{N}{j}, \tag{2.113}$$

so $\pi_0=2^{-N}$. In other words, the stationary distribution for Y is obtained from that of X by summing over the number of micro-states corresponding to a given macro-state.

This model was introduced by Ehrenfest and Ehrenfest (1906) to explain a paradox in thermodynamics, exposed by Loschmidt (1876). The paradox is that although statistical mechanics can be derived from classical mechanics, the laws of classical mechanics are time-reversible while thermodynamics contains irreversible processes: entropy must increase with time. This physical sense of reversibility would require that for given micro-states x and x', with corresponding macro-states y and y' we have both

$$\mathbf{P}(X_k = x \mid X_0 = x') = \mathbf{P}(X_k = x' \mid X_0 = x) \tag{2.114}$$

and

$$\mathbf{P}(Y_k = y \mid Y_0 = y') = \mathbf{P}(Y_k = y' \mid Y_0 = y). \tag{2.115}$$

If now y is small, and y' is nearly $N/2$, (2.114) holds by micro-reversibility, but (2.115) would not hold. Rather, the right-hand side would be much larger than the left-hand side, because of a tendency for the process to veer towards its stationary mean (we are anticipating the results of the next section here, in that the process in the long run tends towards its stationary distribution). The statistical sense of reversibility involves equilibrium behavior, which the classical mechanics laws do not explicitly mention. Our explanation of the Loschmidt paradox, therefore, will be that the process is not micro-reversible at the macro-level. Chandrasekhar (1943, section III.4) and Whittle (1986) contain more material pertinent to this type of question. □

2.5. Long term behavior

As we have discussed before, many physical systems tend to settle down to an **equilibrium state**, where the state occupation probabilities are independent of the initial probabilities. Recall that when we powered up $\hat{\mathbb{P}}$ for the Snoqualmie Falls precipitation model, the rows got more and more similar. Figure 2.2 illustrates this.

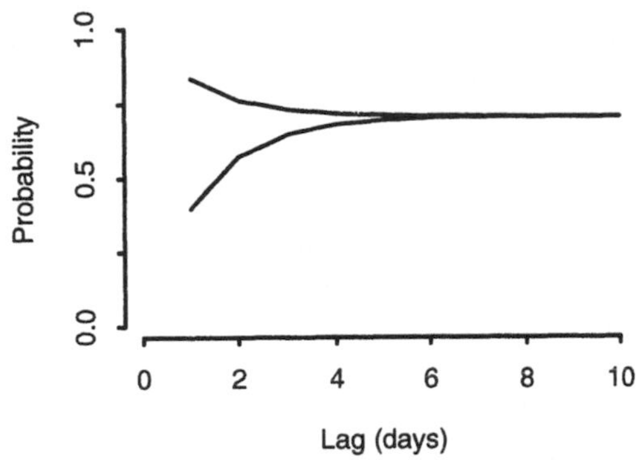

Figure 2.2. n-step transition probabilities for Snoqualmie Falls model. The upper curve is $p_{11}^{(n)}$ while the lower is $p_{01}^{(n)}$.

In fact, under suitable conditions

$$\mathbb{P}^n \to \begin{pmatrix} \boldsymbol{\pi} \\ \boldsymbol{\pi} \\ \cdots \\ \boldsymbol{\pi} \end{pmatrix}. \tag{2.116}$$

We say that the chain has a **limiting distribution**. What this means is that if the chain is left running for a long time, it reaches an equilibrium situation regardless of its initial distribution. In this equilibrium situation the state occupancy probabilities are equal to the stationary distribution. Note namely that

$$\mathbf{p}_n = \mathbf{p}_0 \mathbb{P}_n \to \mathbf{p}_0 \begin{pmatrix} \boldsymbol{\pi} \\ \boldsymbol{\pi} \\ \cdots \\ \boldsymbol{\pi} \end{pmatrix} = \boldsymbol{\pi}$$

regardless of $\mathbf{p}_0$. As the next example shows, there may be a stationary distribution without the chain having a limiting distribution.

Example (A chain without a limiting distribution) Let

$$\mathbb{P} = \begin{bmatrix} 0\,1\,0 \\ 0\,0\,1 \\ 1\,0\,0 \end{bmatrix}. \tag{2.117}$$

Then $\boldsymbol{\pi}=(1,1,1)/3$, but $\mathbb{P}^n$ does not converge. Rather, it cycles through three different matrices. Notice that the period of this chain is 3. This particular $\mathbb{P}$ also is doubly stochastic, and has a uniform stationary distribution (Exercise **6**).

In fact, this trivial example is in many ways typical for what happens in a periodic chain. Assume that the irreducible chain X has period d. Then for every state k we can find integers l,m such that $p_{0k}^{(l)}>0$ and $p_{k0}^{(m)}>0$. Hence $p_{00}^{(l+m)}\geq p_{0k}^{(l)}p_{k0}^{(m)}>0$, so d must divide $l+m$, i.e., $l+m=rd$ for some integer r. Fixing m we see that $l=-m+rd\equiv s+td$ where s is the residue class of m modulo d. Thus we can for every state k find an integer s_k, $0\leq s_k<d$, such that $p_{0k}^{(n)}=0$ unless $n\equiv s_k \bmod d$. Let $G_s=\{k: s_k=s\}$. Then

$$S = G_0+\cdots+G_{d-1}. \tag{2.118}$$

One-step transitions are only possible from states in G_k to states in G_{k+1} (where $G_d\equiv G_0$), and going d steps out of G_k leads back to G_k. Hence for a chain with transition matrix $\mathbb{P}^d$, each G_k is a closed irreducible set. For $d=3$ we have

$$\mathbb{P} = \begin{bmatrix} 0 & A & 0 \\ 0 & 0 & B \\ C & 0 & 0 \end{bmatrix}, \tag{2.119}$$

where A consists of transition probabilities from G_0 to G_1, etc. □

We shall now ascertain the long term behavior of some aspects of a Markov chain. Again we restrict attention to the irreducible case. We start with the asymptotic behavior of n-step transition probabilities.

Theorem 2.9 Let k be an aperiodic state of an irreducible Markov chain with mean recurrence time $\mu_k\leq\infty$. Then

$$\lim_{n\to\infty} p_{kk}^{(n)} = \frac{1}{\mu_k}. \tag{2.120}$$

Proof The transient case is immediate from the corollary to Theorem 2.3, and the null persistent case is Proposition 2.6. To prove the positive persistent case we shall use a technique called **coupling**. Let X and Y be iid copies of the chain, and let $Z=(X,Y)$. Recall from the corollary to Proposition 2.4 that Z is an irreducible Markov chain with transition probabilities $p_{ij,kl}=p_{ik}p_{jl}$. Since X is positive persistent, it has a unique stationary distribution $\boldsymbol{\pi}$. Then Z has stationary distribution $\boldsymbol{\eta}$ with $\eta_{i,j}=\pi_i\pi_j$. Therefore Z is positive persistent by Theorem 2.7. Let $Z_0=(i,j)$, choose $s\in S$, and let $T_{s,s}$ be the hitting time of (s,s) for Z. Since Z is persistent, $\mathbf{P}(T_{s,s}<\infty)=1$. Suppose that $m\leq n$ and that $X_m=Y_m$. Then X_n and Y_n are identically distributed by the strong Markov property. Thus, conditional on $\{T_{s,s}\leq n\}$, the random variables X_n and Y_n have the same conditional distribution. Compute

$$p_{ik}^{(n)} = \mathbf{P}^{i,j}(X_n=k) = \mathbf{P}^{i,j}(X_n=k, T_{s,s}\leq n) + \mathbf{P}^{i,j}(X_n=k, T_{s,s}>n)$$

$$= \mathbf{P}^{i,j}(Y_n=k, T_{s,s}\le n) + \mathbf{P}^{i,j}(X_n=k, T_{s,s}>n) \tag{2.121}$$
$$\le \mathbf{P}^{i,j}(Y_n=k) + \mathbf{P}^{i,j}(T_{s,s}>n) = p_{jk}^{(n)} + \mathbf{P}^{i,j}(T_{s,s}>n).$$

Interchanging i and j, we get a similar result, so that

$$|p_{ik}^{(n)} - p_{jk}^{(n)}| \le \mathbf{P}^{i,j}(T_{s,s}>n) + \mathbf{P}^{j,i}(T_{s,s}>n) \to 0 \text{ as } n\to\infty \tag{2.122}$$

since $\mathbf{P}^{i,j}(T_{s,s}<\infty)=1$ for any i,j. Thus

$$p_{ik}^{(n)} - p_{jk}^{(n)} \to 0 \tag{2.123}$$

as $n\to\infty$ for any i, j, and k. Consequently, if $p_{ik}^{(n)}$ has a limit it does not depend on i. But

$$\pi_k - p_{jk}^{(n)} = \sum_{i\in S}\pi_i(p_{ik}^{(n)} - p_{jk}^{n}) \to 0 \tag{2.124}$$

by bounded convergence. □

Example (The Pólya urn model) Quite a few stochastic processes were originally thought of using colored balls in urns. A paper by Eggenberger and Pólya[1] (1923) dealt with epidemic data for contagious diseases. Given that an individual has a disease, such as smallpox, the probability that other individuals who are in contact with the diseased one themselves become infected is higher than for people who have had no such contact. Hence individuals do not act independently as far as epidemics are concerned.

Eggenberger and Pólya proposed the following urn scheme: consider an urn with N balls, R of which are red and B are black. A ball is pulled out of the urn at random and replaced with $1+d$ balls of the same color. Clearly $d=0$ corresponds to drawing with replacement, and $d=-1$ to drawing without replacement. After the kth replacement the urn has $R+B+dk$ balls. If the draws yield r red and b black balls ($r+b=k$), there are $R+rd$ red and $B+bd$ black balls, whence the probability of a red ball drawn at the $(k+1)$th draw is $(R+rd)/(N+kd)$. Let $X_k=1$(red ball drawn at trial k). Then, with $\mathbf{X}=(X_1,\ldots,X_n)$ and $r_n=\sum_1^n x_i$,

$$\mathbf{P}(\mathbf{X}=\mathbf{x}) = \frac{\prod_{j=1}^{r_n}(R+(r_n-j)d)\prod_{k=1}^{n-r_n}(B+(n-r_n-k)d)}{(N+(n-1)d)(N+(n-2)d)\cdots N}. \tag{2.125}$$

Conditioning on the past yields

$$\mathbf{P}(X_n=x_n \mid X_0^{n-1}=x_0^{n-1}) = \begin{cases} \dfrac{R+r_{n-1}d}{N+(n-1)d} & \text{if } x_n=1, \\ \dfrac{B+(n-1-r_{n-1})d}{N+(n-1)d} & \text{if } x_n=0. \end{cases} \tag{2.126}$$

[1]Pólya, György (1887–1985). Hungarian mathematician. Invented the term "random walk". Perhaps best known for his ideas about general heuristics for solving problems.

Thus X_n is not a Markov chain, unless $d=0$. However, the number of red balls drawn, $R_n=\sum_1^n X_i$, is a Markov chain:

$$\mathbf{P}(R_n=r \mid R_0^{n-1}=r_0^{n-1}) = \begin{cases} \dfrac{R+dr_{n-1}}{N+d(n-1)} & \text{if } r=r_{n-1}+1 \\ \dfrac{B+d(n-1-r_{n-1})}{N+d(n-1)} & \text{if } r=r_{n-1} \end{cases}. \tag{2.127}$$

Note, however, that the transition probabilities for R_n are time-dependent, since they depend explicitly on $n-1$, not only on r_{n-1}. It is not hard (Exercise 7) to derive the marginal distribution of R_n, which is

$$\mathbf{P}(R_n=r) = \binom{n}{r} \times R(R+d)\cdots(R+(r-1)d)$$
$$\times \frac{B(B+d)\cdots(B+(n-r)d)}{N(N+d)\cdots(N+(n-1)d)}. \tag{2.128}$$

Consider the case where n is large and R small relative to N, corresponding to a rare disease. In particular, let $R=hN/n$ and $d=cN/n$. By taking limits in (2.128) we see that

$$\lim_{n\to\infty} \mathbf{P}(R_n=r) = \frac{1}{r!}(1+c)^{-(h/c+r)} h(h+c)\cdots(h+(r-1)c) \tag{2.129}$$

which is a negative binomial distributions with parameters h/c and $c/(1+c)$, and mean h. When $c\to 0$ this limit is just the Poisson approximation to the binomial. □

Corollary For an irreducible aperiodic chain

$$\lim_{n\to\infty} p_{jk}^{(n)} = \frac{f_{jk}}{\mu_k}. \tag{2.130}$$

Proof Recall that $p_{jk}^{(n)}=\sum_{l=0}^n f_{jk}^{(l)} p_{kk}^{(n-l)}$. Taking limits under the summation sign we get

$$p_{jk}^{(n)} \to (1/\mu_k)\sum f_{jk}^{(l)} = f_{jk}/\mu_l. \tag{2.131}$$

To verify that we can take limits under the summation sign, write

$$p_{jk}^{(n)} = \sum_{l=0}^{m-1} f_{jk}^{(l)} p_{kk}^{(n-l)} + \sum_{l=m}^{n} f_{jk}^{(l)} p_{kk}^{(n-l)}. \tag{2.132}$$

Since $p_{kk}^{(n)} \le 1$ for all n we have

$$\sum_{l=0}^{m-1} f_{jk}^{(l)} p_{kk}^{(n-l)} \le p_{jk}^{(n)} \le \sum_{l=0}^{m-1} f_{jk}^{(l)} p_{kk}^{(n-l)} + \sum_{l=m}^{n} f_{jk}^{(l)}. \tag{2.133}$$

Now let $n\to\infty$ to see that

$$\sum_{l=0}^{m-1} f_{jk}^{(l)}/\mu_k \leq \liminf_{n\to\infty} p_{jk}^{(n)} \leq \limsup_{n\to\infty} p_{jk}^{(n)} \leq \sum_{l=0}^{m-1} f_{jk}^{(l)}/\mu_k + \sum_{l=m}^{\infty} f_{jk}^{(l)}, \quad (2.134)$$

and finally let $m\to\infty$ to obtain the result. □

Let $N_k(n)=\sum_{i=1}^{n} 1(X_i=k)$ count the time spent in state k.

Corollary If k is a persistent aperiodic state, then

$$\lim_{n\to\infty} \mathbf{E}N_k(n)/n = \lim_{n\to\infty} \frac{1}{n}\sum_{i=1}^{n} p_{kk}^{(i)} = \frac{1}{\mu_k} \quad (2.135)$$

for any starting state communicating with k.

Remark The limit in the corollary is called a **Cesàro limit** of the $p_{kk}^{(i)}$. The existence of a Cesàro limit is implied by, but does not imply, the existence of a limit of the sequence. □

Proof By Theorem 2.9 we have $p_{kk}^{(i)}\to 1/\mu_k$. By the remark above, this implies that the Cesàro limit is the same. Now notice that

$$\mathbf{E}^k N_k(n) = \sum_{i=1}^{n} \mathbf{P}^k(X_i=k) = \sum_{i=1}^{n} p_{kk}^{(i)}. \quad (2.136)$$

If we instead start at j, communicating with k, we get

$$\mathbf{E}^j N_k(n) = \sum_{i=1}^{n} \mathbf{P}^j(X_i=k) = \sum_{i=1}^{n} p_{jk}^{(i)} \quad (2.137)$$

and by Corollary 1 above $p_{jk}^{(n)}\to f_{jk}/\mu_k$, so the same holds for the Cesàro limit. Since j communicates with k we have $f_{jk}=1$. □

Consider a persistent state k. The **limiting occupation probability** is the proportion of time spent in that state in an infinitely long realization, i.e., $\lim_{n\to\infty} N_k(n)/n$. In Corollary 2 we computed the expected value of this average. The next result yields a law of large numbers.

Theorem 2.10 The limiting occupation probability of an ergodic state is $1/\mu_k$ (with probability 1).

Proof Suppose that the chain starts in state k. Let $T_k(1), T_k(2), \cdots$ be the successive times when the chain reaches k. By the strong Markov property $T_k(1), T_k(2)-T_k(1), T_k(3)-T_k(2), \ldots$ are iid random variables with pgf $F_{kk}(s)$ and mean $\mu_k<\infty$. By the strong law of large numbers we have with probability one that

$$\lim_{r\to\infty}\frac{T_k(1)+(T_k(2)-T_k(1))+\cdots+(T_k(r)-T_k(r-1))}{r}$$

$$=\lim_{r\to\infty}\frac{T_k(r)}{r}=\mu_k. \tag{2.138}$$

Recall that $N_k(n)/n$ is the proportion of time spent in state k up to time n. Thus

$$T_k(N_k(n))\le n\le T_k(N_k(n)+1). \tag{2.139}$$

In addition, $N_k(n)\to\infty$ as $n\to\infty$, again with probability one, since k is revisited infinitely often. Thus

$$\frac{N_k(n)}{n}\le\frac{N_k(n)}{T_k(N_k(n))}\to\frac{1}{\mu_k}\text{ with probability 1} \tag{2.140}$$

and

$$\frac{N_k(n)+1}{n}\ge\frac{N_k(n)+1}{T_k(N_k(n)+1)}\to\frac{1}{\mu_k}\text{ with probability 1} \tag{2.141}$$

so that $N_k(n)/n\to 1/\mu_k$ a.s. The case when the process starts from a state other than k is left as Exercise **8**. □

Example (The Hardy–Weinberg law) Consider a large population of individuals, each of whom possesses a particular pair of genes. We classify each gene as type A or type a. Assume that when two individuals mate, each contributes a randomly chosen gene to the resulting offspring, and assume also that mates are selected at random from the population. Write the proportion of individuals in the population with AA, Aa, and aa genes, respectively, as p, q, and r. Then the proportion of A-genes in the population is $P=p+q/2$ and the proportion of a-genes is $Q=q/2+r$. Under random mating, therefore, an individual will have probability P^2 of receiving the gene combination AA, probability $2PQ$ of receiving Aa, and probability Q^2 of receiving aa. Hence in the next generation the proportion of A-genes is $P^2+PQ=P$, and the proportion of a-genes is Q. We see that the proportions of gene types as well as the proportion of gene pairs remain stable after the first mating. This is called the **Hardy–Weinberg law** (Hardy[1], 1908; Weinberg, 1908). Assume now that we have a population with $P^2{:}2PQ{:}Q^2$ gene pair ratio, and consider the genetic history of a single individual, assuming for simplicity that each individual has exactly one offspring. If X_n is the genetic state of the nth descendant we have a Markov chain with state space {AA,Aa,aa}, and transition matrix

[1]Hardy, Godfrey Harold (1877–1947). English pure mathematician. His main contributions came through his long collaboration with Littlewood on problems in number theory, inequalities, and complex analysis. He was apparently not very fond of this non-theoretical paper, which he published in an obscure American journal.

$$\mathbb{P} = \begin{bmatrix} P & Q & 0 \\ P/2 & (P+Q)/2 & Q/2 \\ 0 & P & Q \end{bmatrix}. \tag{2.142}$$

From the Hardy–Weinberg law it would seem natural that the stationary distribution for this chain, which, by the theorem, is also the proportion of descendants in each genetic state in the long run, should be $(P^2, 2PQ, Q^2)$. This is indeed the case (Exercise **9**). □

It is possible to deduce more general laws of large numbers. The following, which we shall find particularly useful later, is often called the **ergodic theorem for Markov chains.**

Theorem 2.11 Let X be a positive persistent chain. Then, regardless of the initial distribution, if $f:S \to \mathbf{R}$ satisfies $\mathbf{E}^{\pi} |f(X_1)| < \infty$, where π is the stationary distribution, then

$$\frac{1}{n}\sum_{j=1}^{n} f(X_j) \to \mathbf{E}^{\pi} f(X_1) \tag{2.143}$$

in probability.

Remark This result holds with probability one. See Bhattacharya and Waymire (1990, section II.9) for details. □

Proof We divide up the time axis using the random times $T_k(l)$ of successive returns to state k. Write

$$Z_i = \sum_{T_k(l)+1}^{T_k(l+1)} f(X_j) \tag{2.144}$$

where $T_k(0) \equiv 0$. Then $Z_0, Z_1, \ldots$ are independent by the strong Markov property, and $Z_1, Z_2, \ldots$ are also identically distributed. Decompose

$$\sum_{1}^{n} f(X_j) = Z_0 + \sum_{j=1}^{N_k(n)} Z_i - \sum_{j=n+1}^{T_k(N_k^{(n)}+1)} f(X_j) \equiv Z_0 + S_{N_k(n)} - R_n. \tag{2.145}$$

We deal with each of these terms separately. First note that, since the chain is positive persistent, Z_0 is a sum of a finite number of random variables (with probability one). Hence $Z_n/n \to 0$ with probability one, and so in probability.

By persistence we have $\mathbf{P}(N_k^{(n)} \to \infty) = 1$, so using the law of large numbers we deduce, provided that $\mathbf{E}|Z_1| < \infty$, that $S_{N_k(n)}/N_k(n) \to \mathbf{E}Z_1$ in probability. Also, $N_k(n)/n \to \pi_k$ with probability one according to Theorem 2.10. Hence $S_{N_k(n)}/n \to \pi_k \mathbf{E}Z_1$ in probability.

Next note that

$$|R_n| \leq \sum_{j=n+1}^{T_k(N_k^{(n)}+1)} |f(X_j)| \leq \sum_{j=T_k(N_k^{(n)})+1}^{T_k(N_k^{(n)}+1)} |f(X_j)| \equiv \xi_n. \tag{2.146}$$

By the strong Markov property, again, $\xi_1, \xi_2, \ldots$ are iid, and by Markov's inequality $\xi_n/n \to 0$ in probability provided $\mathbf{E}\xi_1 < \infty$. Hence

$$\mathbf{P}(|R_n| > n\varepsilon) \leq \mathbf{P}(\xi_n > n\varepsilon) \to 0. \tag{2.147}$$

Clearly, if $\mathbf{E}\xi_1 < \infty$ then $\mathbf{E}|Z_1| < \infty$. To see that the former holds, note that if ν_i as before is the expected number of visits to i between successive visits to k we have

$$\mathbf{E} \sum_{j=T_k(1)+1}^{T_k(2)} |f(X_j)| = \sum_{i \in S} |F(i)| \nu_i = \sum_{i \in S} |f(i)| \frac{\pi_i}{\pi_k}, \tag{2.148}$$

using the corollary to Theorem 2.7. Finally, compute

$$\mathbf{E}Z_1 = \sum_{i \in S} f(i) \nu_i = \frac{1}{\pi_k} \sum_{i \in S} f(i) \pi_i, \tag{2.149}$$

so that

$$\frac{S_{N_k(n)}}{n} \to \sum_{i \in S} f(i) \pi_i = \mathbf{E}^\pi f(X_1) \tag{2.150}$$

in probability. □

In the next section we shall find an important use of this result. There are central limit type results for ergodic chains as well. We write $\xi_n \sim \mathrm{AsN}(\mu_n, \sigma_n^2)$ if $(\xi_n - \mu_n)/\sigma_n$ converges in distribution to the standard normal distribution. Define $T_k(0) \equiv 0$ and $U_k(m) = T_k(m) - T_k(m-1)$.

Theorem 2.12 Let k be an ergodic state. Suppose that $\sigma_k^2 = \sum_1^\infty (n - \mu_k)^2 f_{kk}^{(n)}$ satisfies $0 < \sigma_k^2 < \infty$, and that the distribution of $U_k(m)$ is non-degenerate. Then

$$N_k(n) \sim \mathrm{AsN}\left(\frac{n}{\mu_k}, \frac{n\sigma_k^2}{\mu_k^3}\right). \tag{2.151}$$

Proof Assume that we start from k. Since the $U_k(l)$ are iid we have by the central limit theorem that $\sum_{l=1}^m U_k(l) \sim \mathrm{AsN}(m\mu_k, m\sigma_k^2)$. Write

$$\{N_k(n) < m\} = \{T_k(m) > n\} \tag{2.152}$$

and choose $m = [n/\mu_k + x(n\sigma_k^2/\mu_k^3)^{1/2}]$, where $[y]$ stands for the integer part of y, and x is an arbitrary real number. Now note that $T_k(m) = \sum_{l=1}^m U_k(l)$, so

$$\mathbf{P}\left[\frac{N_k(n)-n/\mu_k}{(n\sigma_k^2/\mu_k^3)^{½}}<x\right]=\mathbf{P}(\sum_{l=1}^{m}U_k(l)>n)$$

$$=\mathbf{P}\left[\sum_{l=1}^{m}U_k(l)-m\mu_k(m\sigma_k^2)^{-½}>n-m\mu_k(m\sigma_k^2)^{-½}\right] \tag{2.153}$$

$$=\mathbf{P}\left[\sum_{l=1}^{m}U_k(m)-m\mu_k(m\sigma_k^2)^{-½}>\frac{-x}{1+o(n^{-¼})}\right]\xrightarrow{d}\Phi(x),$$

where $\Phi(x)$ is the standard normal cdf. The case when we start from another state only changes the distribution of $U_k(1)$, which is asymptotically negligible. □

So far we have concentrated on aperiodic chains. The periodic case can be dealt with by looking at an imbedded aperiodic chain. Here is a version of Theorem 2.9 for periodic chains.

Theorem 2.13 Let X be an irreducible persistent Markov chain of period d. Then

$$\lim_{n\to\infty} p_{kk}^{(nd)}=\frac{d}{\mu_k} \tag{2.154}$$

and writing $r_{jk}=\min\{r:p_{jk}^{(r)}>0\}$ we also have

$$\lim_{n\to\infty} p_{jk}^{(r_{jk}+nd)}=\frac{df_{jk}}{\mu_k}. \tag{2.155}$$

Proof Let $Y_k=X_{dk}$. Then Y is ergodic with transition matrix $\mathbb{P}_Y=\mathbb{P}^d$. Hence

$$P_{Y;jk}(s)=\sum_n p_{Y;jk}^{(n)}s^n=\sum_n p_{jk}^{(nd)}s^n=P_{jk}(s^{1/d}) \tag{2.156}$$

since $p_{jk}^{(l)}=0$ for $l\neq nd$. Rewriting equation (2.56) we have

$$F_{Y;kk}(s)=\frac{P_{Y;kk}(s)-1}{P_{Y;kk}(s)}=\frac{P_{kk}(s^{1/d})-1}{P_{kk}(s^{1/d})}=F_{kk}(s^{1/d}), \tag{2.157}$$

so by Theorem 2.9

$$p_{Y;kk}^{(n)}\to\frac{1}{\frac{d}{ds}F_{kk}(s^{1/d})\,|_{s=1}}. \tag{2.158}$$

The left-hand side is $p_{kk}^{(nd)}$, while the right-hand side is $(F_{kk}'(1-)/d)^{-1}=d/\mu_k$. The second part follows just as did the first corollary of Theorem 2.9. □

Example (Limiting behavior of a particular chain) Let

$$\mathbb{P} = \begin{bmatrix} 0.5 & 0.5 & 0 & 0 \\ 0.25 & 0.75 & 0 & 0 \\ 0.25 & 0.25 & 0.25 & 0.25 \\ 0 & 0 & 0 & 1 \end{bmatrix}. \tag{2.159}$$

Here S={0,1,2,3} with S_T={2} and S_P={0,1}+{3}. Starting from state 2 where do we go? Let u=$\mathbf{P}^2$(absorption in {0,1}). By partitioning the sum into the possible values of the first step we get

$$\begin{aligned} u &= \sum_{k=0}^{3} \mathbf{P}^2(X_1 = k, \text{ absorption in } \{0,1]) \\ &= \sum_{k=0}^{3} \mathbf{P}^2(Z_1 = k)\mathbf{P}^k(\text{absorption in } \{0,1\}) \\ &= (0.25+0.25)\times 1 + 0.25u + 0.25\times 0 = 0.5 + 0.25u \end{aligned} \tag{2.160}$$

whence u=2/3. The stationary distribution for the subclass {0,1} is (1/3,2/3). Therefore

$$\lim_{n\to\infty} p_{20}^{(n)} = \frac{2}{3}\times\frac{1}{3} = \frac{2}{9}. \tag{2.161}$$

Similarly $\lim_{n\to\infty} p_{21}^{(n)}$=4/9. In summary

$$\lim_{n\to\infty} \mathbb{P}^n = \begin{bmatrix} 1/3 & 2/3 & 0 & 0 \\ 1/3 & 2/3 & 0 & 0 \\ 2/9 & 4/9 & 0 & 1/3 \\ 0 & 0 & 0 & 1 \end{bmatrix}. \tag{2.162}$$

The technique used in this argument, namely conditioning on the first step, often proves very useful. □

2.6. Markov chain Monte Carlo methods

An interesting recent application of the asymptotic theory of Markov chains is to Monte Carlo calculation of complicated integrals. There is a variety of problems that reduce to needing to compute such an integral.

Example (Likelihood) Let $L_{\mathbf{x}}(\boldsymbol{\theta})$ be a likelihood function based on an observation $\mathbf{x}$ of a random vector $\mathbf{X}$. We make no particular assumptions of the structure of $\mathbf{X}$: it could be a sequence of iid random variables, or a realization of a stochastic process. Frequently we can write

$$L_{\mathbf{x}}(\boldsymbol{\theta}) = h(\mathbf{x};\boldsymbol{\theta})/c(\boldsymbol{\theta}) \tag{2.163}$$

where h is known, but the normalizing constant $c(\boldsymbol{\theta})=\int h(\mathbf{x};\boldsymbol{\theta})d\mathbf{x}$ is too complicated to compute explicitly.

Example (Mixture distribution)

Suppose that we have iid observations from a mixture of exponential distributions with density

$$f(x;\boldsymbol{\theta}) = \sum_{j=1}^{k} p_j \lambda_j e^{-x\lambda_j}. \tag{2.164}$$

Here k is assumed known, so the unknown parameter is $\boldsymbol{\theta}=(p_1,\ldots,p_k,\lambda_1,\ldots,\lambda_k)$. One can, of course, write out the likelihood as the product of terms of the form (2.164), but the maximization problem can be unpleasant due to difficulties in the numerical evaluation of some of the terms. We can put it in the form needed for Markov chain Monte Carlo (or MCMC for short) by letting Λ be a random variable, taking on the value λ_i with probability p_i. Then

$$f(x;\boldsymbol{\theta}) = \mathbf{E}\Lambda e^{-\Lambda x}. \tag{2.165}$$

Considering a iid sequence Λ_i, the likelihood can be written

$$L(\boldsymbol{\theta}) = \prod_{i=1}^{n} f(x_i;\boldsymbol{\theta}) = \prod_{i=1}^{n} \mathbf{E}\Lambda_i e^{-\Lambda_i x_i} = \mathbf{E}\prod_{i=1}^{n} \Lambda_i e^{-\Lambda_i x_i}. \tag{2.166}$$

□

Example (Posterior distribution) Suppose that $\boldsymbol{\theta}$, instead of being an unknown constant, is a random variable with a distribution $\pi(\boldsymbol{\theta})$, often called the **prior** distribution. If we have data $\mathbf{x}$ that conditionally upon $\boldsymbol{\theta}$ are drawn from a joint distribution $f(\mathbf{x} \mid \boldsymbol{\theta})$, we can use Bayes' theorem to compute the conditional distribution

$$\pi(\boldsymbol{\theta} \mid \mathbf{x}) = \frac{f(\mathbf{x} \mid \boldsymbol{\theta})\pi(\boldsymbol{\theta})}{\int f(\mathbf{x} \mid \boldsymbol{\theta})\pi(\boldsymbol{\theta})d\boldsymbol{\theta}} \tag{2.167}$$

called the **posterior** distribution, since it is the distribution of $\boldsymbol{\theta}$ after $\mathbf{x}$ was observed. The integral in the denominator is often difficult to compute, as is the ratio of integrals (called **posterior expectation**)

$$\int \boldsymbol{\theta}\pi(\boldsymbol{\theta} \mid \mathbf{x}) = \frac{\int \boldsymbol{\theta} f(\mathbf{x} \mid \boldsymbol{\theta})\pi(\boldsymbol{\theta})d\boldsymbol{\theta}}{\int f(\mathbf{x} \mid \boldsymbol{\theta})\pi(\boldsymbol{\theta})d\boldsymbol{\theta}}. \tag{2.168}$$

□

Example (Monte Carlo testing) Let H_0 be a simple hypothesis about the distribution of a multidimensional random variable X. Suppose that we have a continuous test statistic $T=T(X)$, and we reject H_0 for large observed values t of T. Let f be the density of T, and assume that we can simulate a random sample $t_2, \ldots, t_n$ from f. We base the observed significance level, or P-value, of t on its rank among the n values $t, t_2, \ldots, t_n$. If the rank of t is k, we reject H_0 at the k/n-level, since the rank is uniformly distributed on $1, \ldots, n$ when H_0 is true (Bickel and Doksum, 1977, p. 347). In fact, all that is needed for this to hold is that the T_i have a joint distribution which is invariant under permutations of the indices. Such distributions are called **exchangeable**, and arise, e.g., when the random variables are conditionally independent, given another random variable.

We can extend this procedure to the case of a composite null hypothesis, provided that the problem admits a sufficient statistic. We then merely simulate from the conditional distribution, given the observed values of the sufficient statistics. Of course, this simulation problem can be quite hard. □

Example (The Rasch model of item analysis) Consider r individuals responding to c test items each. Let $X_{ij}=1$(individual i answered item j correctly). Rasch (1960) suggested the model

$$\mathbf{P}(X_{ij}=1) = \exp(\alpha_i+\beta_j)/(1+\exp(\alpha_i+\beta_j)) \tag{2.169}$$

where $\sum_i \alpha_i = \sum_j \beta_j = 0$. The likelihood can be written

$$\prod_{i,j} \frac{\exp(x_{ij}(\alpha_i+\beta_j))}{(1+\exp(\alpha_i+\beta_j))} = \frac{\prod_i \exp(x_{i\bullet}\alpha_i) \prod_j \exp(x_{\bullet j}\beta_j)}{\prod_{i,j}(1+\exp(\alpha_i+\beta_j))}, \tag{2.170}$$

and we see that the totals $x_{i\bullet} = \sum_j x_{ij}$ and $x_{\bullet j} = \sum_i x_{ij}$ are sufficient statistics (cf. Appendix A). Hence, given these totals, all possible binary tables have the same probability. The problem is to device an enumeration scheme for all these tables. It is a very hard combinatorial problem. □

As it happens, it is often possible to construct a Markov chain with limiting distribution proportional to a given function $f(\mathbf{u})$. One can then estimate $\int f(\mathbf{u})d\mathbf{u}$ by running a Monte Carlo simulation of the Markov chain long enough to reach equilibrium. Exactly how long that is depends on the problem at hand.

Example (Likelihood, continued) Let $f(\mathbf{x})=g(\mathbf{x})/c$ be a fixed density, chosen so that $h(\mathbf{x};\boldsymbol{\theta})>0$ implies that $f(\mathbf{x})>0$. The mle of $\boldsymbol{\theta}$ maximizes

$$L_{\mathbf{x}}(\boldsymbol{\theta}) = \frac{h(\mathbf{x};\boldsymbol{\theta})/g(\mathbf{x})}{c(\boldsymbol{\theta})/c}. \tag{2.171}$$

For any $\boldsymbol{\theta}$ we can evaluate $h(\mathbf{x};\boldsymbol{\theta})/g(\mathbf{x})$, but not $c(\boldsymbol{\theta})/c$. Note that

$$\frac{c(\boldsymbol{\theta})}{c} = \int_S \frac{h(\mathbf{x};\boldsymbol{\theta})}{c} d\mathbf{x} = \int_S \frac{h(\mathbf{x};\boldsymbol{\theta})}{g(\mathbf{x})} f(\mathbf{x}) d\mathbf{x} = \mathbf{E}_f \left[h(\mathbf{X};\boldsymbol{\theta})/g(\mathbf{X}) \right]. \tag{2.172}$$

If we can generate samples from f we can estimate the expectation on the right-hand side of (2.172). The classical Monte Carlo method is to draw N observations $\mathbf{x}_i$ iid from f and then compute $(1/n)\sum h(\mathbf{x}_i;\boldsymbol{\theta})/g(\mathbf{x}_i)$. But f is a multivariate distribution, and it may not be easy to generate random samples from this distribution. □

The MCMC method, instead of generating iid observations, generates dependent samples from a Markov chain with stationary distribution $\mathbf{f}=(f(\mathbf{x});\mathbf{x}\in S)$ and uses Theorem 2.11 to obtain the convergence. How can this be done? One approach, the **Gibbs sampler**, was introduced into the statistical literature by Geman and Geman (1984), although it originates in statistical physics where it is called the **heat bath** method. The Gibbs sampler computes successive values of the vector $\mathbf{x}$. At stage t we have a current vector $\mathbf{x}(t)$. At the next stage we update each component of $\mathbf{x}$ in turn. Suppose we have updated $x_1, \ldots, x_{i-1}$ with new values $x_1(t+1), \ldots, x_{i-1}(t+1)$. The new value at component i, $x_i(t+1)$, is drawn at random from $f_i(\bullet \mid \mathbf{x}_1^{i-1}(t+1), \mathbf{x}_{i+1}^{m}(t))$ (recall the notation from section 1.2). At each stage, each component is updated just once. Variants of the Gibbs sampler have the order of updating change from stage to stage, e.g., by going through the components in the order of a random permutation, chosen anew at each iteration.

Proposition 2.7 If $f(\mathbf{x})$ satisfies the positivity condition (1.14), the Gibbs sampler is an ergodic Markov chain with stationary distribution $\mathbf{f}=(f(\mathbf{x});\mathbf{x}\in S)$.

Proof It is clear from the construction that the conditional distribution of $\mathbf{x}(t+1)$ given the past only depends on $\mathbf{x}(t)$, so the process is Markovian. The transition matrix has elements

$$p_{\mathbf{x},\mathbf{y}} = f_1(y_1 \mid \mathbf{x}_2^m) f_2(y_2 \mid y_1, \mathbf{x}_3^m) f_3(y_3, \mathbf{y}_1^2, \mathbf{x}_4^m) \cdots f_m(y_m \mid \mathbf{y}_1^{m-1}). \quad (2.173)$$

The positivity assumption guarantees that $p_{\mathbf{x},\mathbf{y}}>0$ for all $\mathbf{x},\mathbf{y}\in S = \{\mathbf{x}: f(\mathbf{x})>0\}$. Now note that $\mathbb{P}=\mathbb{P}_1\mathbb{P}_2 \cdots \mathbf{P}_m$ where $\mathbb{P}_i$ has $(\mathbf{x},\mathbf{y})$-element

$$p_{i;\mathbf{x},\mathbf{y}} = f_i(y_i \mid \mathbf{x}_{-i}) 1(\mathbf{y}_{-i}=\mathbf{x}_{-i}). \quad (2.174)$$

To see this, it is perhaps easiest to do the case $m=2$, from which the general argument follows by a similar argument. Write

$$\begin{aligned} \left(\mathbb{P}_1\mathbb{P}_2\right)_{\mathbf{x},\mathbf{y}} &= \sum_{\mathbf{z}} p_{1:\mathbf{x},\mathbf{z}}\, p_{2:\mathbf{z},\mathbf{y}} \\ &= \sum_{\mathbf{z}} f_1(z_1 \mid \mathbf{x}_{-1}) 1(\mathbf{z}_{-1}=\mathbf{x}_{-1}) \quad (2.175) \\ &\times f_2(y_2 \mid \mathbf{z}_{-2}) 1(\mathbf{y}_{-2}=\mathbf{z}_{-2}). \end{aligned}$$

The only $\mathbf{z}$'s for which the summands do not vanish have $z_1=y_1$, $z_2=x_2$ and

$\mathbf{z}_3^m=\mathbf{x}_3^m=\mathbf{y}_3^m$. Hence the sum is

$$\left[\mathbb{P}_1\mathbb{P}_2\right]_{\mathbf{x},\mathbf{y}} = f_1(y_1 \mid \mathbf{x}_{-1}) f_2(y_2 \mid y_1, x_3^m) \tag{2.176}$$

as was to be shown. Hence

$$\frac{p_{i;\mathbf{x},\mathbf{y}}}{p_{i;\mathbf{y},\mathbf{x}}} = \frac{f_i(y_i \mid \mathbf{x}_{-i})}{f_i(x_i \mid \mathbf{y}_{-i})} = \frac{f(\mathbf{y})}{f(\mathbf{x})}. \tag{2.177}$$

Recall from section 2.4 that this means that $\mathbb{P}_i$ is a reversible Markov chain with stationary distribution $\mathbf{f}$. Therefore

$$\mathbf{f}\,\mathbb{P} = \mathbf{f}\,\mathbb{P}_1\mathbb{P}_2 \cdots \mathbf{P}_m = \mathbf{f}\,\mathbb{P}_2 \cdots \mathbf{P}_m = \mathbf{f}\,\mathbb{P}_m = \mathbf{f}, \tag{2.178}$$

verifying that the chain has stationary distribution $\mathbf{f}$. By positivity it is irreducible, so the result follows from Theorem 2.7. □

Example (Mixture distribution, continued) The Gibbs sampler draws, given $\boldsymbol{\theta}$, vectors $\boldsymbol{\Lambda}=(\Lambda_1, \ldots, \Lambda_n)$. Since in this very simple case the Λ_i are iid, the Gibbs sampler just repeatedly generates iid $\boldsymbol{\Lambda}^{(i)}$, $i=1, \ldots, N$, and then estimates the likelihood by averaging

$$\hat{L}(\boldsymbol{\theta}) = \frac{1}{N}\sum_{j=1}^{N}\prod_{i=1}^{n}\Lambda_i^{(j)} e^{-\Lambda_i^{(j)} x_i}. \tag{2.179}$$

Rather large values of N may be needed to evaluate the likelihood precisely enough. Of course, in this simple case one can compute the likelihood exactly. Standard optimization routines can be used to find the mle of $\boldsymbol{\theta}$. □

Ideally, in order to obtain observations from the stationary distribution $\boldsymbol{\pi}$ of the Gibbs sampler, we should choose a starting value from $\boldsymbol{\pi}$. But if we knew how to do this there would be no need to run the Gibbs sampler! As outlined in Exercise **14**, the convergence to the stationary distribution is exponentially fast, so we first run the Gibbs sampler for a **burn-in** period in order to get close enough to the stationary distribution. Only after the burn-in period do we actually start to collect observations. The proper length of the burn-in period is a subject of current research.

Example (Monte Carlo testing, continued) Since the Gibbs sampler maintains detailed balance it is reversible. The reverse chain must have the same stationary distribution as the forward chain. We use this to create exchangeable paths. Starting from the observed value $X_0=x$, we run the Gibbs sampler backwards n steps, yielding $X_{-n}=y$, say. The we simulate $N-1$ paths n steps forward in time, all starting from y, yielding observations $X_0^{(i+1)}=x^{(i+1)}$, $i=1, \ldots, N-1$. Since y (at least very nearly) is an observation from $\boldsymbol{\pi}$, the same is true for $x^{(2)}, \ldots, x^{(N)}$. Given y, $X_0, X_0^{(2)}, \ldots, X_0^{(N)}$ are independent, so they form an exchangeable sequence, and the earlier discussion of Monte Carlo testing from exchangeable sequences applies. □

Example (The Rasch model of item analysis, continued) Let S be the set of $r\times c$-tables having marginals $x_{i\bullet}$ and $x_{\bullet j}$. We need to construct a Markov chain having the uniform distribution on S as its stationary distribution. Let $\mathbf{y}=(y_{ij})$ be a configuration in S. Consider any sub-rectangle of $\mathbf{y}$ having ones in two diagonally opposite corners and zeros in the other opposite corners. Exchanging the zeros with ones, and the ones with zeros, does not change the margins, so it yields another $r\times c$-table $\mathbf{z}$ in S. We call this procedure a **switch**, and the sub-rectangle **switchable**. Figure 2.3 shows this concept.

1	0	0	1	0	2
0	1	1	1	0	3
0	0	1	0	0	1
1	0	0	0	1	2
1	1	0	0	0	2
3	2	2	2	1	10

Figure 2.3. Two switchable sub-rectangles in a 5×5 table.

Any table $\mathbf{z}\in S$ can be reached from $\mathbf{y}$ by a series of switches. To produce our Markov chain, pick a non-empty rectangle at random, and switch it if it is switchable. Clearly this preserves the margins. To see that it has a uniform stationary distribution, we just need to check that the transition matrix is symmetric. But that is easy to see: if we can go from $\mathbf{y}$ to $\mathbf{z}$ in one step, we must do that by switching a single rectangle. Thus $P_{\mathbf{y},\mathbf{z}}=P_{\mathbf{z},\mathbf{y}}$, since the opposite switch of the same rectangle brings us back.

In order to apply this procedure to the (very small) table in Figure 2.3, we need to define an appropriate test statistic. This should reflect the type of alternative model we have in mind. Here one may consider the idea that there is an interaction between difficulty and ability. We can rearrange the table so that individuals are ordered by increasing total score (row sum), and questions are ordered by decreasing success rate (column sum). One possible such reordering (ties make it non-unique) is given in Figure 2.4. A measure of interaction could be the χ^2 statistic for independence in the 2×2 table given by summing the two lowest and the two highest scores in the two most solved and the two least solved questions. This statistic is $(N_{11}-c_1r_1/N)^2/(c_1r_1/N)$, which for this table is $(1-2\times3/5)^2/(2\times3/5)=.033$. Simulating this using the Monte Carlo testing method outlined above, by first moving 2,000 steps backwards, and then 99 times move 2,000 steps forward, yields 16 that were larger than and 7 that were

0	0	1	0	0	1
1	1	0	0	0	2
1	0	0	0	1	2
1	0	0	1	0	2
0	1	1	1	0	3
3	2	2	2	1	10

Figure 2.4. Reordering of Figure 2.3.

equal to 0.333, for a P-value between 0.16 and 0.23. We find no evidence against the Rasch model based on this test statistic. □

The first Markov chain Monte Carlo method was developed by Metropolis et al. (1953). The algorithm, called the **Metropolis algorithm**, employs an auxiliary symmetric transition matrix $(q_{x,y})$ (having $q_{x,y}=q_{y,x}$). As before, we want to find a Markov chain with stationary distribution f. The next value of the Markov chain, when the present value is x, is generated by the following update method:

1. Simulate y from the distribution $q_{x,\bullet}$.
2. Calculate the odds ratio $r=f(y)/f(x)$.
3. If $r\geq 1$ the next value is y.
4. If $r<1$ go to y with probability r, and stay at x with probability $1-r$.

It should be clear that the next state only depends on the previous state, so that this is, indeed, a Markov chain. As for the Gibbs sampler, the simplest way to see that it has stationary distribution f is to note that it satisfies detailed balance. Consider a finite state space $\{1,\ldots,K\}$, and order the values so that $f(i)\leq f(j)$ for $i<j$. Then we have $p_{ij}=q_{ij}$, while $p_{ji}=q_{ji}f(i)/f(j)=p_{ij}f(i)/f(j)$, using the symmetry of the auxiliary transition matrix. A generalization of both the Metropolis algorithm and the Gibbs sampler is due to Hastings (1971), and outlined in Exercise **10**. We have demonstrated the following result.

Proposition 2.8 If $f(\mathbf{x})$ satisfies the positivity condition (1.14), the Metropolis algorithm generates an ergodic Markov chain with stationary distribution $\mathbf{f}=(f(\mathbf{x});\mathbf{x}\in S)$.

2.7. Likelihood theory for Markov chains

Given a set of observations from a two-state Markov chain, we saw in section 2.1 how it is possible to estimate the transition matrix, and thus any function thereof, using the method of maximum likelihood. In this section we study the general finite-state Markov chain, and discuss the likelihood theory for both estimation and testing. We will first look at the **nonparametric** case, where the parameter of interest is a point in the space of all transition matrices. Let $N_{ij}(n)=\sum_{l=1}^{n} 1(X_{l-1}=i,X_l=j)$ count the number of i,j-transitions. If $N_{ij}(n)=n_{ij}$, the likelihood (2.9) takes the form

$$L(\mathbf{p}_0,\mathbb{P}) = p_0(x_0)\prod_{l=1}^{n} p_{x_{l-1},x_l}$$

$$= p_0(x_0)\prod_{i\in S}\prod_{j\in S} p_{ij}^{n_{ij}} = p_0(x_0)\prod_{i\in S} L_i(\mathbb{P}) \qquad (2.180)$$

where $L_i(\mathbb{P}) = \prod_{j\in S} p_{ij}^{n_{ij}}$ depends only on the elements in the ith row $\mathbb{P}_{i\bullet}$ of $\mathbb{P}$. In other words, we are estimating $|S|$ independent probability distributions. Let $l(\mathbf{p}_0,\mathbb{P})=\log L(\mathbf{p}_0,\mathbb{P})$. Then (2.180) corresponds, with obvious notation, to

$$l(\mathbf{p}_0,\mathbb{P}) = l_0(\mathbf{p}_0) + \sum_{i\in S} l_i(\mathbb{P}_{i\bullet}). \qquad (2.181)$$

We want to maximize l subject to the constraints that $\mathbf{p}_0\mathbf{1}^T=1$, where $\mathbf{1}$ is a vector of ones, and that $\mathbb{P}_{i\bullet}\mathbf{1}^T=1$. Each of these maximizations can be done separately using Lagrange multipliers by differentiating a term of the form

$$l_i(\mathbb{P}_{i\bullet}) + \lambda(\mathbb{P}_{i\bullet}\mathbf{1}^T - 1) = \sum_{j\in S} n_{ij}\log p_{ij} + \lambda(\sum_{j\in S} p_{ij} - 1). \qquad (2.182)$$

Setting the derivatives equal to zero and writing $n_i = \sum_{j\in S} n_{ij}$ we get

$$\hat{p}_{ij} = \frac{n_{ij}}{n_i} \text{ when } n_i>0 \text{ and } \hat{p}_0(i) = 1(i=x_0). \qquad (2.183)$$

We can think of this as multinomial likelihoods with random sample sizes. The estimates are very reasonable: $\hat{p}_{ij}$ is just the observed proportion of i,j-transitions among all transitions out of i. If $n_i=0$ there are no exits from state i. The likelihood is then flat as a function of p_{ij} for any j in S, and we conventionally set $\hat{p}_{ij}=0$, $i\neq j$.

Let $\hat{S}=\{i\in S:n_i\geq 1\}$ be the observed part of the state space. Obviously, $\hat{S}$ is always finite. We will, for simplicity, ignore the possibility that $T_i=n$, i.e., that state i is reached for the first time at time n, since we then cannot estimate any transitions out of state i (this problem can usually be solved by taking one more observation). Notice that $(\hat{p}_{ij}, i,j\in\hat{S})$ is a stochastic matrix over $\hat{S}$. The class structure of $\hat{S}$ is determined by $\hat{\mathbb{P}}$.

Proposition 2.9 The Markov chain on $\hat{S}$, governed by $\hat{\mathbb{P}}$, has a class of transient states, and precisely one closed class $\hat{S}_P$ of persistent states.

Proof Using Theorem 2.5, we need to show that $\hat{S}_P$ is closed. First note that $x_m \rightarrow x_{m'}$ whenever $m<m'$. Choose m_0 so that $x_{m_0}\in\hat{S}_P$ but $x_m\notin\hat{S}_P$ for $m<m_0$. Then $\{x_{m_0},\ldots,x_n\}$ is closed. □

Remark In particular, $\hat{\mathbb{P}}$ is irreducible on $\hat{S}$ if $\hat{S}_P=\hat{S}$. On the other extreme, $\hat{S}_P$ could be empty, making the estimated chain transient. This would happen if no state entered was ever returned to. □

With only one observation of the initial distribution one cannot learn much about it. There are two possible approaches. One is to condition on $X_0=x_0$, and study the conditional likelihood

$$L^c(\mathbb{P}) = \prod_{i\in S} L_i(\mathbb{P}). \tag{2.184}$$

The conditional mle's are the same as the unconditional ones. The other possibility, appropriate if the chain has been running for a long time, is to use the stationary initial distribution. This is equivalent to maximizing $L(\mathbf{p}_0,\mathbb{P})$ subject to the additional constraint that $\mathbf{p}_0=\mathbf{p}_0\mathbb{P}$. A drawback is that the nice factorization of the likelihood into terms that only depend on rows of $\mathbb{P}$ no longer obtains.

Application (Snoqualmie Falls precipitation, continued) For a two-state chain $\boldsymbol{\pi}=(1-p_{11},p_{11})/(1-(p_{11}-p_{01}))$. If $X_0=0$ we have

$$L(\pi,\mathbb{P}) = \frac{(1-p_{01})^{n_{00}} p_{01}^{n_{01}} (1-p_{11})^{n_{10}+1} p_{11}^{n_{11}}}{1-(p_{11}-p_{01})}. \tag{2.185}$$

Taking logarithms (note that we have parametrized the model so that the rows sum to one), we obtain the likelihood equations

$$\begin{aligned} -\frac{n_{10}+1}{(1-p_{11})} + \frac{n_{11}}{p_{11}} &= -\frac{1}{1-(p_{11}-p_{01})} \\ -\frac{n_{00}}{(1-p_{01})} + \frac{n_{01}}{p_{01}} &= \frac{1}{1-(p_{11}-p_{01})} \end{aligned} \tag{2.186}$$

which are mixed polynomial equations of second order. Clearly, as the n_{ij} increase, the effect of the initial distribution diminishes.

For the Snoqualmie Falls data there were 11 dry and 25 rainy January 1. Hence the likelihood becomes

$$L(\boldsymbol{\pi},\mathbb{P}) = (1-p_{01})^{186} p_{01}^{123+25} (1-p_{11})^{128+11} p_{11}^{643} (1-(p_{11}-p_{01}))^{-36}$$

which is maximized by $\hat{p}_{01}=0.397$ and $\hat{p}_{11}=0.834$, virtually the same estimates as for the conditional method, namely $\hat{p}_{01}=0.398$ and $\hat{p}_{11}=0.834$. □

In terms of long term behavior of the mle's, we cannot hope to estimate $\mathbb{P}$ well if it does not correspond to an irreducible chain, since we need a large number of (i,j)-transitions for all i and j. If there is more than one persistent class we only get to see one of the classes. Therefore we assume that we are dealing with an ergodic chain. It is convenient to introduce the **step chain** $(Y_n,n\geq 0)$ defined by $Y_n=(X_n,X_{n+1})$. If we know Y_n, we know where X_n is going next.

Lemma 2.5 (a) (Y_n) is a Markov chain with state space $\tilde{S}=\{(i,j)\in S^2: p_{ij}>0\}$, initial distribution $\tilde{p}_0$ given by $\tilde{p}_0(i,j)=p_0(i)p_{ij}$ and transition matrix $\tilde{\mathbb{P}}=(\tilde{p}_{ij,kl})$ given by

$$\tilde{p}_{kl,ij} = 1(i=l)p_{ij}. \tag{2.187}$$

(b) If (X_n) is ergodic and $\mathbf{p}_0=\boldsymbol{\pi}$, then (Y_n) is also ergodic with stationary initial distribution $\tilde{\boldsymbol{\pi}}$ given by $\tilde{\pi}(i,j)=\pi(i)p_{ij}$.

Proof

$$\begin{aligned}
&\mathbf{P}(Y_n=(i,j) \mid Y_{n-1}=(k_1,l_1),\ldots,Y_0=(k_n,l_n))\\
&\quad= \mathbf{P}(X_{n+1}=j,X_n=i \mid X_n=l_1,X_{n-1}=k_1,\ldots,X_0=k_n)\\
&\quad= \mathbf{P}(X_{n+1}=j.X_n=i \mid X_n=l_1,X_{n-1}=k_1) \qquad (2.188)\\
&\quad= \mathbf{P}(Y_n=(i,j) \mid Y_{n-1}=(k_1,l_1))
\end{aligned}$$

verifying the Markov property. Furthermore, (2.188) can be evaluated as

$$\begin{aligned}
\mathbf{P}(Y_n=(i,j) \mid Y_{n-1}=(k,l)) &= \mathbf{P}(X_{n+1}=j \mid X_n=i,X_n=l,X_{n-1}=k)\\
&\quad\times \mathbf{P}(X_n=i \mid X_n=l,X_{n-1}=k) \qquad (2.189)\\
&= \begin{cases}\mathbf{P}(X_{n+1}=j \mid X_n=i) & \text{if } i=l\\ 0 & \text{otherwise.}\end{cases}
\end{aligned}$$

The remaining parts are proved using similar computations. □

We now use the step chain and the ergodic theorem to show strong consistency of the estimator $\hat{\mathbb{P}}$. Let $\hat{p}_{ij}(n)$ be the mle of p_{ij} based on observing the chain up to time n.

Theorem 2.14 If (X_n) is an ergodic chain, then $\hat{p}_{ij}(n)\rightarrow p_{ij}$ with probability 1 as $n\rightarrow\infty$ for all $i,j\in S$, regardless of the initial distribution.

Proof If $p_{ij}=0$ then $\mathbf{P}(\hat{p}_{ij}(n)=0)=1$, so we only need to consider $(i,j)\in\tilde{S}$. But

$$N_{ij}(n) = \sum_{k=1}^{n} 1(Y_k=(i,j)) \tag{2.190}$$

and since the step chain is ergodic we have from Theorem 2.10 and Lemma 2.5 (a) that

$$\frac{1}{n}N_{ij}(n) \rightarrow \pi_i p_{ij} \quad \text{with probability 1.} \tag{2.191}$$

Using Theorem 2.10 again we see that

$$\frac{1}{n}N_i(n) \to \pi_i \text{ with probability 1,} \tag{2.192}$$

whence the result follows. □

The mle's are asymptotically normally distributed. To show this, we first need a technical result. Let $W_i^{(m)}=X_{1+T_i(m)}$ be the state entered directly after the mth return to i, and write $Q_{ij}(n)=\sum_{m=1}^{[n\pi_i]} 1(W_i^{(m)}=j)$. Finally let $Q_i(n)=(Q_{ij}(n), j\in S)$.

Lemma 2.6 The vectors $Q_i(n)$ for $i\in S$ are independent having multinomial distributions with sample size $[n\pi_i]$ and success probabilities $\mathbb{P}_{i\cdot}$.

Proof We need to show that the $W_i^{(m)}$ are independent with $\mathbf{P}(W_i^{(m)}=j)=p_{ij}$. But this follows from the strong Markov property, since given that $X_{T_i(m)}=i$ the future and the past are independent. □

We are now able to establish the asymptotic normality of the mle.

Theorem 2.15 Let X_n be an ergodic process. Then, regardless of the initial distribution,

$$\left[N_i(n)^{½}(\hat{p}_{ij}(n)-p_{ij}),\ i,j\in S\right] \xrightarrow{d} \mathrm{N}(\mathbf{0},\Sigma) \tag{2.193}$$

where

$$\Sigma_{ij,kl} = \begin{cases} p_{ij}(1-p_{ij}) & (i,j)=(k,l) \\ -p_{ij}p_{il} & i=k, j\neq l \\ 0 & \text{otherwise} \end{cases} \tag{2.194}$$

Remark The asymptotic covariance has multinomial structure within rows and independence between rows. Note, however, that we have to use a **random norming**, which is quite different from asymptotics for iid sequences. □

Proof Since $n\pi_i/N_i(n)\xrightarrow{\text{a.s.}}1$ we need only show that

$$\left[\frac{N_{ij}(n)-N_i(n)p_{ij}}{(n\pi_i)^{½}},\ i,j\in S\right] \xrightarrow{d} \mathrm{N}(\mathbf{0},\Sigma). \tag{2.195}$$

The basic idea is that $N_{ij}(n)$ is about the same as $Q_{ij}(n)$, and $N_i(n)$ is about $[n\pi_i]$. From the results in Appendix A and the lemma we know that

$$\left[\frac{Q_{ij}(n)-[n\pi_i]p_{ij}}{[n\pi_i]^{½}},\ i,j\in S\right] \xrightarrow{d} \mathbf{N}(\mathbf{0},\Sigma). \tag{2.196}$$

Hence we just need to show that these approximations are adequate, in the sense that

$$D_n = (n\pi_i)^{-1/2}(N_{ij}(n)-N_i(n)p_{ij}-Q_{ij}(n)+[n\pi_i]p_{ij}) \xrightarrow{\mathbf{P}} 0. \tag{2.197}$$

For fixed i,j let $Z_n=1(W_i^{(m)}=j)-p_{ij}$ and $S_n=\sum_1^n Z_i$. The Z_i are iid with mean zero, variance σ^2, and fourth moment κ. We can write D_n from (2.197) in terms of S_n as

$$D_n=(n\pi_i)^{-1/2}(S_{N_i(n)} - S_{[n\pi_i]}). \tag{2.198}$$

Then

$$\mathbf{P}(|D_n|>\varepsilon) \le \mathbf{P}(|D_n>\varepsilon, |N_i(n)-n\pi_i|\le xn^{1/2}) + \mathbf{P}(|N_i(n)-n\pi_i|>xn^{1/2}) \tag{2.199}$$

where x is a number to be chosen below. The first term of the right-hand side of (2.199) can be written as a sum over the possible values M of $N_i(n)$ satisfying the inequality $|m-n\pi_i|\le xn^{1/2}$. Using Chebyshev's inequality twice yields an upper bound of

$$\sum_{m\in M} \frac{1}{n^2\pi_i^2\varepsilon^4}\mathbf{E}(S_m-S_{[n\pi_i]})^4. \tag{2.200}$$

Since $S_m-S_{[n\pi_i]}$ is a sum of $|m-[n\pi_i]+1|$ of the Z_i we have, using $\mathbf{E}(\sum_1^n Z_i)^4 = n\kappa + 3n(n-1)\sigma^4$, that

$$\mathbf{E}(S_m-S_{[n\pi_i]})^4 \le (xn^{1/2}+1)\kappa+3(xn^{1/2}+1)^2\sigma^4. \tag{2.201}$$

The sum in (2.200) has at most $2xn^{1/2}+1$ terms, so we get

$$\mathbf{P}(|D_n|>\varepsilon) \le \frac{2xn^{1/2}+1}{n^2\pi_i^2\varepsilon^4}((xn^{1/2}+1)\kappa+3(xn^{1/2}+1)^2\sigma^4) + \mathbf{P}(|N_i(n)-n\pi_i|>xn^{1/2}). \tag{2.202}$$

The first term on the right-hand side goes to zero, while the second can be made arbitrarily small by making x large and using Theorem 2.10. □

Application (Snoqualmie Falls precipitation, continued) Using the result of Theorem 2.15, we see that $\hat{p}_{01}$ and $\hat{p}_{11}$ are asymptotically independent. Furthermore, $\hat{p}_{11}$ is approximately normally distributed with mean p_{11} and variance $p_{11}((1-p_{11})/n\pi_1$. We estimate that variance using $\hat{p}_{11}=n_{11}/n_1$ and $\hat{\pi}_1=n_1/n$ where $n_{11}=\sum_{i=1}^{36} n_{11}^{(i)}$, etc. Since $n_{11}=643$, $n_1=771$ and $n=1080$, an asymptotic 95% confidence band for p_{11} is (0.808,0.860), while one for p_{01} is (0.343,0.453) using $n_{01}=123$ and $n_0=309$. These are individual confidence bands, and the asymptotic joint coverage probability of the rectangle formed by these intervals is, using the asymptotic independence, $0.95^2 = 0.903$. To find an asymptotic 95% joint confidence set we can use individual 97.5% intervals, which yield the rectangle $(0.775, 0.893) \times (0.272, 0.524)$. □

Sometimes it is natural to look at a smaller model than the full nonparametric model. One may have some relatively simple model in mind, which is parametrized in some fashion.

Application (Russian linguistics) One of Markov's own examples of a Markov chain is in his 1924 probability text. The reconstruction here is using the description by Maistrov (1974). Markov studied a piece of text from Puškin's "Eugen Onegin", and classified 20,000 consecutive characters as vowels or consonants. The data are given below:

Table 2.3 Eugen Onegin characters

	Vowel next	Consonant next	Total
Vowel	1106	7532	8638
Consonant	7533	3829	11362
Total	8639	11361	20000

It is quite clear that the choice of vowel or consonant following a given letter is not independent of the letter. A very simple linguistic model is to assume a constant probability p of switching from one type of character to another. The transition matrix for this hypothesis is

$$\mathbb{P} = \begin{bmatrix} 1-p & p \\ p & 1-p \end{bmatrix}, \tag{2.203}$$

i.e., a one-dimensional subset of the space of stochastic matrices. □

For simplicity we will look only at the case of a finite state space. Assume that the transition probabilities $p_{ij} = p_{ij}(\boldsymbol{\theta})$ depend on an unknown parameter $\boldsymbol{\theta}$, taking values in Θ, an open subset of $\boldsymbol{R}^r$. We will need some regularity conditions:

Conditions A: (i) $D = \{i,j : p_{ij}(\boldsymbol{\theta}) > 0\}$ is independent of $\boldsymbol{\theta}$.
(ii) Each $p_{ij}(\boldsymbol{\theta}) \in C^3$, i.e., each $p_{ij}(\boldsymbol{\theta})$ is three times continuously differentiable.
(iii) The $d \times r$-matrix $\partial p_{ij}(\boldsymbol{\theta}) / \partial \theta_k$, $i,j \in D$, $k = 1, \ldots, r$, where d is the cardinality of D, has rank r.
(iv) For each $\boldsymbol{\theta}$ there is only one ergodic class and no transient states.

We now write the log likelihood from observing $x_1, \ldots, x_n$ (as above we argue conditionally upon the initial state)

$$l(\boldsymbol{\theta}) = \sum_D n_{ij} \log p_{ij}(\boldsymbol{\theta}), \tag{2.204}$$

where as before $n_{ij} = \#\{0 \le k \le n-1 : x_k = i, x_{k+1} = j\}$. Differentiating we get the likelihood equations

$$\frac{\partial}{\partial \theta_k} L_n(\boldsymbol{\theta}) = \sum_D \frac{n_{ij}}{p_{ij}(\boldsymbol{\theta})} \frac{\partial p_{ij}(\boldsymbol{\theta})}{\partial \theta_k} = 0,\ k=1, \ldots, r. \tag{2.205}$$

Let $\boldsymbol{\theta}_0$ be the true parameter value. The following result is due to Billingsley (1961). We will not prove it here.

Theorem 2.16 Assume Conditions A.

(i) There is a consistent solution $\hat{\boldsymbol{\theta}}$ of the likelihood equations.

(ii) $\sqrt{n}\,(\hat{\boldsymbol{\theta}} - \boldsymbol{\theta}_0) \to \mathrm{N}(\mathbf{0}, I^{-1}(\boldsymbol{\theta}_0))$, where I is the **information matrix** with typical element

$$I_{uv}(\boldsymbol{\theta}_0) = \sum_{(i,j)\in D} \frac{\pi_i(\boldsymbol{\theta}_0)}{p_{ij}(\boldsymbol{\theta}_0)} \frac{\partial p_{ij}(\boldsymbol{\theta}_0)}{\partial \theta_u} \frac{\partial p_{ij}(\boldsymbol{\theta}_0)}{\partial \theta_v} \tag{2.206}$$

and $\pi_i(\boldsymbol{\theta}_0)$ is the stationary probability of state i.

(iii) $\mathbf{Var}\sqrt{n}\,(\hat{\theta} - \theta)$ can be consistently estimated by

$$\left[-\frac{n_{ij}}{n} \nabla^2 \log p_{ij}(\hat{\theta})\right]^{-1}. \tag{2.207}$$

The quantity inverted in (2.207) is called the **observed information.**

Application (Russian linguistics, continued) We estimate p by maximizing the log likelihood

$$l(p) = (n_{00} + n_{11}) \log(1-p) + (n_{01} + n_{10}) \log p, \tag{2.208}$$

where 1 denotes a consonant and 0 a vowel. The maximum is obtained at $\hat{p} = (n_{01} + n_{10})/n = (7{,}532 + 7{,}533)/20{,}000 = 0.753$. The second derivative of the log likelihood is

$$l''(p) = -\frac{n_{00} + n_{11}}{(1-p)^2} - \frac{n_{01} + n_{10}}{p^2} \tag{2.209}$$

so from Theorem 2.16 the asymptotic estimated standard error is $(-l''(\hat{p}))^{-\frac{1}{2}} = (\hat{p}(1-\hat{p})/n)^{\frac{1}{2}}$, yielding an asymptotic confidence interval for p of (0.747,0.759). Notice that neither $\hat{p}_{01} = 0.872$ nor $\hat{p}_{10} = 0.663$ fall inside this confidence interval, indicating that the simple one-parameter model is inadequate. □

It is straightforward to develop a likelihood ratio theory of testing hypotheses for Markov chains satisfying Conditions A. Here is a general result, again left without proof (see, e.g., Billingsley, 1961, Theorem 3.1).

Theorem 2.17 Assume Conditions A. Let $\hat{\boldsymbol{\theta}}$ be the mle under the parametric hypothesis H_0. Also, let $\hat{\mathbb{P}}$ be the nonparametric mle, and $\boldsymbol{\theta}_0$ the true value of $\boldsymbol{\theta}$, assuming that H_0 is true. Then

(a) $2(l(\hat{\boldsymbol{\theta}})-l(\boldsymbol{\theta}_0)) \xrightarrow{d} \chi^2(r)$

(b) $2(l(\hat{\mathbb{P}})-l(\hat{\boldsymbol{\theta}})) \xrightarrow{d} \chi^2(d(d-1)-r)$

(c) The statistics in (a) and (b) are asymptotically independent. □

Remark Under conditions similar to Conditions A it is possible to derive a result much like Theorem 2.17 for testing a parametric model against a submodel. That is, in fact, the result given by Billingsley (1961). □

Example (Testing for independence) Suppose we want to study the hypothesis that (x_k) is a sequence of iid random variables, taking values in $\{0, \ldots, K\}$ (so $d=K+1$). In terms of a parametrization this is simply $H_0\colon p_{ij}=\theta_j$ for all $i\in S$ and each $j\in S$. We must compute the maxima of the likelihood under the two models. We already know that $\hat{p}_{ij}=n_{ij}/n_i$. Under the independence assumption we have a multinomial distribution, with $n_{\bullet j}=\sum_i n_{ij}$ observations from the category with probability θ_j. The likelihood is

$$l(\boldsymbol{\theta}) = \sum_{j=0}^{K-1} n_{\bullet j}\theta_j + n_{\bullet K}(1-\sum_{j=0}^{K-1}\theta_j), \tag{2.210}$$

which is maximized by $\hat{\theta}_j=n_{\bullet j}/n$. Hence the log likelihood ratio statistic for testing H_0 is

$$2(l(\hat{\mathbb{P}})-l(\hat{\boldsymbol{\theta}})) = 2\sum_{i,j} n_{ij}\log\frac{n_{ij}/n_i}{n_{\bullet j}/n} \tag{2.211}$$

which asymptotically has a χ^2 distribution with $K(K+1)-K=K^2$ degrees of freedom. In the Snoqualmie Falls rain model we have $K=1$, in accordance with our earlier claim. □

Application (Sedimentology) An important aspect of geology is stratigraphy, the description of sedimental layers. An interesting question is whether or not there is memory in observed sequences of strata, or **facies**. If memory, or perhaps better preference, is present the conditional probability that facies B will be deposited on top of an observed facies A may then be different from the conditional probability given any other underlying facies.

Hiscott (1981) studied the Ordocovian Tournelle Formation in Québec and distinguished, based on field criteria, two different facies associations (or underlying transition probabilities) for six different facies:

Facies	Description
0	Thick shale
1	Interbedded graded siltstones and shale
2	Poorly sorted sandstones with dispersed clasts
3	Interbedded graded sandstones and shales
4	Amalgamated graded sandstones
5	Thick coarse sandstones

Of course, there are no $i \to i$ transitions. The transition counts for one of the facies associations are given in Table 2.4.

Table 2.4 Upward transition counts in the Tournelle Formation

Facies	0	1	2	3	4	5	
0	0	2	0	2	2	0	6
1	5	0	23	31	17	8	84
2	0	21	0	45	27	8	101
3	1	54	44	0	66	25	190
4	0	6	24	81	0	38	149
5	0	5	8	31	32	0	76
	6	88	99	190	144	79	606

Our null hypothesis is independence, i.e., no preference in the sedimentation. The log likelihood statistic works out to 405.9 with $4\times6-5=19$ degrees of freedom, since $K=5$ but the transition matrix is restricted to have zeros on the diagonal. Hence the P-value is 0, and we reject the hypothesis of independence. □

Remark As in the multinomial calculation in Appendix A, the log likelihood statistic is approximately a Pearson χ^2 statistic, in that

$$2\sum n_{ij}\log\frac{n_{ij}}{n_i p_{ij}^0} = \sum\frac{(n_{ij}-n_i p_{ij}^0)^2}{n_i p_{ij}^0} + o_{\mathbf{P}}(1). \tag{2.212}$$

The latter form is sometimes more convenient to compute. □

Application (Russian linguistics, continued) We are testing the hypothesis $H_0: p_{10}=p_{01}$. The expected counts that we need for the χ^2 statistic are computed by multiplying the row sums (n_0,n_1)=(8,638, 11,362) with the transition matrix estimate under H_0

$$\hat{\mathbb{P}} = \begin{bmatrix} 0.247 & 0.753 \\ 0.753 & 0.247 \end{bmatrix} \tag{2.213}$$

yielding

$$\begin{bmatrix} 2131.4 & 6506.6 \\ 8558.4 & 2803.6 \end{bmatrix} \tag{2.214}$$

The χ^2 statistic for testing the one-dimensional null hypothesis given above within the general nonparametric Markov chain model is

$$\chi^2 = \sum \frac{(n_{ij}-n_i\hat{p}^0_{ij})^2}{n_i\hat{p}^0_{ij}} = 1217.7. \tag{2.215}$$

The exact likelihood ratio statistic is also 1217.7, so the approximation is excellent. The statistic has one degree of freedom, since the general nonparametric model has dimension 2 and the null hypothesis has dimension 1. Hence the null hypothesis is rejected, as was suggested by the confidence interval we derived earlier. The test used in this example can also be thought of as a test for stationary distribution (½,½), since this happens if and only if the transition matrix is doubly stochastic (Exercise **6**). □

Application (Stock market pricing) Much effort in finance theory has gone into studying the predictability of the stock market. The **efficient market hypothesis** (see Fama, 1970) implies that the deviations of the overall stock market (or, more precisely, of a portfolio containing all the stocks of a given exchange) from the mean should be independent random variables. This in turn implies that knowledge of the previous behavior of the market does little to help predict future behavior. The stock prices are said to follow a **random walk**. Empirical studies have cast some doubt over this hypothesis. There is some evidence that large deviations from the mean (in essence, highly overpriced or underpriced stocks) tend to be reverting to the mean, leading to negative correlations over long periods of time, and violating the independence assumption.

Several explanations have been proposed to this market behavior. One, proposed by Blanchard and Watson (1982), is called the **rational speculative bubbles** model. In this model, investors realize that prices exceed fundamental values, but they believe that there is a high potential for the bubble to continue to expand and yield a high return. This high return compensates precisely for the risk of a crash, showing the rationality of staying in the marked despite the overvaluation.

McQueen and Thorley (1991) applied a Markov chain model to annual stock returns. This chain took into account the behavior of the market for two successive years, each classified as above or below mean (the mean value used was a running 20-year mean—the results are not sensitive to the choice of mean value). The state space, using 0 to indicate below average prices, is $S=\{(0,0),(0,1),(1,0),(1,1)\}$, where (0,0) denotes two successive below average years. In fact, while it is a regular Markov chain on this state space, it can also be considered a **second order** Markov chain, on the state space (0,1). The term second order indicates that the dependence goes back two steps. We shall study higher order chains in more detail in the next section. The transition matrix for the chain is

$$\mathbb{P} = \begin{bmatrix} 1-p_{001} & p_{001} & 0 & 0 \\ 0 & 0 & 1-p_{011} & p_{011} \\ 1-p_{101} & p_{101} & 0 & 0 \\ 0 & 0 & 1-p_{111} & p_{111} \end{bmatrix} \tag{2.216}$$

where p_{001} is the transition probability from (0,0) to (0,1). Since the second element of the previous state must be the same as the first element of the current state we do not need four binary digits in the subscript for p. In terms of this model, the random walk hypothesis is

$$H_0: p_{001} = p_{011} = p_{101} = p_{111} \tag{2.217}$$

while the rational speculative bubbles hypothesis can be written

$$H_1: p_{001} > p_{111} \tag{2.218}$$

since the probability of state 0 should be larger following (1,1) than following (0,0). Thus, rejection of the hypothesis

$$H'_0: p_{001} = p_{111} \tag{2.219}$$

in the right direction can be taken as evidence in favor of the rational speculative bubbles hypothesis.

The data used by McQueen and Thorley consist of continually compounded returns for a portfolio of all New York Stock Exchange stocks for the calendar years 1947 to 1987. We consider only an equally-weighted portfolio where continually compounded inflation has been subtracted from the nominal rates. The data are given in Table 2.5. In this case we know the initial state: it is (0,0). The mle's are

$$\hat{p}_{001} = 0.750;\ \hat{p}_{011} = 0.818;\ \hat{p}_{101} = 0.5;\ \hat{p}_{111} = 0.1. \tag{2.220}$$

The likelihood ratio test of the random walk hypothesis H_0 yields a test statistic value of 14.0 on 3 degrees of freedom. Nominally, this corresponds to a P-value of 0.003. Since the numbers involved are rather small, McQueen and Thorley performed a simulation study yielding an actual P-value of 0.01. It is quite clear that the random walk hypothesis is untenable. The test of H'_0 is 8.6 on 1 degree

Table 2.5 Transition counts for NYSE portfolio 1947–1987

	0	1
(0,0)	2	6
(0,1)	2	9
(1,0)	5	5
(1,1)	9	1

of freedom, rejected at all levels using the chi-squared distribution, and receiving a P-value of 0.01 in the simulation study. Since $\hat{p}_{111} < \hat{p}_{001}$ we may want to interpret this as evidence in favor of the rational speculative bubbles hypothesis. □

2.8. Higher order chains

In the stock market example at the end of the previous section we saw how the dependence on the past can reach farther than just to the previous time. We define an rth order Markov chain (X_k) on a state space S with d elements by

$$\begin{aligned}\mathbf{P}(X_{n+1}=x_{n+1} \mid X_n=x_n, X_{n-1}=x_{n-1}, \ldots, X_0=x_0) \\ = \mathbf{P}(X_{n+1}=x_{n+1} \mid X_n=x_n,, \ldots, X_{n-r+1}=x_{n-r+1}) \qquad (2.221) \\ = p(x_{n-r+1}, \ldots, x_n; x_{n+1}),\end{aligned}$$

replacing subscripts by function notation for readability. There is no real novelty in an rth order chain. Let namely (Y_k) be a process with state space S^r defined by $Y_k=(X_{k-r+1}, \ldots, X_k)$. Then

$$\mathbf{P}(Y_n=(a_1, \ldots, a_r) \mid Y_{n-1}=(b_1, \ldots, b_r)) = \begin{cases} p(b_1, \ldots, b_r; a_r) \text{ if } a_i=b_{i+1},\ i=1, \ldots, r-1 \\ 0 \text{ otherwise.} \end{cases} \qquad (2.222)$$

Some reflection shows that (Y_k) is a first-order Markov chain with $(d-1)d^r$ states. Hence we can use first-order chain statistical theory to test hypotheses such as the chain being lth order, where $l<r$. We can formulate the hypothesis as

$$p(a_1, \ldots, a_r; a_{r+1}) = p(a_{r-l+1}, \ldots, a_r; a_{r+1}). \qquad (2.223)$$

The mle under this hypothesis is

$$\hat{p}(a_1, \ldots, a_r; a_{r+1}) = \frac{n(a_{r-l+1}, \ldots, a_r, a_{r+1})}{n(a_{r-l+1}, \ldots, a_r)} \qquad (2.224)$$

where $n(a_{r-l+1}, \ldots, a_r) = \sum_l n(_{r-l+1}, \ldots, a_r l)$. The χ^2 statistic is

$$\sum_{a_1,\ldots,a_{r+1}} \frac{(n(a_1,\ldots,a_{r+1})-n(a_1,\ldots,a_r)\hat{p}(a_1,\ldots,a_r;a_{r+1}))^2}{n(a_1,\ldots,a_r)\hat{p}(a_1,\ldots,a_r;a_{r+1})}, \quad (2.225)$$

which asymptotically has a χ^2 distribution with

$$(d-1)d^r-(d-1)d^l = (d-1)d^l(d^{r-l}-1) \quad (2.226)$$

degrees of freedom.

Application (Snoqualmie Falls precipitation, continued) Looking at the Snoqualmie Falls data in more detail, we obtain Table 2.6.

Table 2.6 Data for second-order model

Previous days		Current day		Proportion
Second	First	Dry	Wet	wet
Wet	Wet	100	527	0.841
Dry	Wet	25	94	0.790
Wet	Dry	70	52	0.426
Dry	Dry	109	67	0.381

The 95% asymptotic joint confidence set for (p_{11},p_{01}) from the first-order model was (0.775, 0.893)×(0.272, 0.524). All the observed proportions fall inside this set, indicating that the first-order model is adequate. Note however, that if the previous two days were (dry,wet), the observed proportion 0.790 is outside the individual 95% confidence interval (0.808, 0.860) for p_{11}, showing the importance of using simultaneous rather than individual confidence bands. The χ^2 statistic for testing second order vs. first order is 2.4. Here $r=d=2$, $l=1$ so the statistic has $1\times2^1\times(2^{(2-1)}-1)=2$ degrees of freedom. There are four parameters in the second-order model, and two in the first-order model. The P-value is 0.29, and we see no reason to reject the first-order model. □

In order to test for order of a Markov chain we may use the fact that the test statistics in successively nested hypotheses are asymptotically independent. This creates a multiple decision problem, which is complicated to analyze. Suppose that a chain is really order 1. Then the probability of falsely rejecting the true order is the probability of falsely accepting order 0, and of falsely rejecting order 1 in favor of order 2. These two events are asymptotically independent. The second has asymptotic probability α, but the probability of the first depends on the sample size and on how far the p_{ij} are from the independent case (i.e., on the noncentrality parameter of the χ^2 statistic).

A different possibility is to consider the order as a parameter, and estimate it using maximum likelihood. This does not work, because we can make the likelihood arbitrarily large by having one parameter for each observation. However, if we penalize the likelihood for the number k of independent parameters, by maximizing $2\,l(\hat{\boldsymbol{\theta}})-f\,(k,n)$ for some suitable choice of f, we may be able to offset the increase in the likelihood that is due to an increase in the number of parameters, rather than to an improved fit. Many different choices of f have been suggested. We will use the **Bayes Information Criterion** (BIC), for which $f\,(k,n)=k\log n$. In other words, we look for the model that maximizes

$$\mathrm{BIC}(k) = 2\max_{\Theta_k} l(\boldsymbol{\theta}) - k\log n \tag{2.227}$$

where Θ_k is the parameter space corresponding to k parameters (not all values of k may be possible). An important point is that the sample size used must be consistent with the largest model considered. For example, if we are considering a third-order model, and we have n observations of the chain, we can only use the last $n-3$ observations for the zero order model, the last $n-2$ for the first order, etc. This is because the estimates of the third-order model do not start until the fourth observation, the first three being needed to see what state the chain is leaving. It turns out that for finite state Markov chains, BIC is a consistent estimate of the order of the chain (Katz, 1981). The rules of thumb of Jeffreys (1961, Appendix B) suggest that a difference in BIC of at least 2 log 100 = 9.2 is needed to deem the model with the smaller BIC substantially better.

Application (Snoqualmie Falls precipitation, continued) In order to apply the BIC to the Snoqualmie Falls data we need the maximum likelihood for the chains of order 0, 1, and 2. These are given in Table 2.7, with the value for the model chosen by each criterion shown in boldface.

Table 2.7 BIC for Snoqualmie Falls precipitation

Order	k	$2l(\hat{\theta}_k)$	BIC
0	1	−1259.5	−1266.5
1	2	**−1075.3**	**−1089.2**
2	4	−1073.0	−1100.8

Both the likelihood ratio test and the BIC favor the first-order model. It is usually the case that BIC only has one maximum as a function of k. Generally BIC tends to choose smaller models than the likelihood ratio test. □

High-order Markov chains have rather a lot of parameters: $(d-1)d^l$ for the full lth order chain. This makes the model unsuitable even for relatively small d, as shown in Table 2.8. One would sometimes like a model that allows for high-order dependence, although not using as many parameters as the full lth order

Table 2.8 Number of parameters for different order chains

	Order			
d	1	2	3	4
2	2	4	8	16
3	6	18	54	162
4	12	48	192	768
5	20	100	500	2500

model. Raftery (1985a) proposes a linear model for Markov chain transition probabilities. Let $Q=(q_{ij})$ be a transition matrix, $\boldsymbol{\lambda}$ an l-vector of parameters summing to 1. Assume that

$$\mathbf{P}(X_k=j \mid X_{k-1}=j_1, \ldots, X_{k-l}=j_l) = \sum_{i=1}^{l} \lambda_i q_{j_i j}. \tag{2.228}$$

It is clear that this defines a lth order Markov chain. If Q is ergodic then X is ergodic and has an equilibrium distribution $\boldsymbol{\pi}$, which also is the stationary probability vector for Q. This model is called the **mixture transition distribution** (MTD) model.

It is straightforward to write down the likelihood for an MTD model.

$$L(\boldsymbol{\lambda}, Q) = \sum_{i, i_1, \ldots, i_l=0}^{K} n(i_1, \ldots, i_l, i) \log\left(\sum_{j=1}^{l} \lambda_i q_{i_j i}\right). \tag{2.229}$$

There is no simple closed form for the maximum likelihood estimators. The likelihood must be optimized numerically (Schimert, 1992).

Application (Wind power in Ireland) In order to design turbines for wind power generation in Ireland, data were collected on hourly wind speeds at Belmullet for the first four weeks of July, 1962. The 672 wind speed measurements were grouped into four states:

State 0:	No power produced	0–8 knots
State 1:	Less than full potential	8–16 knots
State 2:	Full capacity	16–25 knots
State 3:	Closed down due to high winds	> 25 knots

Since no transitions to other than neighboring classes were seen, transitions farther away were assumed to have probability zero in order to cut down on the number of parameters in the general Markov chain model (see Table 2.9). The likelihood ratio statistic for testing order 2 vs. 1 is 32.3 on 10 degrees of freedom (P = 0.0003), and that for testing order 3 vs. order 2 is 18.9 on 26 degrees of freedom (P = 0.84). The likelihood ratio test therefore chooses order 2. The

Table 2.9 Order selection for Irish wind power models

Order	Markov chain			MTD chain		
	k	L	BIC	k	L	BIC
0	3	−869.5	−1758.6			
1	6	−417.5	**−874.1**			
2	16	−385.2	−874.5	7	−395.6	−836.7
3	42	−366.3	−1006.1	8	−388.0	**−828.1**
4	110	−347.2	−1410.5	9	−388.0	−834.6

penalty for additional parameters in going to order 3 is simply too large. Note that the maximum likelihood for the MTD chain of order 3 is comparable to that of the general chain of order 2, so not much is lost in reducing from 16 to 8 parameters. The estimated parameter values are

$$\hat{\boldsymbol{\lambda}} = (0.629, 0.206, 0.165) \tag{2.230}$$

and

$$\hat{Q} = \begin{bmatrix} 0.837 & 0.163 & 0 & 0 \\ 0.058 & 0.854 & 0.088 & 0 \\ 0 & 0.133 & 0.847 & 0.040 \\ 0 & 0 & 0.116 & 0.884 \end{bmatrix}. \tag{2.231}$$

The estimated stationary probabilities are

$$\hat{\boldsymbol{\pi}} = (0.148, 0.416, 0.324, 0.111). \tag{2.232}$$

We see that the turbines are expected to be producing power about 3/4 of the time in July, although optimal production only occurs about 1/3 of the time. We interpret $\hat{\boldsymbol{\lambda}}$ as the relative strength of influence of past values. □

2.9. Chain-dependent models

In order to test the goodness of fit of our model of Snoqualmie Falls precipitation, we may, for example, derive the theoretical distribution of some functional of the data, and compare it (using our estimated parameters) to the empirical distribution from our data. Some examples of interesting such functionals are the number of rainy days in a week, the total amount of rainfall in a month, and the maximum rainfall in a month (a number of particular hydrologic importance).

Let $Z_1, \ldots, Z_n$ be a sequence of random variables such that the distribution of Z_k, given that $X_{k-1}=j$ and $X_k=l$, has distribution function F_{jl}. We can think of Z_{k+1} as a score associated with the kth transition. Note that this generalizes the ergodic theory in section 2.5, since we are introducing external randomness (not just looking at a non-random function of the current state).

Conditional on a given sequence of transitions we assume that the Z_i are independent. Let $S_n = \sum_1^n Z_k$. S_n is called an **additive functional** of the Markov chain. Our goal is to compute the distribution of S_n, or equivalently its Laplace transform

$$\phi_n(t) = \mathbf{E}^{\pi}(\exp(-tS_n)), \tag{2.233}$$

assuming that the Markov chain is in equilibrium. Write $\phi_{jk}^{(r)}(t) = \mathbf{E}(\exp(-tS_r) \mid X_0 = j, X_r = k)$. Given that $X_0 = j$, $X_{n-1} = l$, and $X_n = k$, the random variables S_{n-1} and Z_n are conditionally independent, with Laplace transforms $\phi_{jl}^{(n-1)}(t)$ and $\phi_{lk}^{(1)}(t)$, respectively. Thus

$$\mathbf{E}(\exp(-tS_n) \mid X_1 = j, X_{n-1} = l, X_n = k) = \phi_{jl}^{(n-2)}(t)\phi_{lk}^{(1)}(t). \tag{2.234}$$

Averaging over the value of X_{n-1} we get

$$\begin{aligned} \phi_{jk}^{(n-1)}(t) &= \mathbf{E}(\exp(-tS_n) \mid X_1 = j, X_n = k) \\ &= \sum_l \mathbf{P}(X_{n-1} = l \mid X_n = k, X_0 = j) \\ &\quad \times \mathbf{E}(\exp(-tS_n) \mid X_1 = j, X_{n-1} = l, X_n = k) \end{aligned} \tag{2.235}$$

or, since $\mathbf{P}(X_{n-1} = l \mid X_n = k, X_0 = j) = p_{jl}^{(n-1)} p_{lk} / p_{jk}^{(n)}$,

$$p_{jk}^{(n)} \phi_{jk}^{(n)}(t) = \sum_l p_{jl}^{(n-1)} \phi_{jl}^{(n-1)}(t) p_{lk} \phi_{lk}^{(1)}(t). \tag{2.236}$$

Writing $\mathbf{Q}^{(n)}(t) = (p_{jk}^{(n)} \phi_{jk}^{(n)}(t))$, we see that

$$\mathbf{Q}^{(n)}(t) = \mathbf{Q}^{(n-1)}(t)\mathbf{Q}^{(1)}(t) \tag{2.237}$$

so, with $\mathbf{Q}^{(1)} \equiv \mathbf{Q}$, we must have $\mathbf{Q}^{(n)}(t) = (\mathbf{Q}(t))^n$. It now remains to average over X_0 and X_n:

$$\phi_n(t) = \mathbf{E}^{\pi}(\exp(-tS_n)) = \sum_{j,k} \pi(j) p_{jk}^{(n)} \phi_{jk}^{(n)}(t) = \boldsymbol{\pi} Q(t)^n \mathbf{1}^T. \tag{2.238}$$

Example (Semi-Markov chains) One drawback with Markov chains is the inflexibility in describing the time spent in a given state. We have seen (Exercise **2**) that this time must be geometrically distributed with parameter p_{jj}. A generalization is the semi-Markov chain, in which the process moves out of a given state according to a Markov chain with transition matrix $\mathbb{P}$, having $p_{jj} = 0$ for all j, while the time spent in state j is a random variable with distribution function $F_j(l)$, or, more generally, $F_{ij}(l)$ where i is the state the process came from. □

Application (Snoqualmie Falls precipitation, continued) Suppose that we are interested in studying the distribution of the number of days in a week with measurable precipitation. Then $Z_k=1(X_k=1)\equiv X_k$. Thus $\phi_{00}=\phi_{10}=1$ and $\phi_{01}=\phi_{11}=e^{-t}$. In Exercise **11** we show how to compute Q^n using diagonalization. However, for small values of n there is a simple recursive formula for computing the exact distribution of S_n. Let $u_k^{(n)}=\mathbf{P}^0(S_n=k)$ and $v_k^{(n)}=\mathbf{P}^1(S_n=k)$. Then

$$\begin{aligned} u_k^{(n)} &= \mathbf{P}(S_n=k,X_1=0 \mid X_0=0) + \mathbf{P}(S_n=k,X_1=1 \mid X_0=0) \\ &= \mathbf{P}(S_{n-1}=k \mid X_0=0)p_{00} + \mathbf{P}(S_{n-1}=k-1 \mid X_0=1)p_{01}. \end{aligned} \tag{2.239}$$

Performing a similar computation for $v_k^{(n)}$, we have

$$\begin{aligned} u_k^{(n)} &= u_k^{(n-1)}p_{00} + v_{k-1}^{(n-1)}p_{01} \\ v_k^{(n)} &= u_k^{(n-1)}p_{10} + v_{k-1}^{(n-1)}p_{11} \end{aligned}. \tag{2.240}$$

Finally, for a two-state chain in equilibrium, we have

$$\mathbf{P}^{\pi}(S_n=k) = \pi_0 u_k^{(n)} + \pi_1 v_k^{(n)}. \tag{2.241}$$

For the Snoqualmie Falls data the exact distribution was computed for $n=7$, using the estimated transition matrix

$$\hat{\mathbb{P}} = \begin{bmatrix} 0.602 & 0.398 \\ 0.166 & 0.834 \end{bmatrix} \tag{2.242}$$

and compared to data from 144 midwinter weeks (Table 2.10).

Table 2.10 Snoqualmie Falls weekly rainfall

# wet days	Observed frequency	Expected frequency
0	3	1.9
1	5	4.5
2	9	9.0
3	10	15.0
4	21	21.9
5	25	27.6
6	30	30.0
7	41	34.2

The χ^2 statistic for goodness of fit is 4.03 with 4 degrees of freedom, for a P-value of 0.40. Strictly speaking the χ^2 distribution is not applicable, since the observations are from successive quadruples of weeks. However, as we have seen, the dependence between events seven days or more apart is fairly small. More precisely, the correlation between $\sum_1^7 Z_i$ and $\sum_8^{14} Z_i$ is estimated to be 0.074 (Exercise **12**). The adequacy of the χ^2-distribution is the subject of Exercise **C5**. □

In modeling the Snoqualmie Falls precipitation we have so far concentrated on looking at the presence or absence of rainfall. By augmenting the parameter space, as in the Irish wind example in the previous section, we can take into account the amount of precipitation, at least in a rough way. From a forecasting point of view it is important to do better than that. To that end, we will follow Katz (1977) and look at a bivariate process (X_n, Z_n), where $X_n = 1$(precipitation on day n). As before, we assume that X_n is a first-order Markov chain with stationary transition probabilities p_{ij} and stationary distribution $\boldsymbol{\pi}$. The amount of precipitation on the nth day is Z_n, positive precisely when $X_n = 1$. We make the following assumptions:

(i) The distribution of Z_n depends on (X_{n-1}, X_n).

(ii) The Z_i are conditionally independent, given the process (X_n).

It follows from these assumptions that

$$\mathbf{P}(Z_n \leq x \mid X_0, Z_1, X_1, Z_2, X_2, \ldots, Z_{n-1}, X_{n-1} = i) = \mathbf{P}(Z_n \leq x \mid X_{n-1} = i). \tag{2.243}$$

Let $F_i(x) = \mathbf{P}(Z_n \leq x \mid X_{n-1} = i, X_n = 1)$. Suppose that the chain is in equilibrium. Write

$$\mu_i = \mathbf{E}(Z_n \mid X_{n-1} = i, X_n = 1) \tag{2.244}$$

and

$$\sigma_i^2 = \mathbf{Var}(Z_n \mid X_{n-1} = i, X_n = 1). \tag{2.245}$$

By unconditioning

$$\mu = \mathbf{E}Z_n = \sum \mu_i \pi_i p_{i1} \tag{2.246}$$

and

$$\begin{aligned} \rho_0 &= \mathbf{Var}Z_n = \mathbf{E}\,\mathbf{Var}(Z_n \mid X_{n-1}, X_n) + \mathbf{Var}\,\mathbf{E}(Z_n \mid X_{n-1}, X_n) \\ &= \mathbf{E}\sigma_{X_{n-1}}^2 + \mathbf{Var}\mu_{X_{n-1}} = \sum \pi_i \sigma_i^2 p_{i1} + \sum \pi_i \mu_i^2 p_{i1} - \mu^2 \\ &= \pi_0 p_{01}(\sigma_0^2 + \mu_0^2) + \pi_1 p_{11}(\sigma_1^2 + \mu_1^2) - \mu^2. \end{aligned} \tag{2.247}$$

We will now look at the total amount of rainfall in n days. Let $S_n = \sum_1^n Z_i$. Then (O'Brien 1974)

$$\frac{S_n - n\mu}{\sigma n^{1/2}} \xrightarrow{d} \mathrm{N}(0,1) \tag{2.248}$$

where $\sigma^2=\rho_0+2\sum_1^\infty\rho_j$, and $\rho_j=\mathbf{Cov}(Z_n,Z_{n+j})$, provided that $0<\sigma^2<\infty$. To compute σ^2 we do another conditional computation:

$$\mathbf{E}Z_nZ_{n+j} = \mathbf{E}\,\mathbf{E}(Z_nZ_{n+j}\mid X_{n-1},X_n,X_{n+j-1},X_{n+j}) \tag{2.249}$$
$$= \mathbf{E}\mu_{X_{n-1}}\mu_{X_{n+j-1}} = \sum_{i,k}\pi_i\mu_i\mu_k p_{ik}^{(j)}.$$

This is again an additive functional. Adapting equation (2.238), let

$$Q = \begin{bmatrix} 0 & p_{01}\mu_0 \\ 0 & p_{11}\mu_1 \end{bmatrix} \tag{2.250}$$

and $\mathbf{1}=(1,\,1)$. Then we can write

$$\mathbf{E}Z_nZ_{n+j} = \boldsymbol{\pi}Q\mathbb{P}^jQ\mathbf{1}^T. \tag{2.251}$$

Note that $\mu=\boldsymbol{\pi}Q\mathbf{1}^T$. Thus

$$\rho_j=\boldsymbol{\pi}Q(\mathbb{P}^j-\mathbf{1}^T\boldsymbol{\pi})Q\mathbf{1}^T \tag{2.252}$$

and

$$\sigma^2 = \rho_0 + 2\sum_{j=1}^{\infty}\boldsymbol{\pi}Q(\mathbb{P}^j-\mathbf{1}^T\boldsymbol{\pi})Q\mathbf{1}^T. \tag{2.253}$$

The following fact helps in the computation:

Lemma 2.7 $\mathbb{P}^j-\mathbf{1}^T\boldsymbol{\pi} = (\mathbb{P}-\mathbf{1}^T\boldsymbol{\pi})^j$.

Proof We prove this by induction. The case $j=1$ is trivial. Assume that the statement is true for k. Then, since $\boldsymbol{\pi}\mathbb{P}=\boldsymbol{\pi}$

$$\mathbb{P}^{k+1}-\mathbf{1}^T\boldsymbol{\pi} = (\mathbb{P}^k-\mathbf{1}^T\boldsymbol{\pi})\mathbb{P} = (\mathbb{P}-\mathbf{1}^T\boldsymbol{\pi})^k\mathbb{P}$$
$$= (\mathbb{P}-\mathbf{1}^T\boldsymbol{\pi})^{k+1} + (\mathbb{P}-\mathbf{1}^T\boldsymbol{\pi})^k\mathbf{1}^T\boldsymbol{\pi} \tag{2.254}$$
$$= (\mathbb{P}-\mathbf{1}^T\boldsymbol{\pi})^{k+1} + (\mathbb{P}^k-\mathbf{1}^T\boldsymbol{\pi})\mathbf{1}^T\boldsymbol{\pi}$$

using the induction hypothesis twice. Now, since $\mathbf{1}^T$ is a right eigenvector for $\mathbb{P}$ we have $\mathbb{P}\mathbf{1}^T\boldsymbol{\pi} = \mathbf{1}^T\boldsymbol{\pi}$ and by iterating $\mathbb{P}^k\mathbf{1}^T\boldsymbol{\pi}=\mathbf{1}^T\boldsymbol{\pi}$, so $(\mathbb{P}^k-\mathbf{1}^T\boldsymbol{\pi})\mathbf{1}^T\boldsymbol{\pi}=\mathbf{0}$, completing the induction. □

It follows that the infinite sum is $2\boldsymbol{\pi}Q(\mathbb{P}-\mathbf{1}^T\boldsymbol{\pi})^{-1}Q\mathbf{1}^T$, and some algebra shows that

$$\sigma^2 = \rho_0 + \frac{2}{1-(p_{11}-p_{10})}\mu\pi_0(p_{11}\mu_1-p_{01}\mu_0). \tag{2.255}$$

Application (Snoqualmie Falls precipitation, continued) In order to illustrate the theory developed above we will apply it to the Snoqualmie Falls data. In order to fit the precipitation amounts, a gamma distribution was assumed. Figure 2.5 shows the observed and estimated distributions.

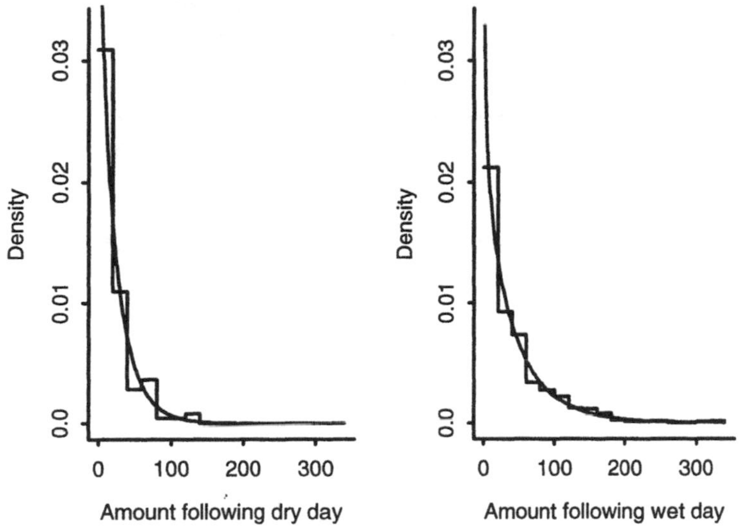

Figure 2.5. Observed and fitted densities of precipitation amounts following wet days (a) and dry days (b).

The parameters were estimated by maximum likelihood. The estimated scale parameters are different for wet days following wet days than for wet days following dry days, while the shape parameters are quite similar. Thus there is a tendency for lower amounts following a dry day. The means are μ_0=22.9 and μ_1=43.5, so μ=28.3 (recall that μ is the overall mean, taking into account the zero rainfall on dry days). The variances are σ_0^2=691.7 and σ_1^2=2351.7, whence σ^2=2339.0. In order to check assumption (ii), the amounts of precipitation on consecutive wet days were plotted on a log–log scale (Figure 2.6). There is no evidence of dependence. In order to study the distribution of the total amounts, the exact distribution, using the fitted gamma distribution, was compared to the limiting normal distribution derived above. Figure 2.7 shows the result for n =20, together with a histogram of observed sums for January 6–25. The fit of the exact distribution to the observed sums is rather bad, the latter showing evidence of bimodality, perhaps corresponding to dry and wet years. In addition, the normal approximation is bad, which is not too surprising since we are only summing up 20 random variables, many of which are zero. □

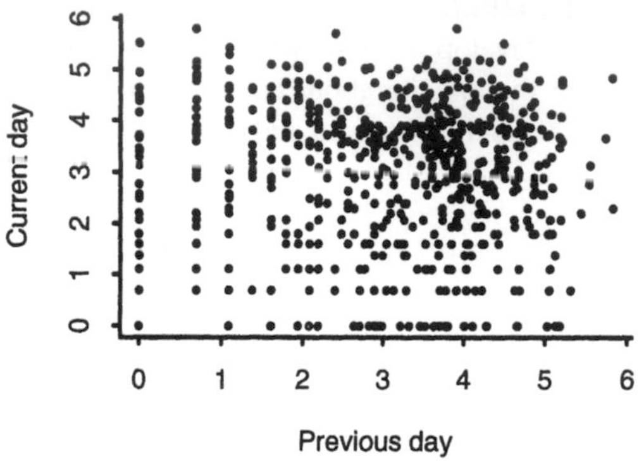

Figure 2.6. Amounts of precipitation on consecutive wet days.

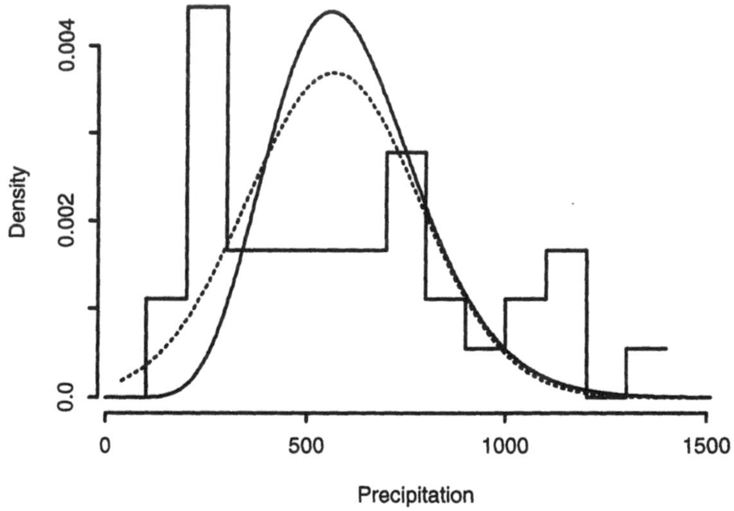

Figure 2.7. Exact theoretical (solid) and asymptotic (dotted) densities of precipitation January 6–20. The solid step function is a histogram of the 36 observed years.

One can easily derive another interesting parameter of the precipitation process, namely the distribution of maximum rainfall. Let $M_n=\max_{i\le n}Z_i$, and $G_n(x;i)=\mathbf{P}^i(M_n\le x)$. Also define $G_n(x)=\pi_0 G_n(x;0)+\pi_1 G_n(x;1)$. Splitting the set

$\{M_n \leq x, X_0 = 0\}$ over the possible values of X_1 we get that

$$G_n(x;0) = \frac{\mathbf{P}(M_n \leq x, X_0=0)}{\mathbf{P}(X_0=0)}$$

$$= \frac{\mathbf{P}(M_n \leq x, X_0=0, X_1=0)}{\mathbf{P}(X_0=0)} + \frac{\mathbf{P}(M_n \leq x, X_0=0, X_1=1)}{\mathbf{P}(X_0=0)}$$

$$= \mathbf{P}(M_{2,n} \leq x \mid X_1=0)\mathbf{P}(X_1=0 \mid X_0=0) \tag{2.256}$$

$$+ \mathbf{P}(M_{2,n} \leq x, Z_1 \leq x \mid X_0=0, X_1=1)\mathbf{P}(X_1=1 \mid X_0=0)$$

$$= p_{00}G_{n-1}(x;0) + p_{01}F_0(x)G_{n-1}(x;1)$$

where $M_{2,n} = \max_{2 \leq k \leq n} Z_k$. Similarly we obtain

$$G_n(x;1) = p_{10}G_{n-1}(x;0) + p_{11}F_1(x)G_{n-1}(x;1). \tag{2.257}$$

Using the initial conditions $G_0(x;0)=G_0(x;1)=1$ these equations can be solved recursively. Assuming that the F_i are cdf's of a gamma distribution, and writing

$$F(x) = \pi_0 + \pi_0 p_{01}F_0(x) + \pi_1 p_{11}F_1(x) \tag{2.258}$$

the following extreme value result is valid (Denzel and O'Brien 1975):

$$\lim_{n\to\infty} G_n\left[u_n + \frac{x}{nF'(u_n)}\right] = \exp(-\exp(-x)), \tag{2.259}$$

where $1-F(u_n)=1/n$. Rewriting (2.259) we see that

$$G_n(y) \approx \exp(-\exp(-nF'(u_n)(y-u_n))). \tag{2.260}$$

The assumption of gamma precipitation distribution is not crucial: Denzel and O'Brien (ibid.) show that the limiting behavior of M_n is the same as for iid observations from F (see Resnick, 1987, for details). Also, the assumption of starting in the stationary distribution is unnecessary: the limiting behavior is the same for all initial distributions.

Application (Snoqualmie Falls precipitation, continued) The fit of the limiting extreme value distribution is substantially better than the fit of the limiting total amounts distribution. Using the mle's determined earlier, and noting that $u_n=114$ for these values, we compute the exact and asymptotic distributions. Figure 2.8 shows them for $n=20$. The asymptotic approximation is excellent. In addition, the observed maximal precipitation (or, more precisely, the empirical distribution function of the 36 years of maxima) for January 6–25 is shown. There is perhaps a slight skewness, with too many small and too few large maximal precipitation values, but the sample size is only 36. □

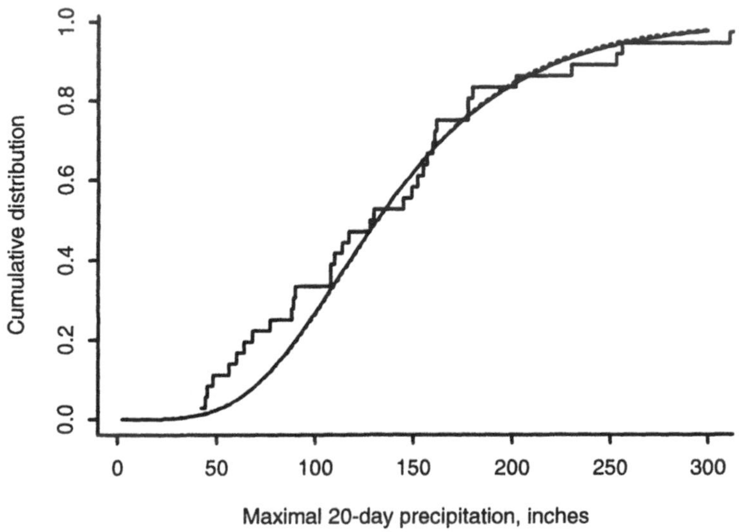

Figure 2.8. Observed (step function), exact (smooth line) and asymptotic (dotted line) distribution functions of maximum precipitation January 6–25.

2.10. Random walks and harmonic analysis

We have encountered the random walk earlier in different contexts. In this section we look at what amounts to an application to mathematics. The harmonic analysis that we shall encounter is very elementary, although it will allow us to deal with first passage problems for finite state space Markov chains in some generality. We begin, however, in the fair coin tossing setup.

Consider a fair, simple random walk on $\{0,\ldots,K\}$, so that $p_{i,i+1}=p_{i+1,i}=½$, $i=0,1,\ldots,K-1$, and all other transition probabilities are 0. An interesting problem is to find

$$f(x) = \mathbf{P}^x(\text{reach } K \text{ before } 0). \tag{2.261}$$

By conditioning on the first step we see that

$$f(x) = ½f(x-1) + ½f(x+1) \quad x=1,2,\ldots,K-1. \tag{2.262}$$

The initial conditions are $f(0)=0$ and $f(K)=1$.

We call $D=\{1,\ldots,K-1\}$ the **interior** points and $B=\{0,K\}$ the **boundary** points. A function $f(x)$ on $S=D+B$ is **harmonic** if for all points in D it satisfies the averaging property

$$f(x) = ½(f(x-1)+f(x+1)). \tag{2.263}$$

The problem of finding a harmonic function given its boundary values is called the **Dirichlet problem**, and the uniqueness principle for Dirichlet problems asserts that there can be no two harmonic functions having the same boundary values. To prove the uniqueness principle, we must first prove the maximum principle:

Theorem 2.18 (Maximum principle) A harmonic function $f(x)$ defined on S takes on its maximum M and its minimum m on the boundary.

Proof Let $M = \max_S f(x)$. If $f(x) = M$ for some $x \in D$, then $f(x-1) = f(x+1) = M$. Continuing left, eventually $f(0) = M$. The same argument works for the minimum. □

Theorem 2.19 (Uniqueness principle) If f and g are harmonic functions on S with $f(x) = g(x)$ for $x \in B$, then $f(x) = g(x)$ for all x.

Proof Let $h(x) = f(x) - g(x)$. Then for $x \in D$

$$\frac{h(x-1)+h(x+1)}{2} = \frac{f(x-1)+f(x+1)}{2} - \frac{g(x-1)+g(x+1)}{2}$$

$$= f(x) - g(x) = h(x), \tag{2.264}$$

so h is harmonic. But h is 0 on B, so by Theorem 2.18 it is 0 everywhere. □

We have now reduced the problem to finding a harmonic function with the given initial conditions. Using the theory in Appendix B we see that the solution is $f(x) = x/K$. Harmonic function theory also gives simple answers to other questions about our random walk. For example, are we certain to reach the boundary eventually? Let

$$h(x) = \mathbf{P}^x(\text{never reach } B). \tag{2.265}$$

Then $h(x) = \tfrac{1}{2}h(x-1) + \tfrac{1}{2}h(x+1)$, so h is harmonic, with boundary values $h(0) = h(K) = 0$. So $h(x)$ must be identically zero.

How long does it take to hit the boundary? The answer to this question is not a harmonic function, but conditioning on the first step yields for an interior x

$$e(x) = \mathbf{E}^x T_B = 1 + \tfrac{1}{2}\mathbf{E}^{x-1} T_B + \tfrac{1}{2}\mathbf{E}^{x+1} T_B. \tag{2.266}$$

The corresponding difference equation then is

$$(E-1)^2 e(x) = -2 \times 1^x \tag{2.267}$$

with initial conditions $e(0) = e(K) = 0$. Using Appendix B again, the solution is $e(x) = (K-x)x$.

Let us now try a more complicated two-dimensional array.

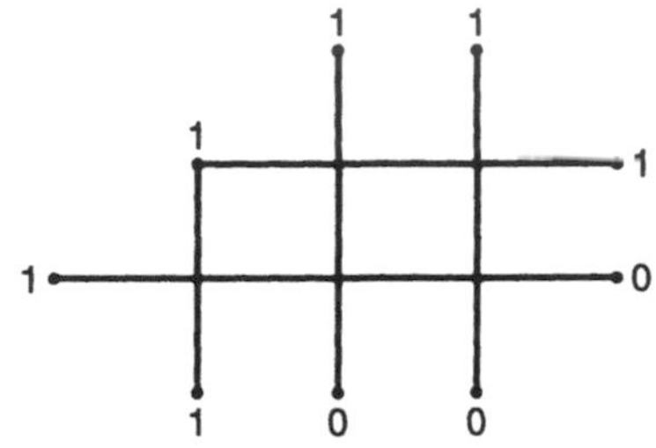

Figure 2.9. A subset of the two-dimensional lattice. Adapted from *Random Walks and Electric Networks* by P. G. Doyle and J. L. Snell, published by the American Mathematical Society.

A **lattice point** is a point with integer coordinates. In Figure 2.9 we see a subset of the two-dimensional lattice, with boundary points divided into two types, marked 0 or 1. Consider a process performing a random walk on the lattice subset. Starting in an interior point, the process moves to each neighboring point with probability ¼. We have two classes of boundary states, B_0 and B_1, and we are interested in the probability of hitting B_1 before B_0.

We describe the general situation as follows. Let $S=D+B$ be a finite set of lattice points, such that each point in D has four neighbors in S, and each point in B has at least one neighbor in D. D consists of the interior points, and B is the set of boundary points. Also assume that S is **connected**, i.e., that there is a route from every point to every other point. Call a function f on S harmonic if

$$f(a,b) = \tfrac{1}{4}(f(a+1,b)+f(a-1,b)+f(a,b+1)+f(a,b-1)). \quad (2.268)$$

As before, our functions of interest, namely the hitting probability $f(x)$ is a harmonic function with boundary values: $f(x)=1(x\in B_i)$, $i=$ 0 or 1. where $B=B_0+B_1$. For a finite lattice the maximum principle and the uniqueness principle are proved just as before. The problem, then, is reduced to solving the difference equation (2.268). Using the ergodic theorem we may run many particles in a random walk on the lattice and note what proportion end up at what boundary point. This must be repeated for each initial interior point. While this approach yields an answer, it is slow and imprecise. The theory of partial difference equations is not very well developed. Let ∇^2 be the **symmetric second difference operator**, so that $\nabla^2 g(x)=g(x+1)-2g(x)+g(x-1)$. Then (2.268) can be written (using a subscript to denote which argument the operator is applied to)

$$\nabla_1^2 f(x,y) + \nabla_2^2 f(x,y) = 0 \quad (2.269)$$

with boundary conditions $f(\mathbf{b})=1(\mathbf{b}\in B_1)$. It is natural to look for guidance to the continuous case. The corresponding partial differential equation is **Laplace's equation**

$$\frac{\partial^2 f(x,y)}{\partial x^2} + \frac{\partial^2 f(x,y)}{\partial y^2} = 0 \tag{2.270}$$

with boundary condition

$$\lim_{(x,y)\to t} f(x,y) = \phi(t),\ t \in \partial D \tag{2.271}$$

where ∂D is the boundary of D. In the continuous case any function satisfying Laplace's equation is called harmonic. The **method of relaxation** was developed to solve the continuous Dirichlet problem. It is based on a result which says that a function is harmonic if its value at (x,y) is equal to its average over any circle inside D centered on (x,y). This suggests a method of successive averaging of function values. Translated to the lattice case, the method of relaxation works as follows. Start with an arbitrary function with the right boundary values, such as

		1	1	
	1	0	0	1
1	0	0	0	0
	1	0	0	

Pick an interior point, such as (2,2) (with (0,0) being at the lower left-hand corner of the minimum rectangle containing all the states). If the function is not equal to its values over the neighbors, adjust it. Run through all the interior points in some order. The new function becomes

		1	1	
	1	0.5	0.625	1
1	0.832	0.328	0.156	0
	1	0	0	

The new function is still not harmonic, in general, but we can repeat the procedure until it converges. After nine iterations we have

		1	1	
	1	0.823	0.787	1
1	0.876	0.506	0.323	0
	1	0	0	

Note that this can be done automatically in a spread sheet program.

There is a third way of solving Dirichlet problems. This method involves the theory of Markov chains directly. Suppose that the chain has state space S, with the boundary states B absorbing, and transition matrix $\mathbb{P}$. Call a function f **harmonic for $\mathbb{P}$** if

$$f(i) = \sum_j P_{ij} f(j) \quad \text{for all } i \in D. \tag{2.272}$$

Writing f as a vector, with the absorbing states first, we have that $f^T = \mathbb{P} f^T$, so $\mathbb{P}^n f^T = f^T$ for all n. Note that

$$\mathbb{P}^n = \begin{bmatrix} \mathbb{I} & 0 \\ R & Q \end{bmatrix}^n = \begin{bmatrix} \mathbb{I} & 0 \\ C_n & Q^n \end{bmatrix}. \tag{2.273}$$

Here $C_{n+1} = R\mathbb{I} + QC_n$. Letting $n \to \infty$ we see that $C = R + QC$. Hence

$$\mathbb{P}^n \to \begin{bmatrix} \mathbb{I} & 0 \\ C & 0 \end{bmatrix}, \tag{2.274}$$

where $C = (\mathbb{I} - Q)^{-1} R$. Then

$$f^T = \begin{bmatrix} f_B^T \\ f_D^T \end{bmatrix} = \begin{bmatrix} \mathbb{I} & 0 \\ C & 0 \end{bmatrix} \begin{bmatrix} f_B^T \\ f_D^T \end{bmatrix}. \tag{2.275}$$

We see that

$$f_D^T = C f_B^T = (\mathbb{I} - Q)^{-1} R f_B^T \tag{2.276}$$

so that f_D is determined by the values of f at the boundary. Numbering the states in Figure 2.9

		1	2	
	9	10	11	3
8	12	13	14	4
	7	6	5	

we have

$$R = \begin{bmatrix} \tfrac14 & 0 & 0 & 0 & 0 & 0 & 0 & 0 & \tfrac14 \\ 0 & \tfrac14 & \tfrac14 & 0 & 0 & 0 & 0 & 0 & 0 \\ 0 & 0 & 0 & 0 & 0 & 0 & \tfrac14 & \tfrac14 & \tfrac14 \\ 0 & 0 & 0 & 0 & 0 & \tfrac14 & 0 & 0 & 0 \\ 0 & 0 & 0 & \tfrac14 & \tfrac14 & 0 & 0 & 0 & 0 \end{bmatrix} \tag{2.277}$$

$$Q = \begin{bmatrix} 0 & \tfrac14 & 0 & \tfrac14 & 0 \\ \tfrac14 & 0 & 0 & 0 & \tfrac14 \\ 0 & 0 & 0 & \tfrac14 & 0 \\ \tfrac14 & 0 & \tfrac14 & 0 & \tfrac14 \\ 0 & \tfrac14 & 0 & \tfrac14 & 0 \end{bmatrix}. \tag{2.278}$$

Since $f_B = (1,1,1,0,0,0,1,1,1)$ we get that

$$(\mathbb{I} - Q)^{-1} R f_B^T = \begin{bmatrix} 0.823 \\ 0.787 \\ 0.876 \\ 0.506 \\ 0.323 \end{bmatrix}; \tag{2.279}$$

the same result as that obtained by the method of relaxation, without the need for iterations, but requiring a matrix inversion which may be difficult if the state space is large.

Application (Airplane fire escape probabilities) To assure profitability, airlines must configure their airplanes to carry the maximum number of seats. However, the maximization is constrained by passenger safety and comfort, as well as by the maximum useful load carrying capacity of the airplane. The Federal Aviation Authority (FAA) of the US requires that all commercial airplanes have a maximum evacuation time of 90 seconds for all seating configurations.

One serious situation requiring evacuation is a fire in an engine with smoke obscuring the exit pathways. For example, in 1985 a Boeing 737 taking off from Manchester developed a fire in the left engine. The pilot first thought that the problem was a blown landing gear, thereby delaying the eventual evacuation of the aircraft. After about a dozen passengers had left the airplane, the interior was filled by thick black smoke. The cabin attendants were unable to see from the left to the right forward door. The 131 passengers were tourists, with many relatively inexperienced flyers. Many crawled over seats looking for exits. 76 passengers survived the fire.

Figure 2.10 shows the seating configuration.

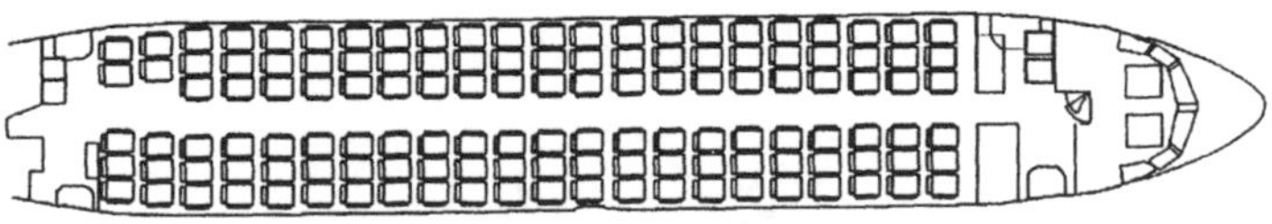

Figure 2.10. Seating arrangement for a Boeing 737.

There are 20 rows of six seats, three on either side of the aisle, and two more rows with three seats to the right and only two on the left. Two seats were unoccupied, but including infants there were 131 passengers. Only two forward exits and the right overwing escape hatch were usable. Except for the front row, escape routes were not generally to the nearest exit, indicating that the behavior was somewhat random.

Assuming the fair random walk model for passengers (this includes the possibility of trying to walk through a wall, corresponding to a zero escape probability, and assumes that individuals reaching an aisle can find an exit) we solved the corresponding harmonic function problem. Escape probabilities for an individual in a window seat on the left varied from 0.101 in row 1 to 0.25 in rows 10–14, while a right window seat occupant had probabilities ranging from 0.101 (row 1) to 0.596 (in the overwing exit row 10). Aisle seats generally had higher escape probabilities, varying from 0.467 (row 1 on left) to 0.808 (row 10 on right). The total expected number escaping the fire from this extremely simple model is 55, corresponding well to the 60–65 passengers who experienced the smoky environment and escaped. We can compute estimates of the change in escape probability if we add exits. Adding an exit amounts to increasing the number of preferred exit states. The same type of computation indicates that each additional exit (up to four additional exits) increases the expected number of survivors by about one. □

Note that a function f which is harmonic for $\mathbb{P}$ satisfies $\mathbb{P}\mathbf{f}^T=\mathbf{f}^T$, so it is, in a sense, a dual concept to the stationary distribution. More precisely, where the stationary distribution is a left eigenvector of $\mathbb{P}$, any harmonic function is a right eigenvector, both corresponding to the eigenvector 1. Recall that the function $f(x)=1$ is always a right eigenvector, and, consequently, so is any constant function. It turns out that this is the unique right eigenvector whenever $\mathbb{P}$ corresponds to an irreducible persistent chain, provided that f is either non-negative or bounded (see Asmussen, 1987, section I.5 for some discussion on how this is useful in the classification of chains). Our application of the concept of harmonic functions to hitting probabilities deals with transient chains, where the behavior is more interesting.

Example (Electric network) Consider a network with five resistors and a unit voltage applied across it (Figure 2.11).

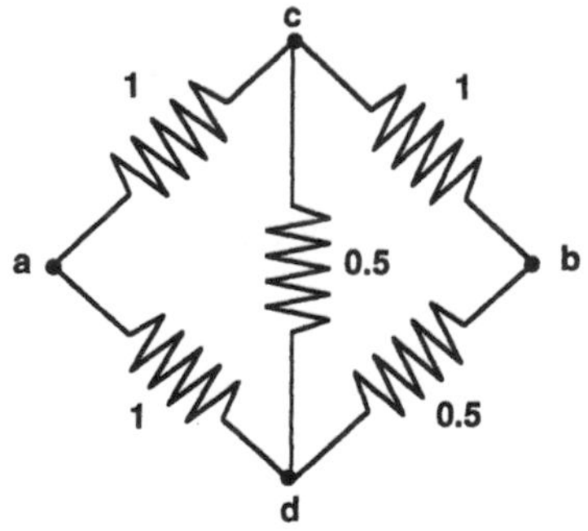

Figure 2.11. A simple Wheatstone bridge with unit voltage to be applied between a and b.

The conductance between two points x and y is the inverse of the resistance between the points. In this case $C_{ac}=C_{ad}=C_{bd}=C_{bc}=1$ while $C_{ab}=C_{cd}=2$. We are interested in determining voltages $v(x)$ at various points in this network. Two laws describe the behavior of the system:

Kirchhoff's law: The current flowing into x is the same as the current flowing out from x.

Ohm's law: If x and y are connected by a resistance R_{xy}, then the current flowing from x to y is

$$i_{xy} = \frac{v(x)-v(y)}{R_{xy}}. \tag{2.280}$$

The voltage at a is 1 and that at b is 0, so if $x \neq a,b$ $\sum_y i_{xy}=0$, and since $i_{xy}=-i_{yx}$ we have

$$v(x)\sum_y C_{xy} = \sum_y C_{xy} v(y) \tag{2.281}$$

or, writing $C_x=\sum_y C_{xy}$,

$$v(x) = \sum_y \frac{C_{xy}}{C_x} v(y). \tag{2.282}$$

Now notice that $\mathbb{P}=(C_{yx}/C_x)$ is a transition matrix. Making a and b absorbing states (i.e., changing the corresponding rows to have 1 on the diagonal and 0 elsewhere) yields a modified transition matrix $\mathbb{P}^*$, and v is harmonic for $\mathbb{P}^*$. We need the modification since (2.282) only holds for $x \neq a$ or b. In this case,

$$\mathbb{P}^* = \begin{bmatrix} 1 & 0 & 0 & 0 \\ 0 & 1 & 0 & 0 \\ 1/4 & 1/4 & 0 & 1/2 \\ 1/5 & 2/5 & 2/5 & 0 \end{bmatrix} \tag{2.283}$$

so

$$(\mathbb{I}-Q)^{-1} R \mathbf{v}_B = \begin{bmatrix} 5/4 & 5/8 \\ 1/2 & 5/4 \end{bmatrix} \begin{bmatrix} 1/4 & 1/4 \\ 1/5 & 2/5 \end{bmatrix} \begin{bmatrix} 1 \\ 0 \end{bmatrix} = \begin{bmatrix} 7/16 \\ 3/8 \end{bmatrix}, \tag{2.284}$$

i.e., $v(c)=0.4375$ and $v(d)=0.375$.

Note that $\mathbb{P}$ is reversible, since $C_{yx}=C_{xy}$, so

$$p_{xy} = \frac{C_{xy}}{C_x} = \frac{C_{yx}}{C_y} \times \frac{C_y}{C_x} = p_{yx} \frac{C_y}{C_x}. \tag{2.285}$$

This reversibility is required by physical theory. In particular, therefore, we must have a stationary distribution π satisfying $\pi_y/\pi_x=C_y/C_x$, whence $\pi_y=C_y/C$ where $C=\sum_y C_y$. A lot more material on the relation between random walks and electric networks can be found in Doyle and Snell (1984). □

Application (Airplane fire escape probabilities, continued) Another Boeing 737 fire with substantial cabin smoke occurred in 1984 in Calgary (this and the Manchester fire are the only instances of this type for the Boeing 737). Again, an engine fire was first interpreted by the pilot as a tire failure on the landing gear. In this case the 114 passengers were frequent flyers, and had all flown with this type of aircraft before. All on board the airplane survived. Here the simple random walk predicts 59 survivors, but the passengers did not act randomly. For example, the passengers in rows 1–7 all exited through the forward doors, those in rows 8–16 through the right overwing exit, and those in rows 17–22 through the right rear exit.

A weighted approach assumes that a passenger in an exit row moves to the exit with probability 5/8, and to the other three adjacent locations with probability 1/8. Passengers in a row adjacent to an exit row have equal probabilities of moving toward the aisle or climbing over seats. All other passengers have probability 5/8 of moving toward the aisle, and 1/8 of moving in any other direction. This simple model of informed behavior yields 101 expected survivors. □

2.11. Bienaymé–Galton–Watson branching processes

The simple **branching process** was first studied by Bienaymé[1] in 1845 in order to find a mathematical (rather than social or genetic) explanation for the fact that a large proportion of the family names, both among nobility and bourgeoisie, seemed to be dying out when viewed over a long period of time. It has been applied to problems of genetics, epidemiology, nuclear fission, queueing theory, and demography, among other areas, and is one of the simplest non-ergodic stochastic models. It is commonly called the **Bienaymé–Galton–Watson** (BGW) process.

Let

$$Z_k = \sum_{i=1}^{Z_{k-1}} X_{k,i} \tag{2.286}$$

where for each k the $X_{k,i}$ are iid random variables with the same distribution $p_l = \mathbf{P}(X_{i,j}=l)$ called the **offspring distribution**. We interpret $\sum_1^0$ as 0. Suppose that the offspring distribution has pgf $P(s)$, with mean m and variance σ^2. Assuming that $Z_0 = 1$, we see that

$$P_k(s) = \mathbf{E}s^{Z_k} = P_{k-1}(P(s)). \tag{2.287}$$

By induction we see that

$$\mathbf{E}Z_k = \mathbf{E}\,\mathbf{E}(Z_k \mid Z_{k-1}) = m\mathbf{E}Z_{k-1} = \cdots = m^k. \tag{2.288}$$

Notice that $\mathbf{E}Z_k \to \infty$ where $m > 1$, while it stays constant when $m = 1$ and goes to zero when $m < 1$. For this reason processes with $m > 1$ are called **supercritical**, those with $m = 1$ are **critical**, and those with $m < 1$ are **subcritical**. Furthermore

---

[1]Bienaymé, Irénée-Jules (1796–1878). Inspector General of the Administration of Finances of France 1834–1848, subsequently independent mathematician. Much of his work had long been overlooked; see Heyde and Seneta (1977).

$$\mathbf{Var}Z_k = \mathbf{E}\,\mathbf{Var}(Z_k \mid Z_{k-1}) + \mathbf{Var}\,E(Z_k \mid Z_{k-1})$$
$$= \sigma^2 \mathbf{E}Z_{k-1} + m^2 \mathbf{Var}Z_{k-1} = \sigma^2 m^{k-1} + m^2 \mathbf{Var}Z_{k-1} \tag{2.289}$$

so that, using the methods in Appendix B,

$$\mathbf{Var}Z_k = \sigma^2 m^{k-1} \sum_0^{k-1} m^i = \sigma^2 m^{k-1} \frac{m^k - 1}{m-1}. \tag{2.290}$$

From (2.286) it is easy to see that the process (Z_k) is a Markov chain, with transition probabilities $p_{ij} = p_j^{*i}$, where p^{*i} is the i-fold convolution of (p_j). Another way of writing the process is

$$Z_n = mZ_{n-1} + e_n, \tag{2.291}$$

where $e_n = Z_n - mZ_{n-1}$ can be thought of as a prediction error, having conditional mean 0, given the past, and conditional variance $\sigma^2 Z_{n-1}$, given the past.

From the construction in (2.286) it is also clear that once a generation is empty, all following generations will be empty as well. Therefore we will compute the probability of eventual extinction. Since this Markov chain has infinite state space, the method used in the previous section to compute hitting probabilities does not apply.

Theorem 2.20 Suppose that $p_0 > 0$, $p_0 + p_1 < 1$, and let $q = \mathbf{P}(Z_k \to 0)$. Then q is the smallest nonnegative root of the equation $P(s) = s$. Furthermore, $q = 1$ iff $m \le 1$.

Proof Extinction will occur in or before the kth generation in one of the following ways: the ancestor has

0 children
1 child whose family becomes extinct in or before the $(k-1)$th generation
2 children, both of whose families become extinct in or before the $(k-1)$th generation, etc.

Let $P_k(s) = \mathbf{E}s^{Z_k}$. Since the probability of the jth of these cases is p_j, the probability $q_k = P_k(0)$ of extinction after k generations is, by conditioning on the first family size,

$$\mathbf{P}(Z_k = 0) = P_k(0) = \sum_{j=0}^{\infty} p_j P_{k-1}(0)^j = P(P_{k-1}(0)). \tag{2.292}$$

Since $p_1 \ne 1$, we note that $P(s)$ is a strictly increasing function. Thus the numbers $q_k = P(q_{k-1})$ form a strictly increasing sequence of positive numbers, all bounded by 1. This sequence therefore has a limit which we denote q, with $p_0 \le q \le 1$. This is the probability of ultimate extinction. By going to the limit in (2.292), we see that $q = P(q)$, whence

$$\frac{P(q) - P(q_k)}{q - q_k} = \frac{q - q_{k+1}}{q - q_k} < 1. \tag{2.293}$$

Letting $k \to \infty$ we see (Figure 2.12) that $P'(q) \le 1$.

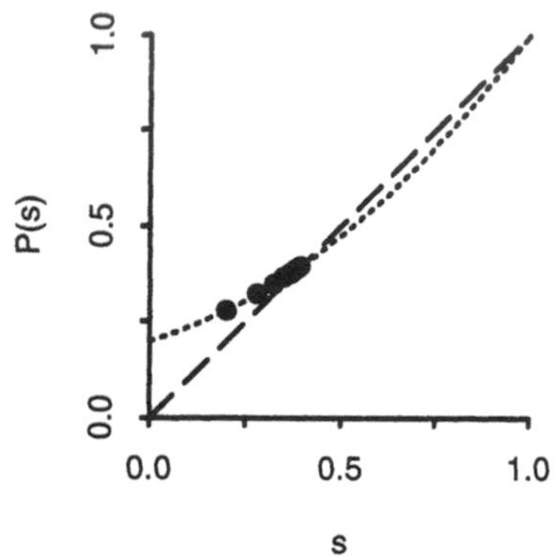

Figure 2.12. The sequence q_k (large dots) for a supercritical branching process. The dotted line is the pgf $P(s)$, and the dashed line is the line $y = s$.

Since $P'(s)$ is a power series with positive coefficients it is an increasing function, i.e., $P(s)$ is convex. If $P'(1) > 1$ we must have $q < 1$. In this case q and 1 are the only positive roots to the equation $P(s) = s$. On the other hand, if $P'(1) \le 1$, we have for $0 \le s < 1$ that $P'(s) - 1 < 0$, so that the smallest zero of the function $P(s) - s$ must be 1, whence $q = 1$.

Finally, if $p_0 = 0$, $Z_k \ge Z_{k-1}$, and extinction is impossible. If $p_1 = 1$, $Z_k = Z_0$, and extinction is again impossible. Upon noting that $P'(1) = m$, the proof is complete. □

Example (Epidemics) The problem of determining the fraction of a community that must be vaccinated in order to prevent major epidemics of a communicable disease is a crucial public health problem. In order to describe an epidemic, the population is divided into three possible health states. An individual can be susceptible to infection by a given disease agent, (s)he may have been infected by the agent and is infectious (possibly after a latent period), or (s)he is removed from the epidemic by death, by isolation, or by immunity or other natural loss of infectiousness. Initially all members of the population are susceptible to infection. The epidemic starts when one or many infectious individuals enter the population and come into contact with its members. A susceptible person is infected if (s)he has adequate contact with an infectious individual. General theory of epidemic models can be found in Bailey (1975), and their statistical inference is discussed, e.g., in Becker (1976) and Rida (1991). A BGW process can be used to approximate the infectious population during the early stages of an epidemic (Becker 1977). Clearly, since the number of susceptible individuals decreases as the epidemic progresses, it is unreasonable to assume that the offspring distribution is the same from generation to generation.

However, for early stages of epidemics in large populations the assumptions underlying the BGW process are not unrealistic.

In order for an epidemic to become serious, a large buildup of cases is needed in the early stage. Consequently we can call an epidemic **major** if the offspring mean is >1, and **minor** if it is ≤1 (so that the extinction probability is 1). In order to prevent major epidemics it is necessary to ensure adequate vaccination in the community so as to make the offspring mean less than one. Suppose that we select a proportion θ of the population at random for vaccination. If the vaccination is effective, the offspring distribution changes to $P^*(s)=\theta+(1-\theta)P(s)$ with mean $m^*=(1-\theta)m$, so that $m^*<1$ only if $\theta>1-1/m$. For a numerical illustration, assume that the offspring distribution is Poisson. Then the probability $1-q$ of a major epidemic is a function of m.

Table 2.11 Probability of a major epidemic

m	1	1.05	1.1	1.4	1.8
$1-q$	0	.09	.18	.51	.73
$1-1/m$	0	.05	.09	.29	.44

The third line of Table 2.11 contains the proportion of vaccination needed to bring the new mean below one. If one accepts this model it becomes of considerable importance to be able to estimate m as accurately as possible. □

In order to estimate m, first suppose that we know all the X_{ij}. Then the likelihood would be $\sum N_k \log p_k$, where $N_k=\#\{X_{ij}=k\}$, and the mle of p_k would be N_k/N, where

$$N=\sum_{k=0}^{\infty} N_k=\sum_{0}^{n-1} Z_k=Y_{n-1}. \tag{2.294}$$

The mle of m is then

$$\hat{m}=\sum k\hat{p}_k=\sum \frac{kN_k}{N}=\frac{\#\text{ children}}{\#\text{ parents}}=\frac{Y_n-1}{Y_{n-1}}. \tag{2.295}$$

Notice that this does not depend on the (N_k), only on the generation sizes. This suggests that $\hat{m}$ may also be the mle based on only observing generation sizes. This happens to be the case (Keiding and Lauritzen, 1978).

It is clear that $\{0\}$ is an absorbing state, and it can be shown that all other states are transient. In fact, either the chain dies out, or it explodes (diverges to infinity). Unless $p_0=0$, extinction has positive probability, and therefore $\hat{m}_n$ has positive probability of converging to $1-1/Y_\infty$. The statistical theory developed in section 2.7 fails to apply, since the process is not ergodic. However, conditional on nonextinction we show below that $\hat{m}_n \to m$ with probability one. It is difficult to determine exact finite-sample properties of $\hat{m}_n$.

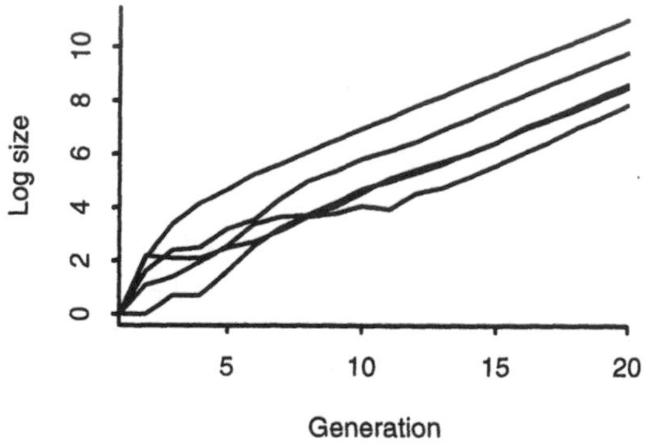

Figure 2.13. Five paths of a BGW process with Poisson offspring distribution. Adapted from P. Guttorp, *Statistical Inference for Branching Processes*, by permission of John Wiley & Sons, Inc. (1991).

The growth of a BGW process, given that it does not become extinct, is geometric. In Figure 2.13 we see the logarithms of five paths from a process with Poisson offspring distribution with $m=2$. After some initial wigglyness, they look quite linear. The figure suggests that if we rescale the generation sizes by their means, which are growing geometrically, we may end up with a limiting constant. This is not quite right: the limit turns out to be a random variable.

Theorem 2.21 Let $m>1$, and define $W_n=m^{-n}Z_n$. Then $W_n \to W$ with probability one, where $\mathbf{E}W=1$, $\mathbf{Var}W=\sigma^2/m(m-1)$, and $\mathbf{P}(W=0)=q$.

A proof of this result can be found in Guttorp (1991, Theorem 1.1). Notice that the set $\{W=0\}$ is precisely the set where the process Z_n becomes extinct.

Corollary $\hat{m}_n \to m$ on the set of nonextinction.

Proof Write

$$m^{-n}Y_n = m^{-n}\sum_1^n Z_k = \sum_1^n W_k m^{-(n-k)} \to \frac{mW}{(m-1)}. \tag{2.296}$$

Hence, whenever $W>0$ we see that

$$\hat{m}_n = (m^{-n}Y_n - m^{-n}\frac{)}{m^{-1}}(m^{-(n-1)}Y_{n-1}) \to m \tag{2.297}$$

with probability one. □

Remark Nonparametric inference for BGW processes is quite different from that for ergodic Markov chains. It can be shown (Guttorp, 1991, section 1.4) that, in essence, only the mean and the variance are consistently estimable from observing a long, non-extinct path. For example, one cannot estimate q or $P(s)$ in general. □

Application (Smallpox epidemics) We now proceed to study in more detail an example of the application of branching processes to smallpox epidemics. This disease, although now completely eradicated (only two laboratory specimens remained by 1993), had a well defined incubation time of about 12 days (rarely outside 9–15 days). Therefore the cases in the early stages of an epidemic tend to be well separated, and it is easy to determine generation sizes. We use data (Table 2.12) from Vila Guarani, a residential district in São Paulo, on *variola minor*, the less lethal form of smallpox. The epidemic was introduced by two travelers, who are not included. They were only responsible for the ancestor.

Table 2.12 Vila Guarani epidemic

Generation	0	1	2	3	4
Size	1	5	3	12	24

After the fourth generation school vacations started, and mass vaccination was introduced. Hence later data does not have similar conditions. The maximum likelihood estimate of the offspring mean is 2.10, indicating that a major epidemic was developing. One may argue, however, that although conditions were relatively constant within generations, they may not be constant between generations. In particular, the weather may be quite different almost two months from the first outbreak. The weather is important, since the disease was spread through close proximity, partly through airborne transmission, and weather affects the social behavior. To investigate this possibility, we need to introduce the idea of **random environment**. A branching process in random environment is obtained by picking an offspring distribution for each generation at random from a set of possible distributions. If there is substantial variability in the successive parent/offspring ratios $m_r = Z_r / Z_{r-1}$, compared to what would be expected from a fixed offspring distribution (constant environment) we would conclude that the environment may be random. These ratios are shown in Table 2.13. Given Z_{r-1}, the conditional variance of m_r is σ^2 / Z_{r-1}. There are several different ways of estimating σ^2. We use a maximum likelihood estimator, based on computing the variance for the maximum likelihood estimator of the offspring distribution (see the Remark below). This yields $\hat{\sigma}^2 = 4.18$. Thus we obtain the results in Table 2.14. These ratios are consistent with the hypothesis of constant environment, in that only ratios larger than 2 in absolute value

Table 2.13 Vila Guarani estimates

r	1	2	3	4
m_r	5	0.6	4	2

Table 2.14 Vila Guarani model evaluation

r	1	2	3	4
$\mathrm{se}(m_r)$	2.04	0.91	1.18	0.59
$(m_r-\hat{m})/\mathrm{se}(m_r)$	1.42	−1.64	1.61	−0.16

would be suspicious. If we want to vaccinate enough of the population to make the epidemic minor, we need to reach $(1-1/\hat{m})$ or 52% of the population. However, since the main path of infection is through the class room, the closing of schools for vacation would already affect the offspring mean considerably. □

Remark The mle of the offspring variance (Guttorp, 1991, section 3.4) does not have a closed form, and is somewhat complicated to compute. A simpler estimate is $\tilde{\sigma}^2=(n-1)^{-1}\sum_{i=1}^{n}(Z_i-\hat{m}Z_{i-1})^2/Z_{i-1}$. This is generally a more variable estimate. In this case $\tilde{\sigma}^2=10.20$, considerably larger than the mle. The corresponding analysis using $\tilde{\sigma}^2$ instead of $\hat{\sigma}^2$ would show even less indication of a random environment. □

Another application of branching process theory is to a very different type of problem. Given the frequency of a particular genetic variant occurring in a population, can we figure out how long ago the mutation first arose? Here the parameter of interest is the age N of a branching process, and our observation is simply Z_N. In order to answer the question, we must specify the offspring distribution. It is not unreasonable to assume that the spread of the variant is quite similar to the population growth in general. A useful and flexible distribution, which has been found to describe human population growth adequately in different situations, is the **modified geometric** distribution:

$$p_k = bc^{k-1},\ k\geq 1 \tag{2.298}$$

$$p_0 = 1-\frac{b}{1-c}$$

with mean $m=b/(1-c)^2$ and variance $\sigma^2=b(1-b-c^2)/(1-c)^4$. We shall think of m as the mean rate of increase of the mutant type, representing a **selective advantage** if it is higher than the overall population mean, and a **disadvantage** if it is lower. It is convenient to reparametrize the distribution in terms of m and $h=(1-c)/c$, so $p_k=mh^2(1+h)^{-(k+1)}$ for $k\geq 1$, and $p_0=1-mh/(1+h)$. Then $q=1-(m-1)h$ if $m>1$. One advantage with the modified geometric distribution is that it is self-reproducing: the distribution of Z_t is of the same form, with

$$\mathbf{P}(Z_t=k) = \begin{cases} m^t M^2(1-M)^{k-1}, & k\geq 1 \\ 1-m^t M, & k=0 \end{cases} \tag{2.299}$$

where $M=(1-s_0)/(m^t-s_0)$ and $s_0=1-(m-1)h$ (Exercise **13**). Furthermore, $\mathbf{P}(Z_t=k \mid Z_t>0)=M(1-M)^{k-1}$, a regular positive geometric distribution, with mean $1/M$ and variance $(1-M)/M^2$. If we observe $Z_N=k>0$ (necessarily), the conditional log likelihood can be written

$$l(N,m,h) = \log M(N,m,h)+(k-1)\log(1-M(N,m,h)). \tag{2.300}$$

We have three parameters, but only one observation. We can reduce the parameter space by assuming that h is a known constant, corresponding to a known value of $c=p_k/p_{k-1}$, assumed to be obtained from population data. This implies that the selective advantage enters only through the parameter b. The likelihood equation becomes $M=1/k$, so the maximum likelihood estimate of N is

$$\hat{N} = \frac{\log(1+(k-1)(m-1)h)}{\log(m)} \tag{2.301}$$

provided that $m\neq 1$. If $m=1$, a similar computation yields $\hat{N}=(k-1)h$.

Application (Yanomama genetics) We apply this theory to the study of rare protein variants in South American Indian populations. A variant is considered rare if it occurs in only one tribe. Such mutations have presumably arisen after tribal differentiation, are probably descended from a single mutant, and because of the low intertribal migration have not spread to the general South American Indian population. We will look at the Yanomama tribe, living in the Andes on the border between Brazil and Columbia. The Yanomama albumin variant Yan-2 is unique to this tribe, although it is fairly frequent, and therefore must be quite old. Extensive sampling of 47 widespread Yanomama villages found 875 replicates of the variant gene in the current adult population. From demographic data for this tribe, a selectively neutral gene would have offspring well described by a modified geometric distribution with $h=1.5$. The offspring mean m is slightly above 1 (although in recent years the mean population increase has been higher, perhaps as high as 1.2). Figure 2.14 shows the likelihood (as a function of m and N) arising from this datum. The middle line is the ridge of maximum likelihood, and the outer contour lines correspond to the maximum likelihood minus 2 and 4, respectively. For $m=1.0$ the maximum likelihood estimate of the age is 1311 generations, or about 30,000 years. This is far longer than the time these tribes have been in the Americas, so it would appear unlikely that the variant would have survived only in this tribe. In fact, this value of m corresponds to a slight selective disadvantage, making the

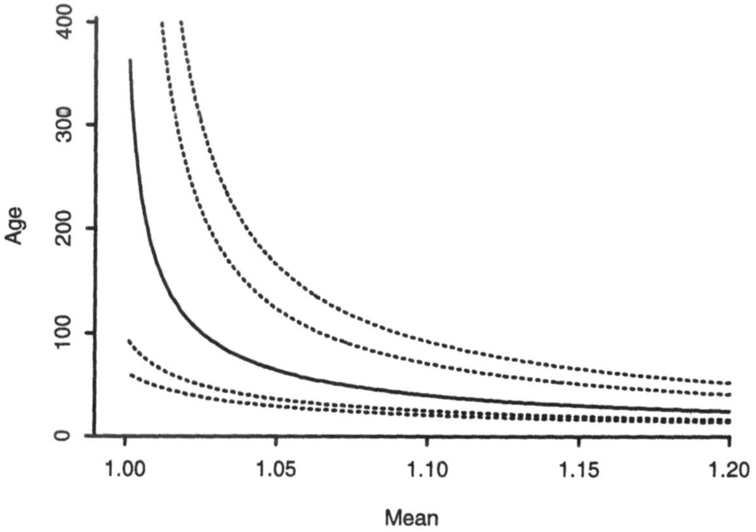

Figure 2.14. Likelihood surface for the Yanomama gene. The solid line is the ridge of maximum likelihood, while the dotted contours are 2 and 4 units of log likelihood below the maximum. Adapted from P. Guttorp, *Statistical Inference for Branching Processes*, by permission of John Wiley & Sons, Inc. (1991).

estimate even less plausible.

An increase rate of 1.02 may have been sustained by the Yanomama over long periods of time. The maximum of the likelihood then occurs at 168 generations, which is a more reasonable value, corresponding to a variant which is perhaps 3,800 years old. This supports the hypothesis that the allele is relatively old, but has arisen after tribal separation. We do not need to assume that the gene has had a selective advantage.

In order to obtain confidence intervals for these values we have to determine the asymptotic properties of the estimator. If $m>1$, given that $Z_N>0$, we have

$$\log m(\hat{N}-N) = \log(m^{-N}(1+(Z_N-1)(m-1)h)) \to \log((1-q)W) \quad (2.302)$$

with probability 1 as $N\to\infty$, since $(m-1)h=1-q$. Here W, as before, is the limit of $m^{-n}Z_n$. For the positive geometric distribution one can determine the distribution of W: it is exponential with parameter $1-q$. Notice that the estimate is not consistent, even in the peculiar sense of this limiting result. We can use the result to compute confidence bands. Note first that $(1-q)W\sim\exp(1)$, and the 97.5 and 2.5 percentiles of the standard exponential distributions are log 1.0256 and log 40, respectively. Thus if $\zeta\sim\exp(1)$ we have

$$\begin{aligned}0.95 &= \mathbf{P}(\log 1.0256 \le \zeta \le \log 40)\\ &= \mathbf{P}(\log 1.0256 \le (1-q)W \le \log 40)\\ &= \mathbf{P}\left(\frac{\log\log 1.0256}{\log m} \le \frac{\log((1-q)W)}{\log m} \le \frac{\log\log 40}{\log m}\right) \qquad (2.303)\\ &\approx \mathbf{P}\left(\frac{\log\log 1.0256}{\log m} \le \hat{N}-N \le \frac{\log\log 40}{\log m}\right)\\ &= \mathbf{P}\left(\hat{N}-\frac{\log\log 40}{\log m} \le N \le \hat{N}-\frac{\log\log 1.0256}{\log m}\right)\end{aligned}$$

which in this case yields the band (102, 353) at $m=1.02$. □

While the BGW branching process is a non-ergodic Markov chain, a slight modification yields an ergodic chain. Assume that at each time n there is a random number I_n of immigrants, each of which behaves like the rest of the population. This is called a **branching process with immigration**. We can write the total population size Z_n as

$$Z_n = \sum_{i=1}^{Z_{n-1}} X_{i,n} + I_n. \qquad (2.304)$$

We assume that I_n is independent of the previous I_k and Z_k, and that the I_n are identically distributed with pgf $Q(s)$ and mean $\lambda<\infty$. If $m=\mathbf{E}X_{i,n}<1$ we have seen that the corresponding branching process becomes extinct with probability one; however, the immigration process ensures that 0 is no longer an absorbing state. It would therefore seem reasonable that the resulting process may be ergodic. The pgf for Z_n satisfies

$$P_n(s) = \mathbf{E}s^{Z_n} = \mathbf{E}(P(s)^{Z_{n-1}})Q(s) = P_{n-1}(P(s))Q(s). \qquad (2.305)$$

If a stationary distribution exists, it must have pgf $\Pi(s)$ satisfying

$$\Pi(s) = Q(s)\Pi(P(s)). \qquad (2.306)$$

Taking derivatives, we see that the stationary mean μ is

$$\begin{aligned}\mu = \Pi'(1) &= Q'(1)\Pi(P(1)) + Q(s)\Pi'(P(1))P'(1)\\ &= \lambda + \mu m \qquad (2.307)\end{aligned}$$

or $\mu=\lambda/(1-m)$, provided that $m<1$.

Remark The equation (2.306) has a solution under the assumption that $\mathbf{E}\log(\max(0,I_1))<\infty$. When $m=1$ the resulting Markov chain can be either null persistent or transient. Asmussen and Hering (1984) give detailed proofs.

Application (Traffic theory) Fürth (1918) counted the number of pedestrians at five-second intervals passing a certain building. Thinking of each pedestrian as arriving as an immigrant, and having offspring 0 or 1 depending on whether or not the pedestrian leaves the observation area between two consecutive observations, we have $P(s)=1-m+ms$ and, assuming Poisson distributed input (some motivation for this will be given in the next chapter), $Q(s)=\exp(\lambda(s-1))$. The stationary distribution satisfies (2.306), i.e.,

$$\Pi(s) = \exp(\lambda(s-1))\Pi(1-m+ms). \tag{2.308}$$

It is easy to check that $\Pi(s)=\exp(\mu(s-1))$ satisfies (2.308), i.e., that the stationary distribution is Poisson with parameter μ.

In order to find the mle for the parameters m and λ, note that

$$p_{ij} = \mathbf{P}(Z_n=j \mid Z_{n-1}=i) = e^{-\lambda}\sum_{k=0}^{\min(i,j)} \frac{\lambda^{j-k}}{(j-k)!}\binom{i}{k} m^k(1-m)^{i-k}. \tag{2.309}$$

It helps to reparametrize in terms of μ and m. Then

$$\log L(\mu,m) = \sum n_{ij}\log p_{ij} \tag{2.310}$$

where

$$p_{ij} = e^{-\mu(1-m)}\sum_{k=0}^{\min(i,j)} \frac{\mu^{j-k}}{(j-k)!}\binom{i}{k} m^k(1-m)^{i+j-2k}. \tag{2.311}$$

To compute the mle we may of course maximize the likelihood function numerically, but it is instructive to manipulate the likelihood equations a little further. Note that

$$\frac{\partial p_{ij}}{\partial \mu} = -(1-m)p_{ij} + \frac{j}{\mu}p_{ij} - \frac{r_{ij}}{\mu}, \tag{2.312}$$

where

$$r_{ij} = ie^{-\mu(1-m)}\sum_{k=1}^{\min(i,j)} \frac{\mu^{j-k}}{(j-k)!}\binom{i-1}{k-1} m^k(1-m)^{i+j-2k}. \tag{2.313}$$

Furthermore,

$$\frac{\partial p_{ij}}{\partial m} = \mu p_{ij} - \frac{i+j}{1-m}p_{ij} + \left[\frac{1}{m}+\frac{2}{1-m}\right]r_{ij}. \tag{2.314}$$

Hence the likelihood equations are

$$\frac{\partial \log L(\mu,m)}{\partial \mu} = \sum n_{ij}\{-(1-m)+\frac{i}{\mu}-\frac{r_{ij}}{\mu p_{ij}}\} = 0 \tag{2.315}$$

$$\frac{\partial \log L(\mu,m)}{\partial m} = \sum n_{ij}\{\mu-\frac{i+j}{1-m}+\left[\frac{1}{m}+\frac{2}{1-m}\right]\frac{r_{ij}}{p_{ij}}\} = 0. \tag{2.316}$$

Let $R=\sum n_{ij}r_{ij}/p_{ij}$, $N_1=\sum in_{ij}$, and $N_2=\sum jn_{ij}$. Notice that R is a function of μ and m, while N_1 and N_2 are not. Then the equations can be written

$$-(1-m)n + \frac{N_1-R}{\mu} = 0$$

$$\mu n - \frac{N_1+N_2-2R}{1-m} + \frac{R}{m} = 0. \tag{2.317}$$

Solving the first equation for μ we have $\mu=(N_1-r)/n\,(1-m)$, and using this in the second we get $m=R/N_2$. Substituting this back into the expression for μ we see that

$$\hat{\mu} = \frac{(N_1-R)N_2}{n(N_2-R)} = N_2/n \tag{2.318}$$

or the mean of the observed path, provided that we ignore edge effects, so $N_1=N_2$. The equation $m=R(\hat{\mu},m)/N_2$ must be solved numerically, and numerical optimization of $\log L(\hat{\mu},m)$ (the **profile likelihood** for m) is just as easy.

The transition counts from Fürth's data are given in Table 2.15.

Table 2.15 Pedestrian counts

	0	1	2	3	4	5	6	7
0	67	24	6	1				
1	23	83	46	11	2			
2	7	41	58	25	5			
3	1	16	18	23	7	5		
4		1	7	8	7	2		1
5			1	1	5		1	
6				1				
7						1		

For these data $N_1=N_2=804$, while $n=505$ so $\hat{\mu} = 804/505 = 1.59$. Numerical optimization yields $\hat{m}=0.69$.

The number $P=1-m$ is called the **probability aftereffect**. It measures the probability that an individual in the system will leave it during the time between observations, but can also be thought of as a measure of dependence. If v is the average speed of a pedestrian, τ the time between observations, and b the width of the building, we see that $P=v\tau/b$. We can estimate P from the data. so the average speed of a pedestrian, using $\tau=5$ seconds and $b=20$ meters (this value is just a guess, as Fürth does not say how wide the building is), can be estimated as $\hat{v}=(1-\hat{m})b/\tau=1.25$ m/s. We can assess the variability of this estimate by producing a likelihood interval for m. Figure 2.15 depicts the likelihood surface. We see that (0.63, 0.74) is a confidence interval for m, so the

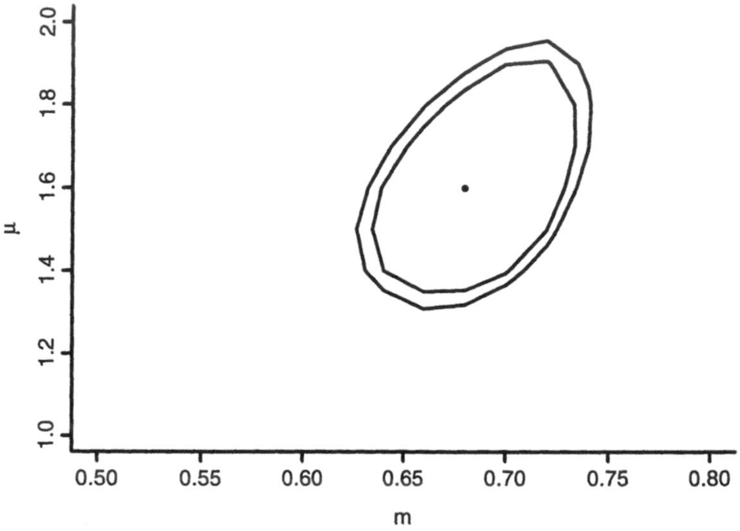

Figure 2.15. Likelihood surface for the pedestrian data. Contours are at 2.3 (approximately 90% coverage) and 3.0 (95% coverage) units of log likelihood below the maximum.

average pedestrian speed confidence interval (using the assumed value of b) is (1.0, 1.5).

In order to assess the fit of this model, Venkataraman (1982) suggests a time series approach, which has been extended and applied to these data by Mills and Seneta (1989). Since we have that

$$\mathbf{E}(Z_k \mid Z_1^{k-1}) = \lambda + mZ_{k-1} \tag{2.319}$$

we look at the fitted residuals

$$\hat{\varepsilon}_k = Z_k - \hat{\lambda} - \hat{m}Z_{k-1}. \tag{2.320}$$

Consider the jth-order autocorrelation function of the $\hat{\varepsilon}_k$, $k=1,\ldots,n$, namely

$$\hat{r}_j = \frac{\sum_{k=j+1}^{n} \hat{\varepsilon}_k\hat{\varepsilon}_{k-j}}{\sum_{k=1}^{n} \hat{\varepsilon}_k^2} \equiv \hat{\psi}_j/\hat{\psi}_0. \tag{2.321}$$

Then Venkataraman shows that

$$N(\hat{r}_2 - \hat{m}\hat{r}_1)^2\hat{v}^2 \xrightarrow{d} \chi_1^2 \tag{2.322}$$

where

$$\hat{\nu}^2 = \frac{(1/N)\hat{\psi}_0}{\hat{r}_2 + \hat{m}^2\hat{r}_1 - 2\hat{m}\sum\hat{\varepsilon}_k^2\hat{\varepsilon}_{k-1}\hat{\varepsilon}_{k-2}/\hat{\psi}_0}. \tag{2.323}$$

Mills and Seneta work this out to be 18.83 for the Fürth data, and the model thus is a poor fit. □

2.12. Hidden Markov models

A criticism frequently raised regarding the Markov chain model of precipitation has been its inability to produce realistic weather data. In particular, the lengths of wet and dry periods do not correspond to observed data. We show an example in Figure 2.16.

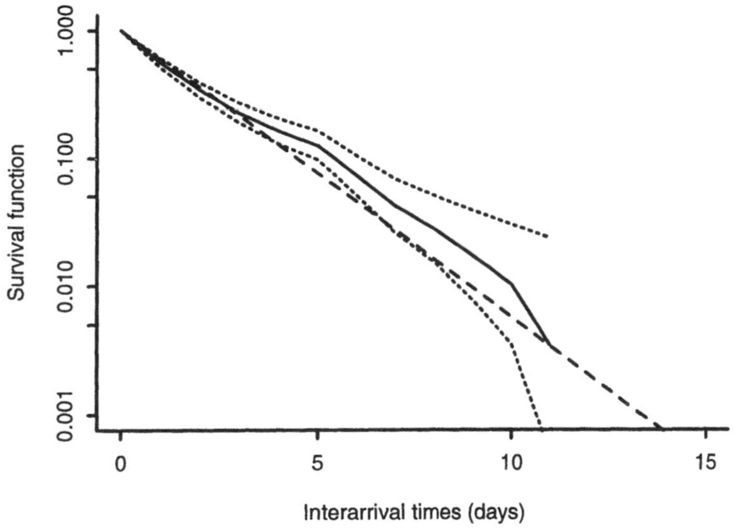

Figure 2.16. Estimated survival function (solid line) for the Snoqualmie Falls dry periods, January–March, together with the Markov chain survival function (short-dash line). The dotted line is an asymptotic 95% pointwise confidence band.

Here the estimated **survival function** (one minus the distribution function) for dry periods is plotted, together with the survival function for the estimated geometric distribution that obtains in the Markov chain situation (see Exercise 2). We see that the Markov chain survival function lies below the observed one, indicating a tendency towards shorter dry spells than what are observed. The dotted lines are asymptotic 95% pointwise confidence bands. The estimation of the survival function has to be done with care, since dry periods at the beginning and the end of the month are incompletely observed. Here we have used methods from survival analysis (Cox and Oakes 1984, ch. 4). Another possibility is to look back and forward in time to get the complete length of dry periods straddling January 1 or 31, assuming that the probability model is stationary over periods longer than exactly one month.

Another criticism of the Markov chain model is that it contains virtually no scientific knowledge (it is a statistical model, in the sense of section 1.2). The dependence observed in actual precipitation data is presumed to be a function of weather systems that pass through an area. It appears reasonable that dependence between rainfall from different weather systems is much smaller than that between rainfall on successive days of the same weather system.

A third problem is that the Markov chain approach has not been very successful when applied to more than one station in a region. While it is straightforward to develop a chain with an N-dimensional state space, representing all possible outcomes of rainfall, the large number (2^N) of parameters does not seem warranted for such a model.

In order to alleviate the problems discussed above, we shall build a model for precipitation at k stations by introducing unobserved states (thought of as somehow summarizing "climate") which account for different distributions of rainfall over the stations. The weather states are assumed to follow a Markov chain, while the pattern of occurrence/non-occurrence of precipitation over the network at any given time, given the weather states, is conditionally independent of the pattern at any other time. This type of model is often called a **hidden Markov model** and is a special case of the general **state space model** approach.

Let $C(t)$ denote the weather state at day t (throughout this section t will denote the discrete time variable, while n will denote the station number). We assume that $C(t)$ is a Markov chain with stationary transition probabilities

$$\gamma_{ij} = \mathbf{P}(C(t)=j \mid C(t-1)=i), \quad i,j=1,\ldots,M \tag{2.324}$$

and equilibrium probabilities

$$\delta = (\delta_1,\ldots,\delta_M) \tag{2.325}$$

so that, writing $\Gamma=(\gamma_{ij})$ for the transition matrix of the weather process we have

$$\delta\Gamma = \delta. \tag{2.326}$$

Let $X_n(t)=1$(rain at site n on day t), $n=1,\ldots,N$, $t=0,1,\ldots$ Write

$$\mathbf{X}(t) = (X_1(t),\ldots,X_N(t)) \tag{2.327}$$

and

$$Y(t) = (2^{N-1},2^{N-2},\ldots,2^0)\mathbf{X}(t)^T \tag{2.328}$$

so that $Y(t)$, which can take on values $l=0,1,\ldots,L\equiv 2^N-1$, is the decimal representation of the binary number $\mathbf{X}(t)$, ordered from all stations dry ($Y=0$, $\mathbf{X}=(0,\ldots,0)$) to all stations wet ($Y=N$, $\mathbf{X}=(1,\ldots,1)$). We assume that, given the weather state, the conditional probability of rain is given by

$$\mathbf{P}(Y(t)=l \mid C(t)=m) = \pi_{lm}. \tag{2.329}$$

The matrix of conditional distributions $\Pi=(\pi_{lm})$ has all columns summing to one. For a square matrix A we write $A_{l\bullet}$ for the lth row vector, and $A_{\bullet l}$ for the vector of elements of the lth column vector.

We assume now that $C(t)$ is in equilibrium, as is therefore $\mathbf{X}(t)$. In the following example we see that in general $\mathbf{X}(t)$ is not a Markov chain.

Example (A hidden Markov model which is not a Markov chain)
Let $N=2$, $M=2$, and assume that weather state 1 means that the two sites each have probability p of rain, independently of each other, while weather state 2 means that the two sites independently have probability q of rain. Then

$$\Pi = \begin{pmatrix} (1-p)^2 & (1-q)^2 \\ p(1-p) & q(1-q) \\ p(1-p) & q(1-q) \\ p^2 & q^2 \end{pmatrix}. \tag{2.330}$$

Suppose now that

$$\Gamma = \frac{1}{3}\begin{pmatrix} 1 & 2 \\ 2 & 1 \end{pmatrix} \tag{2.331}$$

so that $\delta=\frac{1}{2}\mathbf{1}$ where $\mathbf{1}$ is a vector of ones. Then one can compute the conditional probabilities

$$\begin{aligned} &\mathbf{P}(\mathbf{X}(t)=(1,1) \mid \mathbf{X}(t-1)=(1,1),\mathbf{X}(t-2)=(1,1)) \\ &= \frac{1}{3}(p^2+q^2)\left[1+\frac{3p^2q^2}{p^4+4p^2q^2+q^4}\right] \end{aligned} \tag{2.332}$$

and

$$\mathbf{P}(\mathbf{X}(t)=(1,1) \mid \mathbf{X}(t-1)=(1,1), \mathbf{X}(t-2)=(0,0))$$
$$= \frac{1}{3}\left[p^2+q^2 + \frac{3p^2q^2((1-p)^2+(1-q)^2)}{p^2(1-p)^2+2p^2(1-q)^2+2(1-p)^2q^2+q^2(1-q)^2}\right]. \tag{2.333}$$

Setting, for example, $p=0.9$ and $q=0.1$, we see that

$$\mathbf{P}(\mathbf{X}(t)=(1,1) \mid \mathbf{X}(t-1)=(1,1), \mathbf{X}(t-2)=(1,1)) = 0.286 \tag{2.334}$$

while

$$\mathbf{P}(\mathbf{X}(t)=(1,1) \mid \mathbf{X}(t-1)=(1,1), \mathbf{X}(t-2)=(0,0)) = 0.278. \tag{2.335}$$

Hence **X** is not a Markov chain. □

The likelihood of observations $y_1, \ldots, y_t$ can be written using $\lambda(k) = \text{diag}(\Pi_{k\bullet})$, a diagonal matrix of length M with diagonal elements consisting of the kth row of Π, as

$$L_t(\Pi,\Gamma) = \mathbf{P}(Y(1)=y_1, \ldots, Y(t)=y_t)$$
$$= \delta\lambda(y_1)\Gamma\lambda(y_2)\Gamma\lambda(y_3)\cdots\lambda(y_t)\mathbf{1}^T. \tag{2.336}$$

Application (Snoqualmie Falls precipitation, continued) We fit the Snoqualmie Falls precipitation data using a two-state version of the general model described above. Since we have only one site, $Y=X$. In order to have sufficient amounts of data, we use January–March. The model is given by

$$\Pi = \begin{bmatrix} \pi_{01} & \pi_{02} \\ \pi_{11} & \pi_{12} \end{bmatrix} = \begin{bmatrix} 1-\pi_1 & 1-\pi_2 \\ \pi_1 & \pi_2 \end{bmatrix} \tag{2.337}$$

and

$$\Gamma = \begin{bmatrix} \gamma_{11} & \gamma_{12} \\ \gamma_{12} & \gamma_{22} \end{bmatrix} = \begin{bmatrix} 1-\gamma_1 & \gamma_1 \\ \gamma_2 & 1-\gamma_2 \end{bmatrix}. \tag{2.338}$$

The model is determined by the four parameters $(\pi_1, \pi_2, \gamma_1, \gamma_2)$. The steady-state probabilities are

$$\boldsymbol{\delta} = (\delta_1, \delta_2) = \left[\frac{\gamma_2}{\gamma_1+\gamma_2}, \frac{\gamma_1}{\gamma_1+\gamma_2}\right]. \tag{2.339}$$

The following parameter estimates were obtained by numerical maximization of the likelihood for the Snoqualmie Falls data for January through March: $\hat{\gamma}_1=0.326$, $\hat{\gamma}_2=0.142$, $\hat{\pi}_1=0.059$ and $\hat{\pi}_2=0.941$. From (2.339) we see that $\boldsymbol{\delta}=(0.303, 0.697)$. Since $\hat{\pi}_1$ is nearly zero, this corresponds to a mostly dry state, which occurs about 1/3 of the time, while $\hat{\pi}_2$ corresponds to a mostly wet state, occurring 2/3 of the time. If a Markov chain were an appropriate description, we would have $\pi_1=0$ and $\pi_2=1$. The hidden Markov model is a significant improvement. The likelihood under the hidden Markov model is -1790.2, while that under the Markov chain model (the constrained hidden Markov model having $\pi_1=0$ and $\pi_2=1$) is -1796.64. The likelihood ratio test therefore rejects the Markov chain model in favor of the hidden Markov model with a P-value of 0.002. Using BIC, however, the two models are comparable, with the hidden Markov model coming out slightly worse.

In order to compare the survival function for dry periods to that observed, and that obtained for the Markov chain, we need to calculate the theoretical expression for it. Notice that a dry period starts whenever there is a transition from 1 to 0 in the X-process. Denoting the survival function by $G(k)$, we have

$$G(k) = \mathbf{P}(X(1)=0,X(2)=0,\ldots,X(k)=0 \mid X(0)=1,X(1)=0) \qquad (2.340)$$

which can easily be expressed in terms of the likelihood function (2.336). Figure 2.17 shows the survival functions of Figure 2.16, and in addition that of the hidden Markov model. We see that the hidden Markov model is an improvement over the Markov chain model, although still falling short of the observed confidence band for k=5. Remember, though, that the confidence bands are pointwise bands, and that one would therefore not be surprised to see one interval out of ten fail to cover. On the other hand, the bands are appropriate for independent survival times. This assumption is not strictly met for the hidden Markov model (although it does hold for the Markov chain model). □

The advantage of the hidden Markov model over the Markov chain model is more clearly illustrated for a network of sites.

Application (A Great Plains network) A simple two-state version of the general model was fitted to the sequence of wet and dry days at three sites, namely Omaha, Nebraska (site A), Des Moines, Iowa (site B) and Grand Island, Nebraska (site C) using records for the period 1949–1984. Our model is based on the assumption of two weather states which are temporally common to all three sites. Given a particular weather state, the event of rain at any given site is conditionally independent of rain at any other site. The probability of rain at each site varies with the weather state, and can be different from site to site. The matrix Π has entries

$$\pi_{lm} = \prod_{i=1}^{3} \theta_{im}^{x_i^l}(1-\theta_{im})^{1-x_i^l},\ 0 \le l \le 7,\ \ m=1,2 \qquad (2.341)$$

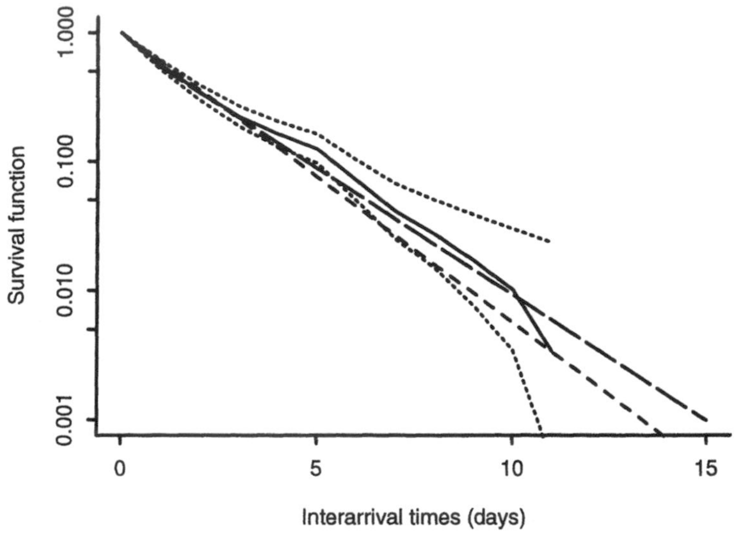

Figure 2.17. Estimated survival function (solid line) for Snoqualmie Falls dry periods in January, together with the survival function for the fitted Markov chain (short-dash line) and for the hidden Markov model (long-dash line). The dotted lines form an asymptotic 95% pointwise confidence band.

where (x_1^l, x_2^l, x_3^l) is the binary representation of l. For example, l=5 corrsponds to the pattern (1, 0, 1) of rain at sites 1 and 3, and no rain at site 2. This model has eight parameters, namely $\boldsymbol{\theta}=(\theta_{11},\theta_{21},\theta_{31},\theta_{12},\theta_{22},\theta_{32})$ and $\boldsymbol{\gamma}=(\gamma_1,\gamma_2)$.

To allow for seasonal changes the year was divided into six seasons, as defined in Table 2.16. Any rain occurring on February 29 was added to that of March 1. The parameter estimates, computed by numerical maximization of the likelihood are also given in the table.

The two weather states which result can again be described as ''mostly wet'' and ''mostly dry'', i.e., $\hat{\theta}_{l1}$ are fairly close to one and $\hat{\theta}_{l2}$ close to zero. The estimates vary quite smoothly with respect to changes in season, as does the estimated unconditional probability of being in a given state.

Table 2.17 gives the observed frequencies for season 1 of various events of interest together with the frequencies derived from the fitted model. For comparison we also give the fitted pattern from the Markov chain model, obtained by fitting an 8-state chain and computing the expected pattern under the stationary distribution. Thus, for example, there were 866 days on which it rained at both sites A and B in season 1, while the hidden Markov model predicted 862

Table 2.16 Parameter estimates for Great Plains network

Season	Days	$\hat{\gamma}_1$	$\hat{\gamma}_2$
1	001–060	0.352	0.335
2	061–120	0.358	0.350
3	181–180	0.333	0.381
4	181–240	0.389	0.354
5	241–300	0.389	0.248
6	301–365	0.359	0.269

Season	$\hat{\theta}_{11}$	$\hat{\theta}_{21}$	$\hat{\theta}_{31}$	$\hat{\theta}_{12}$	$\hat{\theta}_{22}$	$\hat{\theta}_{23}$
1	0.927	0.874	0.762	0.039	0.211	0.139
2	0.944	0.875	0.785	0.041	0.196	0.163
3	0.908	0.806	0.782	0.042	0.185	0.208
4	0.851	0.745	0.720	0.010	0.141	0.187
5	0.880	0.804	0.775	0.030	0.126	0.062
6	0.891	0.865	0.707	0.015	0.175	0.066

Table 2.17 Observed and fitted frequencies of rainfall patterns

	Dry	A	B	C	AB	AC	BC	ABC
Obs	718	1020	1154	957	866	752	728	657
HMM	725	1019	1153	956	862	749	728	657
MC	722	942	1076	1031	789	750	727	655

and the Markov chain predicted 789. The hidden Markov model has preserved the spatial dependence structure of the records very well, while the Markov chain model fails badly at that task. □

An important feature of the hidden Markov model is that it is possible to estimate the underlying Markov chain. In the precipitation model above this implies that it is possible to relate the hidden weather states to atmospheric observations (this has been carried out, with promising results, by Hughes, 1993). We shall see how it is done in a different setting below.

Application (Neurophysiology) Cell membranes contain several types of ion channels which allow selected ions to pass between the outside and the inside of the cell. These channels, corresponding to the action of a single protein molecule in the cell membrane, can be open (allowing passage of the selected ion) or closed at any one time. Some channels respond to changes in electric potential, while others are chemically activated. Among the latter are acetylcholine activated channels at so-called post-synaptic membranes of neurons, the basic cells of the nervous system. A more detailed description of neurons is in section 3.8, where we discuss the modeling of neural activity.

Patch clamp recordings provide the principal source of information about channel activity. A glass micro-pipette is positioned at the cell membrane and sealed to it by suction. Currents through the channels in the tip of the pipette are measured directly. The technique was developed by Sakmann and Neher, who were awarded the 1991 Nobel prize in medicine for it.

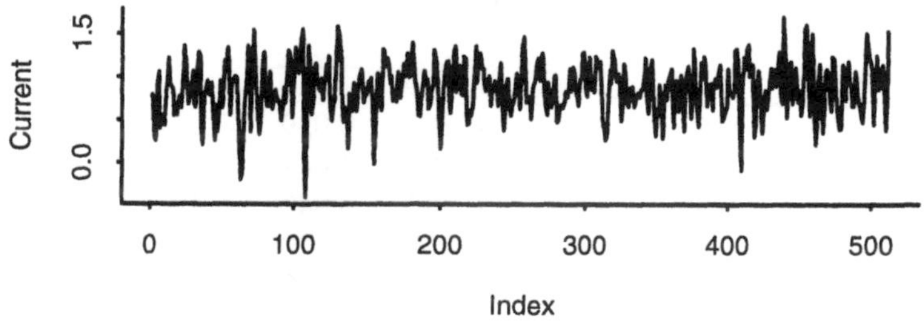

Figure 2.18. Current measured in a single channel of an acetylcholine receptor in a rat.

Figure 2.18 depicts a single channel recording from an acetylcholine receptor. Disregarding the noise, the current appears to move between two levels, corresponding to the open and closed states. The sampling interval is of the order of 10^{-4} seconds, while openings typically last some microseconds.

A very simple model for these observations are independent random variables, $X_1, \ldots, X_n$, distributed $N(\theta_i, \sigma^2)$ where θ_i is 0 or 1, in appropriate units. The likelihood for each n-digit binary sequence $\boldsymbol{\theta}^{(i)}$, $i=1,\ldots,2^n$, is (ignoring irrelevant constants)

$$L(\boldsymbol{\theta}^{(i)}) = \sigma^{-n}\exp(-\frac{1}{2\sigma^2}\sum_{j=1}^{n}(x_j-\theta_j^{(i)})^2). \tag{2.342}$$

Clearly, the mle of $\boldsymbol{\theta}$ must be $\hat{\theta}_j = 1(x_j > ½)$. This method of estimating the channel status is called the **threshold method**.

In order to analyze data such as those in Figure 2.18 we must restore the underlying quantized signal, i.e., the sequence of zeros and ones. If we, as is commonly done, use the threshold method, we tend to get a very ragged reconstruction, since this method does not take into account the fact that nearby signals are more likely to be similar than different.

To avoid the raggedness we need somehow to force the estimate to be smoother. Similarly to the approach in section 2.7, where we estimated the order of a Markov chain, we ensure smoothness by penalizing the likelihood for ragged reconstructions. Technically we shall use what is called **Bayesian statistics**. The way it works is that we assume that $\boldsymbol{\theta}$ is a realization of a stochastic process $\Theta_1, \Theta_2, \ldots$; in this case a Markov chain. Recall Bayes' theorem

$$\mathbf{P}(\boldsymbol{\Theta}=\boldsymbol{\theta} \mid \mathbf{X}=\mathbf{x}) = \frac{\mathbf{P}(\mathbf{X}=\mathbf{x} \mid \boldsymbol{\Theta}=\boldsymbol{\theta})\mathbf{P}(\boldsymbol{\Theta}=\boldsymbol{\theta})}{\sum_{\boldsymbol{\theta}} \mathbf{P}(\mathbf{X}=\mathbf{x} \mid \boldsymbol{\Theta}=\boldsymbol{\theta})\mathbf{P}(\boldsymbol{\Theta}=\boldsymbol{\theta})} \tag{2.343}$$

where the sum on the right-hand side is taken over all the 2^n possible binary n-digit numbers $\boldsymbol{\theta}$. We call $\mathbf{P}(\boldsymbol{\Theta}=\boldsymbol{\theta})$ the **prior** distribution of $\boldsymbol{\Theta}$ (it is obtained prior to observing $\mathbf{X}$), and $\mathbf{P}(\boldsymbol{\Theta}=\boldsymbol{\theta} \mid \mathbf{X}=\mathbf{x})$ the **posterior** distribution. One way of writing (2.343), with $\pi(\boldsymbol{\theta} \mid \mathbf{x})=\log(\mathbf{P}(\mathbf{X}=\mathbf{x} \mid \boldsymbol{\Theta}=\boldsymbol{\theta}))$ and $\pi(\boldsymbol{\theta})=\log \mathbf{P}(\boldsymbol{\Theta}=\boldsymbol{\theta})$, is

$$\pi(\boldsymbol{\theta} \mid \mathbf{x}) = c(\mathbf{x}) + l_{\mathbf{x}}(\boldsymbol{\theta}) + \pi(\boldsymbol{\theta}) \tag{2.344}$$

where $l_{\mathbf{x}}$ is the log likelihood function. Thus, maximizing $\pi(\boldsymbol{\theta} \mid \mathbf{x})$ is equivalent to maximizing the penalized log likelihood $l_{\mathbf{x}}(\boldsymbol{\theta})+\pi(\boldsymbol{\theta})$. The effect is to discount outcomes that are unlikely under the prior distribution. One way to think about this is that the threshold method maximizes $\pi(\theta_i \mid x_i)$ for each i, while the Bayesian approach maximizes $\pi(\boldsymbol{\theta} \mid \boldsymbol{x})$, simultaneously for all i. The two methods are the same whenever the θ_i are conditionally independent given the data, but if there is dependence (smoothness) the results, as we shall see shortly, can be quite different.

If we assume that the x_i are independent, conditionally upon the θ_i, and that the θ_i are a realization of a Markov chain, we must maximize

$$-n \log \sigma^2 + \sum_{i=1}^{n} \log p_{\theta_{i-1},\theta_i} - \frac{1}{2\sigma^2}\sum (x_i-\theta_i)^2. \tag{2.345}$$

In the case where $p_{01}=p_{10}=p<½$, this is equivalent to minimizing

$$J \log \frac{1-p}{p} + n \log \sigma^2 + \frac{1}{2\sigma^2}\sum (x_i-\theta_i)^2 + n \log p \tag{2.346}$$

where $J=\sum_1^n 1(\theta_i \neq \theta_{i-1})$ is the number of jumps in the sequence $\boldsymbol{\theta}$. Equivalently we can write (2.346)

$$\log \frac{1-p}{p} \sum (\theta_i-\theta_{i-1})^2 + n \log \sigma^2 + \frac{1}{2\sigma^2}\sum (x_i-\theta_i)^2. \tag{2.347}$$

We see explicitly how the penalty term (the first term in (2.347)) penalizes adjacent dissimilar θ-values.

A naive minimization of (2.344) entails evaluating $\pi(\boldsymbol{\theta} \mid \boldsymbol{x})$ for all the 2^n possible values of $\boldsymbol{\theta}$, once the transition matrix $\mathbb{P}$ is known. There are two problems with this: 2^n is a gigantic number for a typical data set, and the transition matrix is unknown. In order to avoid the first problem we employ a dynamic programming algorithm due to Viterbi (1967), while for the second we iterate the Viterbi algorithm to reconstruct θ given our current estimate of $\mathbb{P}$, and use the nonparametric mle of $\mathbb{P}$ from the reconstruction in the next iteration of the Viterbi algorithm.

The basic idea of the Viterbi algorithm is as follows. At time t, suppose that we have computed the most likely sequence up to this time for both the possible values of θ_t. We now proceed to determine the most likely history up to $\theta_{t+1}=j$. If this history has $\theta_t=i$ it must include that of the previously calculated most likely histories which ends in $\theta_t=i$. In this way one moves forward through the data, maintaining and updating two possible most likely histories at each time step. At the end, we compare the likelihood of the two histories to choose between them. Formally, write for $\theta_t=0$ or 1

$$H_t(\theta_t) = \{\hat{\theta}_0,\hat{\theta}_1,\ldots,\hat{\theta}_{t-1}\} \tag{2.348}$$

where $\hat{\theta}_0,\hat{\theta}_1,\ldots,\hat{\theta}_{t-1}$ maximizes (for the given θ_t)

$$L_t(\theta_0,\ldots,\theta_t) = \pi(\theta_0)f(x_0 \mid \theta_0)\prod_{k=0}^{t-1} p_{\theta_k,\theta_{k+1}}f(x_{k+1} \mid \theta_{k+1}). \tag{2.349}$$

Here π is the stationary distribution of the chain with transition matrix $\mathbb{P}$, and f is the conditional density of x given θ. In this application we take f to be Gaussian with mean 0 or 1 and common variance σ^2. Then L_t is just the likelihood of the hidden Markov chain $(X_0,\ldots,X_t)$, conditional upon the hidden sequence $\theta_0,\ldots,\theta_t$. Now consider the maximum $\hat{L}_t(\theta_t)=L_t(\hat{\theta}_0,\ldots,\hat{\theta}_{t-1},\theta_t)$. This satisfies the recursive relation

$$\hat{L}_{t+1}(\theta_{t+1}) = \max_{\theta_t}\hat{L}_t(\theta_t)p_{\theta_t,\theta_{t+1}}f(x_{t+1} \mid \theta_{t+1}). \tag{2.350}$$

Letting $\hat{\theta}_t$ be the maximizing value in the right-hand side of (2.350) we can write a recursion for the history H_t as

$$H_{t+1}(\theta_{t+1}) = \{H_t(\hat{\theta}_t),\hat{\theta}_t\}. \tag{2.351}$$

In practice it often happens that $\hat{\theta}_t$ is the same for both values of θ_{t+1}. Then the two most likely histories have merged at time t, and we no longer need to keep them in active memory.

Before we can apply the procedure to actual data, we need to estimate the noise variance σ^2. The exact value is not very critical; a simple approach is to use the threshold reconstruction of assigning all values above ½ to 1, and all below ½ to zero, and then compute the variance of the residual from the respective means. For the data in Figure 2.18 this yields $\sigma=0.067$. The reconstruction as outlined above was then applied to the data three times (each time updating the variance estimate and the transition matrix estimate based on the previous reconstruction), at which point there was no change in the updated state vector $\hat{\boldsymbol{\theta}}$. The resulting transition probabilities were $p_{01}=0.64$ and $p_{10}=0.06$, and the final $\sigma^2=0.073$. The restoration is shown, together with the maximum likelihood reconstruction, in Figure 2.19. Clearly, for this quite noisy sequence it is not reasonable to use the threshold method.

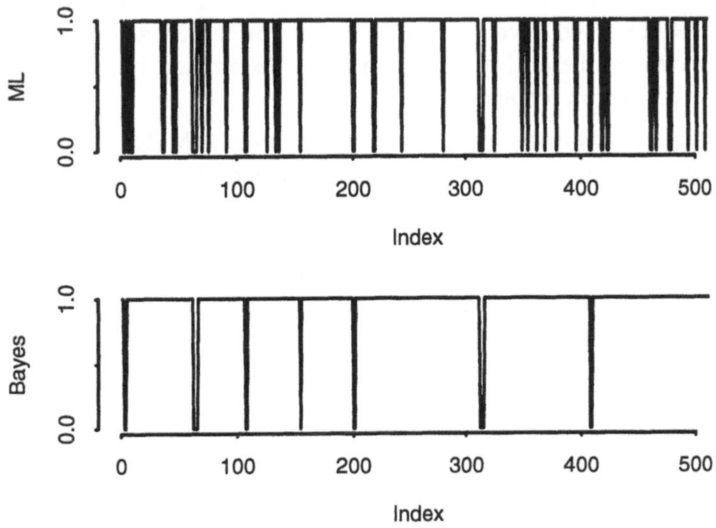

Figure 2.19. Maximum likelihood (upper panel) and Bayesian reconstruction (lower panel) of the signal in Figure 2.18.

□

2.13. Bibliographic remarks

Most of the material in sections 2.1–2.5 is standard. Among my sources have been Bhattacharya and Waymire (1990), Cox and Miller (1965), Feller (1968), Freedman (1983), Grimmett and Stirzaker (1982), and Whittle (1992), each of which is (in its own way) an excellent introduction to the behavior of Markov chains.

The material in section 2.6 owes much to discussions with and papers by Julian Besag, Elizabeth Thompson, and Charlie Geyer. Geyer (1991) and Tierney (1991) are nice presentations of the procedures; the former concentrating on likelihood evaluation, and the latter on computing posterior distributions. Discussion of implementation can be found in Geyer (1992). Some recent advances are in the dissertations by Lin (1993) and Higdon (1994). The Monte Carlo test examples come from Besag and Clifford (1989).

Section 2.7 follows largely Jacobsen and Keiding (1990). The standard source is Billingsley (1961), but it can be difficult to follow at times.

The linear model for higher order chains was developed by Raftery (1985a,b; Raftery and Tavaré, 1994) in a series of papers. A dissertation by Le (1990) developed these ideas further, and the dissertation by Schimert (1992) deals with estimation problems.

Section 2.9 borrows much theory from Cox and Miller (1965) while the application is inspired by Katz (1977). The recursion (2.238) was pointed out by Julian Besag. Section 2.10 is based on the presentation by Doyle and Snell (1984). The material in section 2.11 is from my monograph (Guttorp, 1991). The application to the Amazon indian tribe follows Thompson (1976).

Finally, section 2.12 comes from two main sources, namely Zucchini and Guttorp (1991) for the precipitation model, and Fredkin and Rice (1992) for the application to neurophysiology. General material on hidden Markov models can be found in Juang and Rabiner (1991), albeit from the point of view of speech recognition, and Bayesian image analysis is described in Besag (1989).

2.14. Exercises

Theoretical exercises

1. Prove that the Markov property (2.5) is equivalent to each of the following statements:

(a) Let T_1 be a set of times later than n, and T_0 a set of times less than or equal to n. Let $t_0=\max T_0$. Then

$$\mathbf{P}(X_k=x_k,\, k\in T_1 \mid X_l=x_l, l\in T_0) = \mathbf{P}(X_k=x_k,\, k\in T_1 \mid X_{t_0}=x_{t_0}).$$

(b) Let T_1 be a set of times later than n, and T_0 a set of times prior to n. Then

$$\mathbf{P}(X_k=x_k, k\in T_1,\, X_l=x_l, l\in T_0 \mid X_n=x_n)$$
$$= \mathbf{P}(X_k=x_k, k\in T_1 \mid X_n=x_n)\mathbf{P}(X_l=x_l, l\in T_0 \mid X_n=x_n).$$

2. Show that the time spent in state k upon each return to it for a Markov chain has a geometric distribution with mean $1/(1-p_{kk})$.

3. Show that $\lambda_0=1-2^{-k}$ in the Reid–Landau model of radiation damage.

4. Prove that for a fair simple random walk on the integer lattice in 2 dimensions, $\mathbf{0}$ is persistent.

5. Prove the time invariance result given by equation (2.80).

6. Show that a transition matrix $\mathbb{P}$ on a finite state space has a uniform stationary distribution iff it is doubly stochastic.

7. Derive the marginal distribution of the number of red balls after n draws from a Pólya urn.

8. Verify that Theorem 2.7 holds when the process starts from any state leading to the ergodic state k.

9. Show that the Hardy–Weinberg proportions are the stationary distribution for the chain with transition matrix (2.142).

10. A slight variant of the Metropolis algorithm, which is actually a generalization of both the Gibbs sampler and the Metropolis algorithm, was proposed by Hastings (1970). Let Q be a transition matrix, not necessarily symmetric, and assume for simplicity that the target function $f(x)>0$ for all x. Define

$$\alpha(x,y) = \min\left\{\frac{f(y)Q(y,x)}{f(x)Q(x,y)}, 1\right\}$$

when $f(x)Q(x,y)>0$, and $\alpha(x,y)=1$ otherwise. Let $X_n=x$, and generate a candidate point $Y=y$ from the distribution $Q(x,\bullet)$. With probability $\alpha(x,y)$ we set $X_{n+1}=y$, and with the complementary probability $X_{n+1}=x$.

(a) Show that the process (X_n) is a Markov chain with stationary distribution f.

(b) Show that the Metropolis algorithm results for symmetric Q.

11. Let X_k be a 0-1 process, and consider the distribution of $S_n=\sum_{i=1}^{n} X_k$. In order to evaluate (2.238) we must compute $Q(t)^n$. This can be done by diagonalization: $Q(t)=E\,D\,E^{-1}$ where E contains the right eigenvectors of Q and D is the diagonal matrix of the corresponding eigenvalues.

(a) Show that the eigenvalues solve

$$\lambda^2 - \lambda(p_{11}e^{-t}+p_{00}) + \det(\mathbb{P})e^{-t} = 0.$$

(b) Show that the eigenvectors are given by

$$\mathbf{q}_0(t) = (p_{01}e^{-t}, \lambda_1(t)+p_{00})$$

and

$$\mathbf{q}_1(t) = (p_{01}e^{-t}, \lambda_2(t)+p_{00}).$$

(c) Using that $\lambda_0(t)+\lambda_1(t)=p_{00}+p_{11}e^{-t}$ and $\lambda_0(t)\lambda_1(t)=(p_{00}-p_{01})e^{-t}$, show that (2.238) simplifies to

$$\phi_n(t) = \mathbf{E}^{\pi}(e^{-tS_n}) = \frac{\lambda_1^{n+1}-\lambda_2^{n+1}-(p_{11}-p_{01})(\lambda_1^n-\lambda_2^n)}{\lambda_1-\lambda_2}.$$

(d) Deduce that

$$S_n \sim \mathrm{AsN}\left[\frac{n(p_{01}^2+p_{10}^2)}{(p_{01}+p_{10})^2}, \frac{np_{01}p_{10}(p_{00}+p_{11})}{(p_{01}+p_{10})^3}\right].$$

Hint: Argue that $\log\phi_1^{(n)}(t)\approx n\log\lambda_1(t)$ and do a Taylor expansion of the eigenvalue.

12. Let X_n be a 0-1 chain. Derive the correlation between $\sum_k^l X_i$ and $\sum_m^n X_j$, where $l<m$. Apply the result to the Snoqualmie Falls data with $k=1$, $l=7$, $m=8$, and $n=14$.

13. Let $p_k=bc^{k-1}$, $k\geq 1$ and $p_0=1-\sum_{k\geq 1} p_k$.

(a) Write $P(s)$ in the form

$$P(s) = \frac{\alpha+\beta s}{\gamma+\delta s},\ \alpha\delta-\beta\gamma\neq 0.$$

This is called a **fractional linear** generating function.

(b) Consider a BGW branching process with offspring distribution given by P(s). Show by induction that $P_k(s)$ is a fractional linear generating function.

14. Suppose that $p_{ij}>0$ for all $i,j\in S$, where S is a finite set, so that the chain X_n has a limiting distribution. Show that

$$|p_{ij}^{(n)} - \pi_i| \leq (1-d\delta)^n$$

where $d=|S|$ and $\delta=\min p_{ij}$.

Hint: Divide the terms in the equation $\sum_j (p_{ij}-p_{kj})=0$ into those with positive and negative values. The sum of the positive terms is bounded by $1-n\delta$. Now bound $\max_i p_{ij}^{(n+1)}-\min_i p_{ij}^{(n+1)}$ using Chapman–Kolmogorov.

15. Using the parametric result in Theorem 2.16 and the parametrization $p_{ij}(\mathbf{\theta})=\theta_{ij}$ verify the nonparametric result in Theorem 2.14.

16. Prove or disprove the following statement: The Markov property is equivalent to the property that for any class of (measurable) sets $A_1,\ldots,A_n$ we have that

$$\mathbf{P}(X_n\in A_n \mid X_1\in A_1,\ldots,X_{n-1}\in A_{n-1}) = \mathbf{P}(X_n\in A_n \mid X_{n-1}\in A_{n-1}).$$

17. Let S_n be a simple random walk. Show that $|S_n|$ is a Markov chain, and determine its transition probabilities.

Hint: What can you say about the distribution of S_n given the values of $|S_n|,\ldots,|S_1|$?

18. Show that all two-state chains are reversible.

19. Let (X_k) be a K-state ergodic Markov chain with transition matrix P, such that all $P_{ij}>0$. Let $p=(p_1,\ldots,p_K)$ be a probability distribution. Develop a likelihood ratio test for the hypothesis that p is a stationary distribution for (X_k).

Remark: You may not be able to get an explicit solution (except if K=2), but you should develop a system of equations that the constrained estimates have to satisfy.

20. It is not very difficult (at least in principle) to allow for a Markov chain with continuous state space. Let $X_1, X_2, \ldots$ be continuous random variables such that the density of X_k given that $X_{k-1}=x$ is given by $f(x,\bullet)$. Similar to the discrete state case, we say that the process is a Markov chain if the conditional density of X_n, given $X_{n-1}, \ldots, X_1$, depends only on the value of X_{n-1}.

(a) Let $f_n(x,\bullet)$ denote the conditional density of X_n, given that $X_0=x$. Show that the Chapman–Kolmogorov equation holds in the form

$$f_{n+m}(x,y) = \int_{\mathbb{R}} f_m(x,z) f_n(z,y)\,dz.$$

(b) Show that the stationary distribution (when it exists) has density $\pi(y)$ satisfying

$$\pi(y) = \int_{\mathbb{R}} \pi(x) f(x,y)\,dx.$$

21. The hidden Markov model originated in work in engineering, where a system described by the **state variable** $\mathbf{X}_k$ of dimension s was assumed to develop according to the **state equation**

$$\mathbf{X}_k = \mathbf{X}_{k-1}\mathbf{A} + \varepsilon_k$$

where ε_k is assumed $\mathrm{N}(\mathbf{0},\sigma_\varepsilon^2 \mathbb{I}_{s\times s})$. However, the system cannot be observed directly. Instead, one observes a related random vector $\mathbf{Y}_k$, of dimension o, satisfying the **observation equation**

$$\mathbf{Y}_k = \mathbf{X}_k\mathbf{B} + \delta_k$$

where δ_k is $\mathrm{N}(\mathbf{0},\sigma_\delta^2 \mathbb{I}_{o\times o})$. The parameters $\mathbf{A}$, $\mathbf{B}$, σ_ε^2 and σ_δ^2 were originally taken as known, although modern applications allow for estimation of these parameters. The **Kalman filter** (Kalman, 1960) estimates the value of $\mathbf{X}_k$, given observations $\mathbf{Y}_1^k$. Show that $\mathbf{m}_k=\mathbf{E}(\mathbf{X}_k \mid \mathbf{Y}_1^k)$ and $\Gamma_k=\mathbf{Var}(\mathbf{X}_k \mid \mathbf{Y}_1^k)$ satisfy the recursions

$$\mathbf{m}_{k+1} = \mathbf{m}_k\mathbf{A} + (Y_k - \mathbf{m}_k\mathbf{B})\mathbf{K}_k$$

and

$$\Gamma_{k+1} = \mathbf{A}^T\Gamma_k(\mathbb{I} - B(\mathbf{B}^T\Gamma_k\mathbf{B} + \sigma_\delta^2\mathbb{I})^{-1}\mathbf{B}^T)\mathbf{A} + \sigma_\varepsilon^2\mathbb{I}$$

where the **gain** $\mathbf{K}_k$ is given by

$$\mathbf{K}_k = (\mathbf{B}\Gamma_k\mathbf{B}^T + \sigma_\delta^2\mathbb{I})^{-1}\mathbf{B}^T\Gamma_k\mathbf{A}.$$

Hint: Use the formula for conditional expectation of jointly Gaussian random variables, i. e., normal regression.

22. Let (X_k) be a random walk, but suppose that it is observed with independent measurement error.

(a) Write this process in state space form.

(b) Derive the Kalman filter for it.

Computing exercises

C1. Given a subroutine that generates multinomial random vectors from input values for sample size and probability vector, how would you build a subroutine that generates a Markov chain with given (finite-dimensional) transition matrix $\mathbb{P}$?

C2. (a) How long a stretch of data from an ergodic chain do you need to estimate $\mathbb{P}$ accurately? You choose $\mathbb{P}$ and what you mean by "accurately".

(b) How long a stretch do you need to estimate the stationary distribution accurately?

C3. (a) Let (X_k) be a Markov chain with transition matrix

$$P = \begin{bmatrix} 0.1 & 0.2 & 0.3 & 0.3 & 0.1 \\ 0.5 & 0.1 & 0.1 & 0.1 & 0.2 \\ 0.1 & 0.5 & 0.1 & 0.1 & 0.2 \\ 0.1 & 0.2 & 0.5 & 0.1 & 0.1 \\ 0.2 & 0.2 & 0.1 & 0.4 & 0.1 \end{bmatrix}.$$

Starting from state 1, find $\mathbf{E}^1(T_5)$.

(b) Suppose that you do not know P, but have a realization $(x_k, k \leq 100)$ of (X_k). Find a way of estimating the quantity in (a). Perform the estimation on a simulated realization.

(c) The expected value in part (a) is a certain function of

$$F_{15}(t;P) = \mathbf{P}^1\{T_5 \leq t\}.$$

It is complicated to determine this distribution analytically, even when P is known. If P is unknown, it is necessary to apply simulation methods. The **bootstrap** (Efron, 1978) can be modified to do this. Let $\hat{P}$ be the estimated transition matrix. Generate a path from $\hat{P}$, and use that to estimate the transition matrix the usual way, yielding a matrix $\hat{P}^*$. The bootstrap idea is to say that the distribution of

$$\sqrt{n}\,(F_{15}(t;\hat{P}) - F_{15}(t;P))$$

is well approximated by that of

$$\sqrt{n}\,(F_{15}(t;\hat{P}^*) - F_{15}(t;\hat{P})).$$

In practice, one would use repeated samples from $\hat{P}$ to get realizations of the hitting time, and then use the empirical distribution function of these hitting times to estimate $F_{15}(t;P)$ as well as whatever function of it you are interested in. The approximation argument above is the theoretical basis for this procedure. Compute a bootstrap estimate using the sample generated in part (b).

Use a bootstrap sample of size 100. Compare the mean of the resulting estimate to the answer in (b). Can you use the bootstrap distribution to estimate the variance of the estimated mean?

C4. An electrical network consists of nine nodes, labeled a to i. A battery is connected between nodes a and i such that the potential difference $v(a)-v(i)$ is 1. [Without loss of generality $v(a)=1$, $v(i)=0$.] The network consists of the following eighteen wires, with the given resistances:

10-ohm resistance: *cf*

5-ohm resistance: *ab*, *bc*, *bh*, *ce*, *hi* and *fi*

2-ohm resistance: *ac*, *bd*, *cd*, *de*, *df*, *eg* and *gi*

1-ohm resistance: *ad*, *ef*, *eh* and *fg*.

Find the transition matrix for the corresponding Markov random walk on the nodes a to i, and hence find the potentials of the nodes in each of the following ways:

(a) Use the method of relaxation to solve the equation $v = \mathbb{P}v$.

(b) Use the exact method (2.276) to find the solution; be careful about how you put in the boundary conditions, and remember to make a and i absorbing states.

(c) Simulate the random walk to find $v(e)$. See how large a sample you need to obtain an accurate estimate. See if you can think of a way of estimating all of v without having to do a lot of separate simulations.

C5. Simulate the χ^2 statistic on p. 76 (based on the Snoqualmie Falls Markov model) and compare its distribution under this model to the χ^2-distribution suggested by standard (iid) goodness-of-fit theory.

C6. Generate a data set of 100 observations from a mixture of three exponentials with means 48.7, 5.83, and 0.65 with weights 0.0032, 0.1038, and 0.8930, respectively. From these data, using a MCMC method (Gibbs sampler, Metropolis algorithm, or otherwise), estimate the parameters of the underlying distribution. Compare the results to a regular optimization of the likelihood. How many iterations do you need to estimate the likelihood accurately?

Data exercises

D1. Data set 1 contains the Snoqualmie Falls precipitation occurrence data. Extract the January precipitation (the first 31 numbers on each line), and recode it to 0 or 1, where 1 denotes measurable precipitation.

(a) Test the hypothesis that the years are identically distributed.

(b) Current research in precipitation modeling (Woolhiser, 1992) indicates that the El Niño-Southern Oscillation phenomenon (an unusual pattern of surface pressure and sea surface temperature in the south Pacific) may have a profound effect on precipitation occurrence in North America. Table 2.18 contains the Southern Oscillation Index (SOI), defined as the difference in mean January sea level pressure between Tahiti and Darwin, for each of the years of the Snoqualmie Falls data. Assess the relationship between transition probabilities and the Southern Oscillation Index for January.

Table 2.18 January SOI 1948–1983

38	29	55	79	25	49	57	33	68	56
9	26	45	38	80	62	36	35	19	74
53	16	22	50	52	38	88	33	69	37
38	36	51	50	64	−20				

D2. Gabriel and Neumann (1962) analyzed precipitation data from the rainy season (December through February) in Tel Aviv, Palestine/Israel, from 1923 to 1950. Table 2.19 contains the data.

Table 2.19 Tel Aviv precipitation

Previous days		Current day		
Second	First	Wet	Dry	Total
Wet	Wet	439	249	688
Dry	Wet	248	192	350
Wet	Dry	87	261	348
Dry	Dry	263	788	1051

We may contemplate a 3-parameter submodel of the general second-order Markov chain, based on the idea of relatively short-term fronts passing through the area. Then the only long-term influence should be in cases where it rained the previous day. This suggests the following transition matrix:

$$\mathbb{P} = \begin{pmatrix} p_{ww} & 1-p_{ww} \\ p_{dw} & 1-p_{dw} \\ p_d & 1-p_d \\ p_d & 1-p_d \end{pmatrix}.$$

Evaluate this model using the BIC.

D3. A common assumption in sociology is that the social classes of successive generations in a family follow a Markov chain. Thus, the occupation of a son is assumed to depend only on his father's occupation and not on his grandfather's. Suppose that such a model is appropriate, and that the transition matrix is given by

$$\mathbb{P} = \begin{pmatrix} 0.4 & 0.5 & 0.1 \\ 0.05 & 0.7 & 0.25 \\ 0.05 & 0.5 & 0.45 \end{pmatrix}.$$

Here the social classes are numbered 1 to 3, with 1 the highest. Father's class are the rows, son's class the columns. A sociologist has some data that is supposed to illustrate this model. However, he got these data from a colleague, who did not say whether the counts had father's class along rows or columns. The counts are

$$\begin{bmatrix} 7 & 6 & 5 \\ 9 & 207 & 60 \\ 2 & 64 & 60 \end{bmatrix}.$$

Try to determine the more likely labeling.
Hint: How are the counts from a chain run backwards related to those from the same chain run forwards?

D4. In the early 1950s, the Washington Public Opinion Laboratory in Seattle carried out "Project Revere" which was intended to study the diffusion of information (in particular, since the project was funded by the US Air Force, information contained in leaflets dropped from the air). A subexperiment took place in a village with 210 housewives. Forty-two of these were told a slogan about a particular brand of coffee. Each housewife was asked to pass the information on. As an incentive, participants were told that they would get a free pound of coffee if they knew the slogan when an interviewer called 48 hours later. It was possible to trace the route by which each hearer had obtained the slogan, so that they could be classified by generations. The data are given in Table 2.20.

Table 2.20 Spread of slogan

Generation	1	2	3	4	5
Size	69	53	14	2	4

Using a branching process approach, estimate the mean number of offspring in the first generation. How about the final generation?

D5. The data in Table 2.21 show the state of health (1 = self-care, 2 = intermediate care, 3 = intensive care) for 47 patients suffering from chronic bronchial asthma during five different asthmatic seasons (Jain, 1986). For each season we give the observed transitions between the three health states. Test the hypothesis of stationarity, i.e., that the transition matrix does not depend on season.

D6. Table 2.22 contains second-order transition counts between high (1) and low (0) return weeks, relative to mean weekly returns in the past, for weekly nominal returns on a value-weighted portfolio of stocks (McQueen and Thorley, 1991) from April, 1975, through December, 1987. Test the hypothesis that the portfolio returns perform a random walk.

Table 2.21 Severity of asthma by season

Season	State of health 1	2	3
	19	1	2
Winter	2	9	4
	1	2	7
	15	2	2
Trees	3	10	4
	1	1	9
	17	1	2
Grass	3	10	5
	1	1	7
	13	2	3
Ragweed	3	12	3
	1	1	9
	12	2	3
Fall	6	6	7
	1	2	8

Table 2.22 High and low return weeks for a stock portfolio

Previous		0	1
0	0	63	75
0	1	74	90
1	0	75	88
1	1	89	109

D7. The data in data set 2 are bi-daily counts of the number E_k of emerging blowflies (*Lucilia cuprina*) and the number D_k of deaths in a cage maintained in a laboratory experiment by the Australian entomologist A. J. Nicholson. The flies were supplied with ample amounts of water and sugar, but only limited amounts of meat, which is necessary for egg production. Let $\mathbf{A}_t = (A_{1,t}, \ldots, A_{m,t})^T$ be the age distribution vector, so $A_{x,t}$ is the number of individuals aged x at time t. Let $p_{k,t}$ be the proportion of individuals aged k at time t that survive to age $k+1$, and let P_t be the survival matrix having $p_{k,t}$ on the sub-diagonal, and zeroes elsewhere. A model for the population dynamics

(Brillinger et al., 1980) is given by

$$\mathbf{A}_{t+1} = P_t(H_t) + \mathbf{B}_{t+1} + \boldsymbol{\varepsilon}_t$$

where $\boldsymbol{\varepsilon}_t$ is an age- and time-dependent noise sequence and H_t denotes the history of the population sizes N_t up to time t.

(a) Write this model, using $\mathbf{A}_t$ as the state, in a state space model like that in Exercise 21 (except it will have random time-dependent coefficients).

(b) Deduce that the state vector can be estimated by

$$a_{1,t+1} = E_{t+1}$$

$$a_{k,t+1} = \frac{p_{k-1,t}\, a_{k-1,t}}{\sum_{l\geq 1} p_{l,t}\, a_{l,t}} (N_{t+1} - E_{t+1}), \quad k > 1$$

either using a Kalman filter approach, or by using an intuitive argument.

(c) For the model

$$p_{i,t} = (1-\alpha_i)(1-\beta N_t)$$

estimate the parameters α_i and β.
Hint: One possibility is to compare observed and predicted deaths for the next time interval, in a (weighted) least squares fashion. Another is to try a likelihood approach.

D8. The data in data set 3 consists of four years of hourly average wind directions at the Koeberg weather station in South Africa. The period covered is 1 May 1985 through 30 April 1989. The average is a vector average, categorized into 16 directions from N, NNE, . . . to NNW. Analyze the data using a hidden Markov model, with 2 hidden states, and conditionally upon the state a multinomial observation with 16 categories. Does the fitted model separate out different weather patterns? Also estimate the underlying states, and look for seasonal behavior in the sequence of states.

D9. The data in Table 2.23 are numbers of movements by a fetal lamb observed by ultrasound and counted in successive 5-second intervals (Wittman et al., 1984; given by Leroux and Puterman, 1992). Changes in activity may be due to physical changes in the uterus, or to the development of the central nervous system. Assume that there is an underlying unobserved binary Markov chain Z_i, such that if $Z_i = k$, the observed counts have a Poisson distribution with parameter λ_k, $k = 0, 1$. Fit this model using maximum likelihood.

D10. Christchurch aerodrome is one of two international airports in New Zealand. The runways at Christchurch are prone to fog occurrence, and fog forecasting is difficult, particularly because of the sheltering of the area by the Southern Alps. Renwick (1989) reports hourly data on weather type for 1979–1986, a total of 70,128 observations. Table 2.24 show the transition frequencies for observations between 2 hours before and 3 hours after sunrise. The

Table 2.23 Fetal lamb movements

0	0	0	0	0	1	0	1	0	0	0	0	0	0	1
0	1	0	0	0	0	2	2	0	0	0	0	1	0	0
1	1	0	0	1	1	1	0	0	1	0	0	0	0	0
0	0	0	0	0	0	0	0	0	0	0	0	0	2	0
1	0	0	0	0	0	0	0	0	0	0	0	0	0	0
0	0	0	1	0	0	0	0	0	7	3	2	3	2	4
0	0	0	0	1	0	0	0	0	0	0	0	1	0	2
0	0	0	0	1	0	0	0	0	1	0	0	0	0	0
0	0	0	0	0	0	0	0	0	1	0	0	0	0	0
2	1	0	0	1	0	0	0	1	0	1	1	0	0	0
1	0	0	1	0	0	0	1	2	0	0	0	1	0	1
1	0	1	0	0	2	0	1	2	1	1	2	1	0	1
1	0	0	1	1	0	0	0	1	1	1	0	4	0	0
2	0	0	0	0	0	0	0	0	0	0	0	0	0	0
0	0	0	0	0	0	0	0	0	0	0	0	0	0	0

Table 2.24 Fog and mist occurrence at Christchurch aerodrome

		Next		
		Clear	Mist	Fog
	Clear	13,650	245	30
Now	Mist	480	427	71
	Fog	11	171	198

overall distribution of weather type has clear weather 95.1% of the time, mist 3.4% of the time, and fog 1.5% of the time. Test whether the hours around sunrise are reasonably described by these stationary probabilities.

CHAPTER 3

Continuous time Markov chains

A natural generalization of the Markov chain in discrete time is to allow the times between transitions to be continuous random variables. The motivating example concerns multiplicative showers of cosmic radiation. The difference between deterministic and stochastic models is illustrated using linear and nonlinear birth processes. The relation to the previous chapter is brought out in the discrete skeleton and the jump chain. Some queueing models are presented, and the statistical inference for both nonparametric and parametric models of continuous time Markov processes is discussed. Partially observed processes are used to model the activity of synaptic nerve cells. The blood production in cats is described using a hidden Markov process.

3.1. The soft component of cosmic radiation

High-energy particles, many near the speed of light, continuously bombard the atmosphere from all directions. This cosmic radiation, discovered by V. F. Hess during a balloon flight in 1912, consists primarily of nuclei of the most abundant elements. It has two clearly distinguishable parts, the **soft** and **hard** components. The former, on which we shall concentrate, consists of very energetic electrons together with gamma ray photons, while the latter consists of mesons.

When soft component cosmic rays pass through the atmosphere, they collide with atmospheric atoms, producing showers of additional particles. If the number of shower particles is measured as a function of height above the ground, it is found to have a maximum at about 16 km. Very few of the primary rays reach the surface of the earth. When an energetic photon travels a distance Δt through some substance, such as lead or air, it has a certain chance of being absorbed by an atom and emitting a pair of particles consisting of one positron and one electron. By doing so it loses all its energy. This process is called **pair production**. The probability is proportional to Δt for small distances, and the factor of proportionality is a function of the energies of the parent photon and the resulting electron offspring. A fast electron (or positron) radiates high energy photons when it is absorbed in the nucleus of an atom (**Bremsstrahlung**). Again, the probability is proportional to the distance traveled, and the factor of proportionality is again a function of the energies involved.

In order to measure the cosmic radiation one places a number of particle counters (such as Geiger–Müller tubes) in a pattern around an absorber, such as a block of lead. The absorber is expected to produce showers as described above, and if the counters are wired to register only simultaneous registration at several counters, one can be fairly confident that a shower phenomenon has been observed. The number of coincidences per unit time is measured and plotted against the thickness of the lead. Figure 3.1 illustrates the results of an experiment of Schwegler, with curve III corresponding to the soft component and curve II to the hard component.

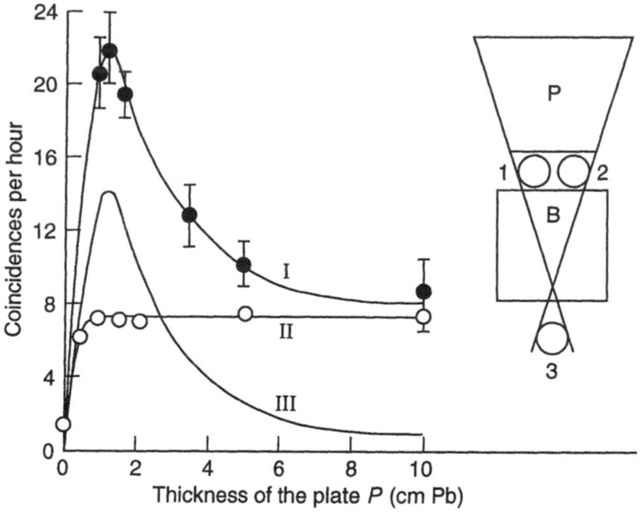

Figure 3.1. Coincidence counts of cosmic radiation as a function of lead absorber thickness. Curve I is obtained without the lead block B between the counters, curve II with B placed between the counters. Curve III is the difference between the two curves. This figure is an adaptation of one published in Arley (1943). Reproduced with permission from L. Janossy: *Cosmic Rays*, published by Oxford University Press, 1948; and with the permission of the estate of Dr. Arley.

We start with a very simple model that was proposed by Furry (1937). He disregarded the photons, and assumed that one electron by traveling a distance Δt could be converted into two electrons, with a probability asymptotically (as $\Delta t \to 0$) proportional to Δt. All electrons were assumed to act independently. Let $P_{1n}(t)$ denote the probability, starting from a single electron, of finding n electrons at depth t, where depth 0 is the top of the atmosphere. Then Furry's model says that the conditional probability of finding $n+m$ electrons at depth $t+\Delta t$, given that there were n at depth t, is given by

$$\frac{P_{1,n+m}(t+\Delta t)}{P_{1n}(t)} = \begin{cases} n\lambda\Delta t+o(\Delta t) & \text{if } m=1 \\ o(\Delta t) & \text{if } m>1 \\ 1-n\lambda\Delta t+o(\Delta t) & \text{if } m=0. \end{cases} \tag{3.1}$$

Rewriting this slightly, we see that

$$\begin{cases} P_{10}(t) \equiv 0 \\ P_{11}(t+\Delta t) = P_{11}(t)(1-\lambda\Delta t) + o(\Delta t) \\ P_{1n}(t+\Delta t) = P_{1n}(t)(1-n\lambda\Delta t) + P_{1,n-1}(t)(n-1)\lambda\Delta t + o(\Delta t), \quad n\geq 2. \end{cases} \tag{3.2}$$

By manipulating these equations and letting $\Delta t \to 0$ we see that

$$\begin{cases} P_{10}(t) \equiv 0 \\ \dfrac{dP_{11}(t)}{dt} = -\lambda P_{11}(t) \\ \dfrac{dP_{1n}(t)}{dt} = \lambda(n-1)P_{1,n-1}(t) - \lambda n P_{1n}(t). \end{cases} \tag{3.3}$$

Adding the initial condition $P_{1n}(0)=1(n=1)$ it is easy to solve these equations recursively, getting

$$\begin{cases} P_{10}(t) \equiv 0 \\ P_{1n}(t) = e^{-\lambda t}(1-e^{-\lambda t})^{n-1}. \end{cases} \tag{3.4}$$

In other words, the number of electrons at depth t has a positive geometric distribution with parameter $\exp(-\lambda t)$. The mean number of electrons therefore is $\exp(\lambda t)$, which is increasing in t, and therefore does not describe the observed mean values very well. This process is an example of a **pure birth process** (a process that only moves upwards in unit steps), and was first introduced in a paper by McKendrick (1914), and rediscovered by Yule[1] (1924). The process is now often called the Yule–Furry process, or the linear birth process.

[1] Yule, George Udny (1871–1951). English statistician. Student of Karl Pearson. Introduced the multiple correlation coefficient.

There are at least three features of the actual shower process that are disregarded in Furry's model. First, the absorption of electrons is disregarded, second, the alternating character of the successive generations (photons $\rightarrow$ electrons $\rightarrow$ photons $\cdots$) is not taken into account, and, third, the energy decrease resulting from the production of photons is neglected. Later we will modify the model to produce something closer to the observed data.

3.2. The pure birth process

Now consider a generalization of the Yule–Furry model, by letting the conditional probability of a birth in the time interval $(t, t+\Delta t)$, given n births at a given time t, be $\lambda_n \Delta t$, with an error of smaller order in Δt. This is called the **general birth process**. The linear model (3.1) has $\lambda_n = \lambda n$. Arguing as when deriving (3.3), and starting from $X(0)=1$, we obtain the system of equations

$$\begin{cases} P_{10}(t) \equiv 0 \\ \dfrac{dP_{11}(t)}{dt} = -\lambda_1 P_{11}(t) \\ \dfrac{dP_{1n}(t)}{dt} = \lambda_{n-1} P_{1,n-1}(t) - \lambda_n P_{1n}(t) \end{cases} \tag{3.5}$$

From the second equation in the system (3.5) we see that $P_{11}(t)=\exp(-\lambda_1 t)$, whence

$$\frac{dP_{12}(t)}{dt} = \lambda_1 \exp(-\lambda_1 t) - \lambda_2 P_{12}(t). \tag{3.6}$$

It turns out to be convenient to use Laplace transforms (much as we needed generating functions in the previous chapter). The Laplace transform f^* of a function $f: \mathbf{R}_+ \rightarrow \mathbf{R}$ is defined by

$$f^*(\theta) = \int_0^\infty e^{-\theta x} f(x)dx \tag{3.7}$$

which exists for $\theta > 0$ whenever f is integrable. Under some regularity conditions involving continuity, the Laplace transform determines f uniquely. In fact, f can be computed from f^* using integration in the complex plane. Here we are mainly interested in the following properties:

Lemma 3.1

(a) If $h(x) = f(x) + g(x)$, then $h^*(\theta) = f^*(\theta) + g^*(\theta)$.

(b) If $h(x) = \int_0^\infty g(x-y) f(y)dy$, then $h^*(\theta) = g^*(\theta) f^*(\theta)$.

(c) $\left[f'(t) \right]^*(\theta) = \theta f^*(\theta) - f(0)$.

(d) $\left[\int_0^\infty f(s)ds \right]^*(\theta) = f^*(\theta)/\theta$.

The proofs are straightforward computations using integration by parts, and are left as Exercise 1.

Example (Some Laplace transforms) If $f(x)=1$ then $f^*(\theta)=1/\theta$, while if $f(x)=x^\alpha \exp(-\beta t)$, then $f^*(\theta)=\Gamma(\alpha+1)(\beta+\theta)^{-(\alpha+1)}$. Also if $f(x)=g(\alpha x)$, we have $f^*(\theta)=g^*(\theta/\alpha)/\alpha$. □

The Laplace transform (over t) of the third equation in (3.5) is (for $n>1$)

$$(\theta+\lambda_n)P_{1n}^*(\theta) = \lambda_{n-1}P_{1,n-1}^*(\theta). \tag{3.8}$$

This is a difference equation in n with nonconstant coefficients, so the theory in Appendix B does not apply. However, the solution is easily obtained recursively as

$$P_{1n}^*(\theta) = \frac{\lambda_1}{\theta+\lambda_1}\frac{\lambda_2}{\theta+\lambda_2}\cdots\frac{1}{\theta+\lambda_n}. \tag{3.9}$$

Notice that the first $n-1$ terms in (3.9) are Laplace transforms of exponential densities with parameter λ_i, while the last term is the Laplace transform of the function $\exp(-\lambda_n x)$, which is the probability that an Exp(λ_n) distributed random variable exceeds x. Using Lemma 3.1(b), and recalling that convolution of densities corresponds to a sum of the corresponding independent random variables, we see that the process reaches n in time t provided the sum of $n-1$ exponential random variables with parameters λ_i is less than t, and that adding an independent Exp(λ_n)-random variable makes the sum greater than t. We shall return to this construction of the process later.

Summing (3.5) over n we get

$$\sum_{n=1}^{N}\frac{dP_{1n}(t)}{dt} = -\lambda_N P_{1N}(t), \tag{3.10}$$

so by Laplace transforming and using (3.9) we obtain

$$\theta\sum_{n=1}^{N}P_{1n}^*(\theta) = 1-\lambda_N P_{1N}^*(\theta) = 1-\prod_{n=1}^{N}\left[1+\frac{\theta}{\lambda_i}\right]^{-1}. \tag{3.11}$$

By letting $N\to\infty$ we see that $\sum P_{1n}^*(\theta)\to 1/\theta$ if and only if the product diverges to zero, i.e., iff $\sum 1/\lambda_i=\infty$. In other words, $\sum P_{1n}(t)=1$. This proves the following result.

Proposition 3.1 For any $t>0$ $\sum_{n=1}^{\infty}P_{1n}(t) = 1$ iff $\sum\frac{1}{\lambda_i}=\infty$.

The interpretation of this result is that if the λ_i grow too fast, so that $\sum 1/\lambda_i < \infty$, then there is positive probability that the process at time t is not anywhere in the state space! In other words, it has gone off to infinity, or **exploded**. Such processes are called **dishonest**, and create a lot of problems in the theory. In particular, what ought to happen after an explosion?

Example (Linear birth process) In the case of a linear birth process we have $\lambda_i = i\lambda$, so that $\sum 1/\lambda_i = 1/\lambda \sum 1/i$ which is infinite. Hence the linear birth process does not explode in finite time. □

Example (The Pólya process) Consider the Pólya urn model described in section 2.5. Fix $\tau > 0$, and assume that the urn scheme is carried out every τ time units. Let R_t be the number of red balls drawn at time $[t/\tau]$. Recall from equation (2.129) that

$$\mathbf{P}(R_{(n+1)\tau} = i+1 \mid R_{n\tau} = i) = \frac{R+di}{N+dn}. \tag{3.12}$$

Let $p(\tau) = R/N$, $c(\tau) = d/N$, and $t = n\tau$. Then $p(\tau)/\tau$ is the rate of red balls drawn per unit time, and we may set things up so that $p(\tau)/\tau \to \lambda$ as $\tau \to 0$. Assume further that $c(\tau)/\tau \to \alpha > 0$, and consider the limit of R_t as $\tau \to 0$, which we, abusing notation, also call R_t. This birth process has transition probabilities

$$P_{i,i+1}(t, t+dt) = \lambda_i(t)dt + o(dt) \text{ as } dt \to 0 \tag{3.13}$$

where $\lambda_i(t) = \lambda(1+i\alpha)/(1+t\alpha)$. We see that the rate of births is decreasing with time, but for any fixed time roughly linear in population size. Starting from $R_0 = 0$ it is straightforward to verify (Exercise 2) that

$$\mathbf{P}^0(R_t = n) = \frac{(\lambda t)^n \prod_{k=1}^{n-1}(1+k\alpha)}{n!\,(1+\alpha\lambda t)^{n+1/\alpha}} \tag{3.14}$$

where, as before, $\mathbf{P}^x$ denotes probability starting from $X(0) = x$. The Pólya process can be used as a model of neutron showers, which takes into account the loss of energy in particles as the shower proceeds. From (3.14) we see that $\mathbf{E}^0(R_t) = \lambda t$, which is again an increasing function of t, albeit growing much slower than the mean of a Yule–Furry process. Consequently the Pólya process is also an inadequate model of cosmic radiation. □

The deterministic counterpart to a linear birth process satisfies the differential equation

$$\frac{dn_t}{dt} = \lambda n_t \tag{3.15}$$

with solution $n_t = n_0 \exp(\lambda t)$. One may expect that the deterministic equation would describe the average behavior of the stochastic model. Since

X_t~Geom(exp($-\lambda t$)), as shown in section 3.1, it has mean exp(λt) when starting from one individual. Since we can think of the process starting from n_0 individuals as the sum of n_0 iid processes, the mean then is $n_0$exp(λt). Hence, in this case, the deterministic equation describes the average behavior of the system. The next example shows, however, that this is not generally the case.

Example (Sociology) Consider a social group of N individuals. Following Bartholomew (1973, Ch. 9) we shall build a simple model for the transmission of information through the group. The information originates from a source such as a television commercial, and is spread by word of mouth or by direct contact with the source. We assume that any individual in the group encounters the source at a constant rate α, and that the group is homogeneous, so that any pair of individuals has the same rate β of exchanging information.

Let $X(t)$ be the number of individuals that have received the information by time t, starting with $X(0)=0$. Presumably individuals will not forget the information, so $X(t)$ should be a birth process. The intensity of transition from state n to state $n+1$ is

$$\lambda_n = (N-n)\alpha + n(N-n)\beta = (N-n)(\alpha+n\beta) \tag{3.16}$$

since the $N-n$ individuals who are not yet informed either can receive information from the source of from any of the n informed individuals. Using (3.9) and Lemma 3.1 we see that

$$\mathbf{P}(X(t)=n) = (-1)^n \left[\prod_{i=0}^{n-1}\lambda_i\right] \sum_{j=0}^{n} \frac{e^{-\lambda_j t}}{\prod_{j\neq k}(\lambda_j-\lambda_k)}, \quad n\leq N. \tag{3.17}$$

A quantity of interest is the expected value of $X(t)$. Its derivative is (in a different application) often called the **epidemic curve** and depicts the average rate of growth of the process. While it is possible to compute $\mathbf{E}X(t)$ directly from (3.17) (see Haskey, 1954), we will take a roundabout route to obtain an approximate expression for it. Let T_n be the time when the nth person becomes informed. We can write $T_n=\sum_0^{n-1}\tau_i$ where the τ_i~Exp(λ_i) are independent. Hence

$$\begin{aligned}\mathbf{E}T_n &= \sum_0^{n-1}\frac{1}{\lambda_i} = \sum_0^{n-1}\frac{1}{(N-i)(\alpha+i\beta)} \\ &= \frac{1}{\alpha+N\beta}\sum_{i=0}^{n-1}\frac{1}{N-i} + \frac{\beta}{\alpha+N\beta}\sum_{i=0}^{n-1}\frac{1}{\alpha+i\beta} \\ &= \frac{1}{\beta(N+\delta}(\psi(n)-\psi(N-n)-\psi(\delta-1)+\psi(n+\delta-1))\end{aligned} \tag{3.18}$$

where $\delta=\alpha/\beta$ and ψ is the **digamma function**, given by

$$\psi(x) = \sum_{i=1}^{\infty}\left[\frac{1}{i} - \frac{1}{i+x}\right]. \tag{3.19}$$

For large x we can approximate $\psi(x) \approx \log x + \gamma$ where γ is Euler's constant 0.5772... (Abramowitz and Stegun, 1965, p. 259; our $\psi(x)$ is their $\psi(x+1)+\gamma$). Consider large values of N, and let $n=Np$ for some $0<p<1$. Then we have the approximation

$$\mathbf{E}T_{Np} \approx \frac{1}{\beta(N+\delta)} \left\{ \log \left[\frac{\beta N+\delta-1}{1-p} \right] + \gamma - \psi(\delta-1) \right\}. \tag{3.20}$$

Now let $F(t)$ be the distribution function for the time until a randomly chosen individual receives the information. From the times $T_1, \ldots, T_N$ of exposures we would estimate F by $F_N(t)=\#\{T_i \le t\}/n$. In particular, we can write $T_{Np}=F_N^{-1}(p)$. Assuming that N is large we have $\mathbf{E}F_N^{-1}(p) \approx F^{-1}(p)$, so $\mathbf{E}T_{Np} \approx F^{-1}(p)$, where the approximation is quite good for values of p away from 0 or 1. Hence

$$F(p) \approx \frac{\exp(\beta(N+\delta)(p-\gamma+\psi(\delta-1)) + 1 - \delta}{\exp(\beta(N+\delta)(p-\gamma+\psi(\delta-1)) + N}. \tag{3.21}$$

Finally $\mathbf{E}X(t) = NF(t)$.

A deterministic counterpart to this problem would assume that (3.16) holds exactly in each infinitesimal time interval $(t, t+dt)$, so

$$\frac{dx(t)}{dt} = (N - x(t))(\alpha + x(t)\beta). \tag{3.22}$$

The solution, with $x(0)=0$, is

$$x(t) = N \frac{\exp(\beta(N+\delta)t - 1}{\exp(\beta(N+\delta)t) + N/\delta} \tag{3.23}$$

which is different from the stochastic counterpart $\mathbf{E}X(t)$. Figure 3.2 shows the two curves in one case.

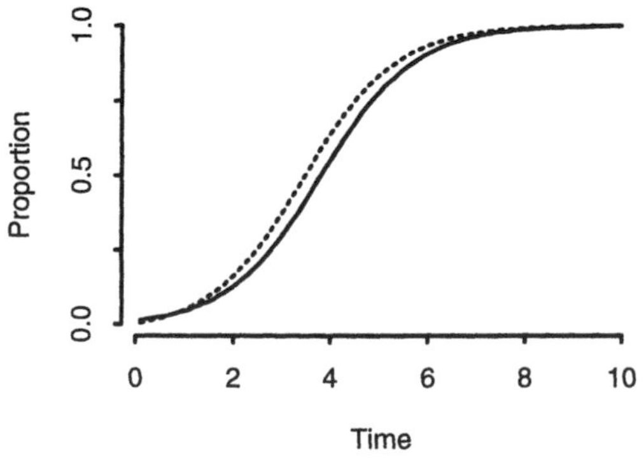

Figure 3.2. Deterministic description (dotted line) and mean of stochastic description (solid line) for the information transmission model. Here N=50, α=0.0292, and β=0.02.

Thus we see that it is not generally the case that a deterministic counterpart to a stochastic model is an expression for the mean of the stochastic model. This is, in fact, only the case for models that are linear in population size. □

3.3. The Kolmogorov equations

In this chapter we are concerned with Markovian stochastic processes with continuous time and discrete state space S, represented as the integers (or a subset of the integers) as for the discrete time chains in the previous chapter. We define the Markov property of a process $X(t)$ by

$$\mathbf{P}(X(t) = k \mid X(t_1) = k_1, \ldots, X(t_n) = k_n) = P_{k_n,k}(t_n,t) \tag{3.24}$$

for $0 \le t_1 \le \cdots \le t_n \le t$, and any integers n, k, $k_1, \ldots, k_n$. If $P_{ij}(s,t) = P_{ij}(|t-s|)$ the process has **stationary transition probabilities**. Unless otherwise stated this property will be assumed henceforth. As we saw in the previous section, there is a possibility that these processes may reach infinity in finite time. We will assume that if this happens the process stays at infinity for ever (infinity is then called a **coffin state**). This is called the **minimal construction** (see, e.g., Asmussen, 1987, section II.2, for a discussion). The following facts are easily established.

Proposition 3.2

$$0 \le P_{ij}(t) \le 1. \tag{3.25}$$

$$\sum_j P_{ij}(t) \le 1. \tag{3.26}$$

$$P_{ik}(s+t) = \sum_j P_{ij}(s) P_{jk}(t). \tag{3.27}$$

$$P_{ij}(0) = 1(i=j). \tag{3.28}$$

Proof The first and fourth statements are trivial, while the second follows from

$$\sum_j P_{ij}(t) = \mathbf{P}(0 \le X(t) < \infty \mid X(0)=i). \tag{3.29}$$

If the inequality is strict, the process is dishonest. To show the third equation, the **Chapman–Kolmogorov equation**, we compute

$$\begin{aligned} P_{ik}(t+s) &= \mathbf{P}^i(X(t+s)=k) = \sum_j \mathbf{P}^i(X(s)=j)\mathbf{P}^i(X(t+s)=k \mid X(s)=j) \\ &= \sum_j P_{ij}(s) P_{jk}(t), \end{aligned} \tag{3.30}$$

noticing that the coffin state is ruled out by our construction. □

As in the case of discrete time it is convenient to express things in matrix notation. Let $\mathbb{P}(t)=(P_{ij}(t), i,j\in S)$. Then (3.27) can be written

$$\mathbb{P}(s+t) = \mathbb{P}(s)\mathbb{P}(t). \tag{3.31}$$

Incidentally, $(\mathbb{P}(t); t\geq 0)$ is therefore a **semigroup**. It is **stochastic** if there is equality in (3.26), and **substochastic** otherwise. In order to proceed we need to assume some regularity. We call the process (or the semigroup) **standard** if the transition probabilities are continuous at 0, i.e., if

$$\lim_{t\downarrow 0} P_{ij}(t) = P_{ij}(0). \tag{3.32}$$

We will restrict attention to standard stochastic semigroups. Even then pathologies, such as a process which never stays in any state, can happen (e.g., Blackwell, 1958).

Lemma 3.2 Let $(\mathbb{P}(t))$ be a standard stochastic semigroup. Then $P_{ij}(t)$ is a continuous function for all i,j.

Proof We show that for any j

$$|P_{ij}(t+h) - P_{ij}(t)| \leq 1-P_{ii}(h), \quad h>0. \tag{3.33}$$

From the Chapman–Kolmogorov equation (3.27) we have that

$$P_{ij}(t+h) = \sum_k P_{ik}(h)P_{kj}(t) \tag{3.34}$$

so

$$P_{ij}(t+h)-P_{ij}(t) = (P_{ii}(h)-1)P_{ij}(t) + \sum_{k\neq i} P_{ik}(h)P_{kj}(t). \tag{3.35}$$

But $P_{kj}(t)\leq 1$, so the second term on the right-hand side is bounded by $\sum_{i\neq k} P_{ik}(h)=1-P_{ii}(h)$. Hence

$$P_{ij}(t+h)-P_{ij}(t) \leq (1-P_{ii}(h))(1-P_{ij}(t)), \tag{3.36}$$

whence the claim follows. □

In fact, it is possible to prove more (although we omit the proof here; see, e.g., Freedman, 1983, Theorem 21 in Chapter 5).

Proposition 3.3 For a standard stochastic semigroup $(\mathbb{P}(t))$ we have

(i) $P_{ii}'(0)$ exists and is nonpositive (but not necessarily finite);

(ii) $P_{ij}'(0)$ exists and is finite for $i\neq j$.

Let $\mathbf{Q}=(Q_{ij})=(P_{ij}'(0))$. $\mathbf{Q}$ is sometimes called the **generator** of $(\mathbb{P}(t))$. The following result is left as Exercise 3.

Lemma 3.3 If S is finite then $\sum_j Q_{ij}=0$, while if S is countably infinite we have $\sum_j Q_{ij}\leq 0$.

For the moment, let us concentrate on the case of finite S. We will look at extensions to countable state space a little later. Using Proposition 3.3 and Lemma 3.3 we can define $q_i=\sum_{j\neq i} Q_{ij}$, so $q_i=-Q_{ii}$. For any t write the Taylor expansions

$$P_{ij}(h) = P_{ij}(t,t+h) = Q_{ij}h + o(h) \tag{3.37}$$

and

$$P_{ii}(h) = P_{ii}(t,t+h) = 1-q_i h + o(h). \tag{3.38}$$

It follows that

$$\mathbf{P}(\text{transition } i\to j \text{ in } (t,t+h)\,|\,X(t)=i\text{ ; transition occurred}) \tag{3.39}$$
$$= \frac{P_{ij}(h)}{1-P_{ii}(h)} = \frac{Q_{ij}h+o(h)}{q_i h+o(h)} \to \frac{Q_{ij}}{q_i} \quad \text{as } h\to 0,$$

provided that $q_i>0$. We call Q_{ij} the **intensity** of the transition $i\to j$.

Example (The Yule–Furry process) We have $P_{j,j+1}(\Delta t)=j\lambda\Delta t+o(\Delta t)$ from the development in section 3.1, and $P_{jk}(\Delta t)=o(\Delta t)$ for $k\neq j+1$. Hence

$$Q_{j,j+1} = j\lambda \tag{3.40}$$

and

$$q_j = \sum_{k\neq j} Q_{jk} = j\lambda. \tag{3.41}$$

The intensity of new events increases with the number of events that have happened in the past. In other words, the probability of an event happening in the next time interval is proportional to how many events have occurred in the lifetime of the process. □

The following result is central to the theory of continuous time Markov chains.

Theorem 3.1 The transition probabilities of a finite state space, continuous time Markov chain satisfy

$$\frac{dP_{ij}(t)}{dt} = \sum_{k\in S} P_{ik}(t)Q_{kj} = \sum_{k\in S} Q_{ik}P_{kj}(t). \tag{3.42}$$

Proof By the Chapman–Kolmogorov equations

$$P_{ij}(t+h) = \sum_k P_{ik}(t)P_{kj}(h). \tag{3.43}$$

Hence

$$P_{ij}(t+h) = P_{ij}(t)P_{jj}(h)+\sum_{k\neq j} P_{ik}(t)P_{kj}(h) \tag{3.44}$$

so

$$\frac{P_{ij}(t+h)-P_{ij}(t)}{h} = -P_{ij}(t)\left(\frac{1-P_{jj}(h)}{h}\right) + \sum_{k\neq j} P_{ik}(t)\frac{P_{kj}(h)}{h}. \tag{3.45}$$

The right-hand side of (3.45) converges to

$$-P_{ij}(t)q_j+\sum_{k\neq j} P_{ik}(t)Q_{kj} = \sum_k P_{ik}(t)Q_{kj}. \tag{3.46}$$

The left-hand side therefore converges to the derivative of P_{ij}. This system of equations is called Kolmogorov's **forward equations**. To get the second equality in the statement of the theorem we proceed in a similar fashion, but now writing

$$P_{ij}(t+h) = \sum_{k\in S} P_{ik}(h)P_{kj}(t). \tag{3.47}$$

The resulting set of equations is called the **backward equations**. □

Remark

(i) The theorem is true also for a large class of processes with infinite state space. The necessary assumptions have to do with assuring the smoothness of the functions $P_{ij}(t)$. Call the semigroup $(\mathbb{P}(t))$ **uniform** if $P_{ii}(t)\to 1$ as $t\downarrow 0$ uniformly in i. There is an easy criterion for uniformity (for a proof, see Freedman, 1983, Theorem 29 in Chapter 5).

Lemma 3.4 $(\mathbb{P}(t))$ is uniform if

$$\sup_{i\in S} q_i < \infty. \tag{3.48}$$

In particular, if S is finite, the semigroup is uniform. Theorem 3.1 is true for uniform semigroups. In addition, $(\mathbb{P}(t))$ is uniform whenever $\sum_j Q_{ij}=0$ (this is part of the result in Freedman, 1983) so, in essence, the process behaves just like the finite state space case.

(ii) In matrix notation, writing

$$\frac{d\mathbb{P}(t)}{dt} = \left[\frac{dP_{jk}(t)}{dt}\right], \tag{3.49}$$

we can restate Theorem 3.1 in the following fashion. The forward equation becomes

$$\frac{d\mathbb{P}(t)}{dt} = \mathbb{P}(t)Q, \tag{3.50}$$

and the backward equation becomes

$$\frac{d\mathbb{P}(t)}{dt} = Q\mathbb{P}(t). \tag{3.51}$$

□

Example (General birth process) Note that equations (3.3) and (3.5) are the forward equations for the Yule–Furry and the general birth processes, respectively. For the general birth process the intensity matrix is

$$Q = \begin{bmatrix} -\lambda_1 & \lambda_1 & 0 & 0 & \cdots \\ 0 & -\lambda_2 & \lambda_2 & 0 & \cdots \\ \cdots & \cdots & \cdots & \cdots & \cdots \end{bmatrix} \tag{3.52}$$

and $q_i = \lambda_i$. If $\sum \lambda_i^{-1} < \infty$, i.e., the process is dishonest, then $\lambda_i^{-1} \to 0$ so $\sup q_i = \infty$ and $(\mathbb{P}(t))$ is not uniform. An honest birth process (such as the Yule–Furry process) may or may not be uniform; in fact, the Yule–Furry process is not uniform since $q_i = i\lambda$. Nevertheless, the Yule–Furry process satisfies both the backward and forward equations. A general honest birth process has a unique solution to the forward equation, but may have many solutions to the backward equation.

□

We often prefer to define a process in terms of its intensities. For a very simple example, assume that we have a birth process which has constant intensity of births. That means that

$$Q_{j,j+1} = \lambda, \quad Q_{jj} = -\lambda \tag{3.53}$$

with all other entries being zero. This is a uniform semigroup (provided that $\lambda < \infty$), and the forward equation yields

$$\frac{dP_{jk}(t)}{dt} = -\lambda P_{jk}(t) + \lambda P_{j,k-1}(t). \tag{3.54}$$

In particular, if $j = 0$ we get

$$\frac{dP_{0k}(t)}{dt} = -\lambda P_{0k}(t) + \lambda P_{0,k-1}(t). \tag{3.55}$$

The initial conditions are taken to be $P_{00}(0) = 1$ and $P_{0i}(0) = 0$ for $i \geq 1$, so that the process starts in state 0. In order to solve this differential equation, we will attempt to convert it into a partial differential equation for the probability generating function

$$G(s;t) = \mathbf{E}s^{X(t)} = \sum_i P_{0i}(t)s^i. \tag{3.56}$$

Multiplying both sides of (3.55) by s^k and summing over k we compute

$$\frac{\partial G(s;t)}{\partial t} = -\lambda G(s;t) + \lambda s G(s;t) \tag{3.57}$$

and $G(s;0)=1$. For a fixed value of s we see that

$$\frac{\partial G(s;t)}{\partial t} = -\lambda(1-s)G(s;t) \tag{3.58}$$

so that

$$G(s;t) = A(s)e^{-\lambda(1-s)t}. \tag{3.59}$$

From the initial condition we must have $A(s)=1$, so $X(t)\sim\text{Po}(\lambda t)$ (i.e., $X(t)$ follows a Poisson distribution with mean λt). In general

$$P_{ij}(s,s+t) = \frac{(\lambda t)^{j-i}}{(j-i)!}e^{-\lambda t}, \quad j\geq i. \tag{3.60}$$

The process we just derived is called the **Poisson**[1] **process**. Because the intensity of events in this process is constant, regardless of which state the process is in, it is sometimes called a **totally random** process. It has the property that the numbers of events in disjoint intervals of time are independent. To see that, first note from (3.60) that $P_{j,j+l}(s)=P_{0,l}(s)$ for all j. Write $X(t,t+s]=X(t+s)-X(t)$, so that $X(t,t=s]$ counts the number of births in $(t,t+s]$. Then

$$\begin{aligned}\mathbf{P}(X(t,t+s]=l \mid X(t)=j) &= \mathbf{P}(X(t+s)=j+l \mid X(t)=j)\\ &= P_{j,j+l}(s) = P_{0,l}(s)\end{aligned} \tag{3.61}$$

independent of j. Thus the number of events in $(0,t]$ is independent of the number of events in $(t,t+s]$. The argument can be extended to any disjoint time sets. We say that the Poisson process has **independent increments**. Hence it has no memory: the chance of something happening at any instant in time is independent of what has happened in the past. It has therefore found uses in describing events such as alpha-particle emissions from radioactive substances, large earthquakes, volcanic eruptions, or arrival of phone calls to a large exchange.

Example (Subsampling Poisson process events) Consider events occurring according to a Poisson process of rate λ, and recorded using a device which detects each event with probability p, independently from event to event.

[1] Poisson, Siméon Denis (1781–1840). French mathematical physicist. Generalized Bernoulli's law of large numbers.

Thinking of $Q_{i,i+1}$ as the instantaneous rate of events, it is clear that the recorded process Y_t will be a birth process with intensity $Q^Y_{i,i+t}=p\lambda$, i.e., a Poisson process with rate $p\lambda$. The case where p varies with time occurs sometimes when using historical records of earthquakes (Lee and Brillinger, 1979) or volcanic eruptions (Guttorp and Thompson, 1991). □

We can develop a formal solution to the forward equation in matrix form. Since $d\mathbb{P}(t)/dt=\mathbb{P}(t)\mathbf{Q}$, an analogy with the one-dimensional case suggests that

$$\mathbb{P}(t) = e^{t\mathbf{Q}} \equiv \sum_0^\infty t^n \mathbf{Q}^n/n\,!. \tag{3.62}$$

Indeed, this can be shown to be the unique stochastic solution to the forward equation satisfying $\mathbb{P}(0)=\mathbb{I}$ and $\mathbb{P}(u+t)=\mathbb{P}(u)\mathbb{P}(t)$, the Chapman–Kolmogorov equations. This way of solving the equation suggests a way of developing properties of the model (see, e.g., Feller, 1971, ch. X and XIV). Using a spectral decomposition of Q one can implement this solution in practice, but here we will concentrate on solving some special cases using the method of probability generating functions.

Define a **linear death process** by the intensities

$$Q_{j,j-1} = \mu j; \quad Q_{jj} = -\mu j. \tag{3.63}$$

We see that the intensity of a decrease by one (i.e., a death) is linear in current population size, while all other changes have intensity zero. The forward equation yields, starting from $X(0)=N$,

$$dP_{Nk}(t)/dt = -\mu k P_{Nk}(t) + \mu(k+1)P_{N,k+1}, \ k<N. \tag{3.64}$$

Note that $P_{NN}(t) = e^{-\mu N}$. Writing $G(s;t)=\sum s^k P_{Nk}(t)$ we derive the partial differential equation

$$\frac{\partial G(s;t)}{\partial t} = -\mu\sum k s^k P_{Nk}(t) + \mu\sum(k+1)s^k P_{N,k+1}(t) \tag{3.65}$$

or

$$\frac{\partial G(s;t)}{\partial t} = \mu(1-s)\frac{\partial G(s;t)}{\partial s} \tag{3.66}$$

which is a Lagrange equation (Appendix C contains solution methods for such equations) with solution

$$G(s;t) = (1-(1-s)e^{-\mu t})^N, \tag{3.67}$$

the probability generating function of a binomial random variable. Thus

$$P_{Nk}(t) = \binom{N}{k} e^{-k\mu t}(1-e^{-\mu t})^{N-k}. \tag{3.68}$$

Note that if $T\sim\text{Exp}(\mu)$ we have that $\mathbf{P}(T\geq t)=\exp(-\mu t)$. In other words, the death process is in state k if $N-k$ independent exponential random variates have occurred at time t. This is like having N independent individuals, each dying at an exponentially distributed time. The time to extinction has the distribution of the maximum of N exponential random variates (see Figure 3.3).

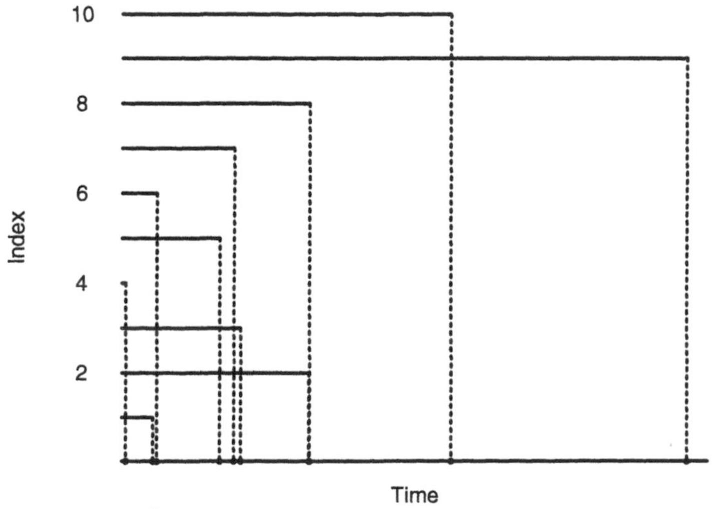

Figure 3.3. The relation between the exponential random variates and the death process.

For a more general death process, with death intensity μ_k, a similar result will hold: the next event at stage k will happen after an time which is exponentially distributed with parameter μ_k.

3.4. A general construction

The following result describes the structure of finite state space continuous time Markov chains.

Theorem 3.2 Let $(X(t), t \geq 0)$ be a finite state space, continuous time Markov chain.

(i) If $q_j=0$, and $X(t_0)=j$, then

$$\mathbf{P}(X(t)=j, t>t_0 \mid X(t_0)=j) = 1. \tag{3.69}$$

(ii) If $q_j>0$ and $X(t_0)=j$, then with probability one there is a jump discontinuity at some time $t>t_0$.

(iii) Let $0<\alpha<\infty$, $q_j>0$, and $X(t_0)=j$. Given that there is a discontinuity of $X(t)$

in $[t_0, t_0+\alpha)$, then the conditional probability that the first discontinuity is a jump to k is Q_{jk}/q_j.

(iv) $\mathbf{P}(X(u)=j, t\le u\le t+\alpha \mid X(t)=j) = \exp(-q_j\alpha),\ \alpha>0$.

Proof (i) The backward equation is

$$\frac{dP_{jj}(t)}{dt} = \sum Q_{jr}P_{rj}(t) \equiv 0 \tag{3.70}$$

since $q_j=\sum_{r\neq j}Q_{jr}=0$. (ii) follows, if we can prove (iv), by letting $\alpha\to\infty$. Property (iii) was argued as (3.(Pn). To verify (iv), let τ be the time spent in state j after time t until it makes a transition to a different state. By the Markov property, the probability that the process stays in j during the time interval $(u,u+v]$, given that it stays at least u time units, is precisely the unconditional probability that it stays in state j for v time units. In equation form

$$\mathbf{P}(\tau>u+v \mid \tau>u) = \mathbf{P}(\tau>v). \tag{3.71}$$

The result follows if we can prove that this equation only has the solution $\mathbf{P}(\tau>v) = \exp(-\alpha v)$. To see that $\alpha=q_j$, notice that by the backward equation $P_{jj}'(0)=-q_j$, and that $\mathbf{P}(\tau>t)=P_{jj}(t)$. It just remains to show that the only solution to the equation

$$g(u+v) = g(u)g(v) \tag{3.72}$$

is the exponential function. In order to verify this fact, we first compute $g(2/n)=g(1/n+1/n)=g(1/n)^2$. By repeating this calculation, we obtain $g(m/n)=g(1/n)^m$. Also $g(1)=g(1/n)^n$, so $g(m/n)=g(1)^{m/n}$. Since g is a distribution function, it is right continuous. Therefore (taking limits through the rational numbers) $g(x)=g(1)^x$. But $g(1)=g(½)^2\ge 0$, so we can write $g(x)=\exp(-\alpha x)$ where $\alpha=-\log g(1)$. □

Remark This result is true in some generality, although care is needed to deal with the possibility of explosion for infinite state spaces. A careful statement and proof is in Freedman (1983, section 5.6). □

We can use Theorem 3.2 to construct the process $(X(u), 0\le u\le t)$ with initial distribution $\mathbf{p}^0$ and intensities Q_{ij}. The procedure is illustrated in Figure 3.4.

(1) Choose i_0 from $\mathbf{p}^0$. Let $X(0)=i_0$.

(2) If $q_{i_0}=0$, i.e., if i_0 is absorbing, we are done: $X(u)=i_0, 0\le u\le t$.

(3) If $q_{i_0}>0$, draw an $\exp(q_{i_0})$-distributed random variable τ. If $\tau\ge t$ we are done: $X(u)=i_0, 0\le u\le t$.

(4) If $\tau<t$, choose i_1 according to the distribution (R_{i_0k}), where $R_{jk}=Q_{jk}/q_j$.

(5) If $q_{i_1}=0$ the resulting path is $X(u)=i_0, 0\le u\le\tau$ and $X(u)=i_1, \tau<u\le t$.

(6) If $q_{i_1}>0$ go to (3).

The same algorithm applies in the case of infinite state space, as long as we stick to the minimal construction.

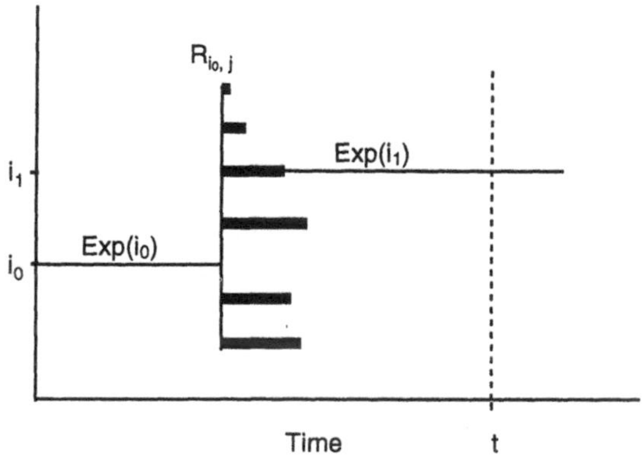

Figure 3.4. The construction using Theorem 3.2.

There are several important discrete time Markov chains associated with a given continuous time Markov chain. Let $Y_n=X(nh)$ for a fixed $h>0$. This process is called the **discrete skeleton** of X (to scale h).

Proposition 3.4 The discrete skeleton (Y_n) is a Markov chain, with transition probabilities $P_{ij}(h)$.

Proof Using the Markov property of X we have

$$\mathbf{P}(Y_n=i \mid Y_{n-1}=j, Y_{n-2}=j_2, \ldots, Y_0=j_n)$$
$$= \mathbf{P}(X(nh)=i \mid X((n-1)h)=j) = P_{ij}(h). \tag{3.73}$$

□

Define

$$R_{jk} = \begin{cases} (1-\delta_{jk})Q_{jk}/q_j & q_j>0 \\ \delta_{jk} & q_j=0 \end{cases} \tag{3.74}$$

where $\delta_{jk}=1(j=k)$. Then (R_{jk}) is a stochastic matrix, called the **jump matrix** of $(X(t))$. The jump matrix governs where the process goes when it leaves a state. Let $\tau_1, \tau_2, \ldots$ be the successive jump times of the process. Define the **jump**

chain $Z_k=X(\tau_k+)$. From the construction following Theorem 3.2 we see that this is a Markov chain with transition matrix R.

Example (The Yule–Furry process) The discrete skeleton to scale h of a Yule–Furry process has transition probabilities

$$P_{ij}(h) = \binom{j-1}{i-1} p^i (1-p)^{j-i} \tag{3.75}$$

where $p=\exp(-\lambda h)$. The jump matrix has $R_{j,j+1}=1$ and $R_{jk}=0$ for $k\neq j+1$. This describes the process as a pure birth process, only moving upwards one step at the time. The jump chain is therefore very dull. □

The discrete skeleton $Y_n=X(nh)$ is useful in determining properties of the process $X(t)$. Since for Y_n we have $p_{jj}^{(n)}=P_{jj}(nh)$ the state j is persistent (in the discrete time sense) iff $\sum P_{jj}(nh)=\infty$. Looking at this sum as a Riemann sum, an equivalent condition is

$$\int P_{jj}(t)dt = \infty. \tag{3.76}$$

We say that a state j is persistent for a continuous time Markov chain if

$$\mathbf{P}^j(\sup\{t\geq 0:X(t)=j\} = \infty) = 1. \tag{3.77}$$

Clearly this implies that $Y_n=j$ infinitely often, so j is persistent in the discrete time sense. If, on the other hand, j is persistent in the discrete time sense, we know that Y_n returns to j infinitely often, so $\mathbf{P}^j(\sup\{n:X(nh)=j\} = \infty)=1$ for all h, and hence in the limit as $h\to 0$.

We define transience similarly: j is transient if

$$\mathbf{P}^j(\sup\{t\geq 0:X(t)=j\} < \infty) = 1, \tag{3.78}$$

and again a state is transient in the continuous time sense iff it is transient in the discrete time sense for the skeleton chain.

There is a major difference between discrete and continuous time chains: there are no periodic states in continuous time. In fact, we have $P_{ii}(t)>0$ for all $t\geq 0$. To see this, recall that (since we only consider standard semigroups) $P_{ii}(t)\to 1$ as $t\to 0$, so there is a $h>0$ with $P_{ii}(s)>0$ for all $s\leq h$. For any real t choose n large enough so that $t\leq hn$. By Chapman–Kolmogorov $P_{ii}(t)\geq (P_{ii}(t/n))^n > 0$, since $t/n<h$. We have shown the following result.

Theorem 3.3 (a) A state i is persistent (transient) for $X(t)$ iff it is persistent (transient) in the discrete skeleton.

(b) A state i is persistent iff $\int_0^\infty P_{ii}(t)dt = \infty$.

(c) $P_{ii}(t)>0$ for all $t\geq 0$.

Remark For any two states i and j the **Lévy dichotomy** holds: either $P_{ij}(t)>0$ for all $t>0$, or $P_{ij}(t)=0$ for all $t>0$. A proof can be found in Bhattacharaya and Waymire (1990, p. 304). As in the discrete time case we call a chain **irreducible** if $P_{ij}(t)>0$ for all i,j. □

We can use the discrete skeleton to assess stationarity as well. As in Chapter 2 we restrict attention to the irreducible case. Then if Y_n is non-null persistent we have $p_{ij}^{(n)}=P_{ij}(nh)\to\pi_j$; otherwise it goes to zero. Looking at two rational values h_1 and h_2 we see that the limit for each of these skeletons must be the same, and since the $P_{ij}(t)$ are continuous, we have a unique stationary distribution. Here is how you compute it: π_k is a solution to the equation

$$\sum \pi_j Q_{jk} = 0 \tag{3.79}$$

since, for any value of t, $\boldsymbol{\pi}$ solves

$$\sum_j \pi_j P_{jk}(t) = \pi_k \tag{3.80}$$

or, equivalently,

$$\pi_k(P_{kk}(t)-1) + \sum_{j\neq k} \pi_j P_{jk}(t) = 0. \tag{3.81}$$

Divide by t and let $t\to 0$ to get (3.79). The stationary distribution of the jump chain of $X(t)$ satisfies a different equation (Exercise 4).

Another way of obtaining (3.79) is to look at the forward equation $d\mathbb{P}(t)/dt=\mathbb{P}(t)Q$. We can think of the stationary distribution as one in which there is no probability flux, so $\mathbb{P}(t)$ is constant equal to $\boldsymbol{\pi}$, and the time derivative is zero, whence $0=\boldsymbol{\pi}Q$. In the special case of reversible chains the law of detailed balance holds for the discrete skeleton. It translates into the requirement that

$$\pi_i Q_{ij} = \pi_j Q_{ji} \tag{3.82}$$

for all i and j, using the same argument as in deriving the equation for the stationary distribution.

Example (Birth and death process) Consider a process with intensities

$$\begin{cases} Q_{j,j+1} = \lambda_j \\ Q_{j,j-1} = \mu_j \\ Q_{jk} = 0 \text{ for } |j-k|\geq 2 \end{cases} \tag{3.83}$$

This is called a **birth and death** process. It was introduced by McKendrick (1925; in a special case in 1914), used by Feller (1939) to describe biological population growth, and studied in detail by Kendall (1948). In order to determine the stationary distribution, the equation $\pi Q=0$ yields

$$\begin{cases} -\lambda_0\pi_0 + \mu_1\pi_1 = 0 \\ \lambda_0\pi_0 - (\mu_1+\lambda_1)\pi_1 + \mu_2\pi_2 = 0 \\ \cdots \\ \lambda_{k-1}\pi_{k-1} - (\mu_k+\lambda_k)\pi_k + \mu_{k+1}\pi_{k+1} = 0 \end{cases} \tag{3.84}$$

with solution

$$\begin{cases} \pi_1 = \dfrac{\lambda_0}{\mu_1}\pi_0 \\ \pi_2 = \dfrac{\lambda_1}{\mu_2}\pi_1 = \dfrac{\lambda_1\lambda_0}{\mu_2\mu_1}\pi_0 \\ \cdots \\ \pi_k = \dfrac{\lambda_{k-1}\cdots\lambda_0}{\mu_k\cdots\mu_1}\pi_0 \\ \cdots \end{cases} \tag{3.85}$$

Since $\sum\pi_k=1$ we obtain

$$\pi_0 = \left(1+\sum_{k=1}^{\infty}\frac{\lambda_0\cdots\lambda_{k-1}}{\mu_1\cdots\mu_k}\right)^{-1} \tag{3.86}$$

and

$$\pi_k = \frac{\lambda_0\cdots\lambda_{k-1}}{\mu_1\cdots\mu_k\left(1+\sum_{n=1}^{\infty}\dfrac{\lambda_0\cdots\lambda_{n-1}}{\mu_1\cdots\mu_n}\right)}. \tag{3.87}$$

In order for these equations to be well defined, we must have

$$\sum_{n=1}^{\infty}\frac{\lambda_0\cdots\lambda_{n-1}}{\mu_1\cdots\mu_n} < \infty. \tag{3.88}$$

This condition is also sufficient for the existence of a stationary distribution, which can easily be seen from the results for birth and death chains in the previous chapter upon noticing that the jump chain for a birth and death process is a birth and death chain. □

Example (The linear BIDE process) While the birth and death process can be a reasonable description of many closed populations, a natural generalization allows for immigration and emigration as well. We restrict attention to the linear BIDE (birth, immigration, death, and emigration) process. As before, a birth or a death occurs with intensity proportional to the population size, while immigration occurs at constant rate, and emigration is similar to death. Thus the BIDE process is a nonlinear birth and death process with birth intensity $\lambda_n = \lambda n + \nu$ and death intensity $\mu_n = (\mu + \eta)n$. Unless we have detailed information about which events are deaths and which are emigrations, we are unable to separate μ and η from data. By adding a constant to μ and subtracting it from η the death intensity remains the same. Thus we say that μ and η are not **identifiable** from the BIDE process $X(t)$. We may write $\mu_n = \kappa n$, where $\kappa = \mu + \eta$.

Applying the same type of argument as before we see that the pgf $G(s,t) = \sum P_{ij}(t)s^j$ satisfies the partial differential equation

$$\frac{\partial G(s,t)}{\partial t} = G(s,t)\nu(s-1) + \frac{\partial G(s,t)}{\partial s}(s-1)(\lambda s - \kappa). \tag{3.89}$$

Using the method of auxiliary equations the solution corresponding to the initial condition $X(0) = a$ is

$$G(s,t) = \frac{(\lambda-\kappa)^{\nu/\lambda}(\kappa(1-s)e^{(\lambda-\kappa)t} + \lambda s - \kappa)^a}{(\lambda(1-s)e^{(\lambda-\kappa t)} + \lambda s - \kappa)^{a+\nu/\lambda}}. \tag{3.90}$$

Letting $t \to \infty$, assuming that $\lambda < \kappa$, we get the pgf for the stationary distribution

$$G(s,\infty) = \left[\frac{\lambda-\kappa}{\lambda s - \kappa}\right]^{\nu/\lambda}, \tag{3.91}$$

which is the pgf of a negative binomial distribution with parameters ν/λ and $(\kappa-\lambda)/\kappa$, the latter being the success probability. Thus the mean of the stationary distribution is $\nu/(\kappa-\lambda)$ and its variance is $\nu\kappa/(\kappa-\lambda)^2$.

If we know which events are of what type, a more detailed analysis is possible. Let $(B(t), I(t), D(t), E(t))$ be the number of births, immigrants, deaths, and emigrants, respectively, during the time interval $(0,t]$. These processes are called **counting processes**, and we shall consider them in more detail in Chapter 5. Note that $X(t) = X(0) + B(t) + I(t) - D(t) - E(t)$, and

$$\mathbf{P}(B(t+dt) - B(t) \mid X(t) = x) = x\lambda dt + o(dt), \tag{3.92}$$

so $B(t)$ is very similar to a linear birth process, except that its intensity does not only depend on the current state of itself, but also on the current state of other processes. Clearly, the four-dimensional process is a Markov process with discrete state space. As we shall see later, μ and η are both identifiable if we observe the four-dimensional process. □

3.5. Queueing systems

Waiting in line is a common way of spending (some would say wasting) one's time in modern society. Queues occur in grocery stores, in multi-user computers, in satellites providing intercontinental telephone service, in hospitals, and in countless other activities. Queueing theory is one of the largest subfields of applied probability. Asmussen (1987) is a good source for a theoretical development. Here we shall concentrate on some very simple queueing models.

A simple queue consists of an **arrival process** of customers, a set of **service facilities** dealing with the customers, and a **queue discipline**, determining the order in which customers are served. Following Kendall (1953) a queueing system is written A/B/k, where A describes the interarrival time distribution (most queues are assumed to have independent interarrival times), B the service time distribution (again, service from different servers and for different customers are often assumed independent), and k the number of servers. There are many possible choices of A and B; we restrict attention to the exponential distribution, denoted M (for Markovian), a deterministic distribution denoted D, and a general (unspecified) distribution, written G. Thus M/M/1 is a single server queue with exponential interarrival times and service times, while M/G/∞ is a queue with exponential interarrival times, general service time, and no waiting (each arriving customer is immediately served).

The M/M/m processes are simply birth and death processes. These are the only Markovian queues. If $m=1$ we have $\lambda_n=\lambda$ and $\mu_n=\mu 1(n>0)$. The stationary distribution is easily seen to be geometric with parameter $\rho=\lambda/\mu$, provided that $\lambda<\mu$ (Exercise 6). The parameter ρ is called the **traffic intensity** and measures the ratio between input rate and output rate at full capacity. From the equilibrium distribution we see that

$$\mathbf{P}(\text{server idle}) = \pi_0 = 1-\rho \tag{3.93}$$

and the mean queue length in equilibrium is

$$\mathbf{E}^{\pi}X(t) = \frac{\rho}{1-\rho}. \tag{3.94}$$

As $\rho\uparrow 1$ the server is busy with high probability, and the mean queue length is large, as would be expected. Let W be the waiting time before service in equilibrium. Since each individual in the queue, at the arrival (which we take to be time 0) of the customer whose waiting time we are determining, will take an exponentially distributed time τ_i to be served, we have

$$(W \mid X(0)=i) = \tau_1 + \cdots + \tau_i \sim \Gamma(i,\mu) \tag{3.95}$$

with conditional Laplace transform $(1+t/\mu)^{-i}$. Hence

$$\mathbf{E}(e^{-W\theta}) = \sum(1-\rho)\rho^i(1+t/\mu)^{-i} = \frac{1-\rho}{1-\rho\mu/(\mu+t)} \tag{3.96}$$

which is the Laplace transform of a random variable which is 0 with probability $1-\rho$ and Exp($\mu(1-\rho)$) with probability ρ.

Next consider the M/M/∞ queue. Here $\lambda_n=\lambda$, $\mu_n=n\mu$, so we have an immigration-death process, whence the process is ergodic with equilibrium distribution Po(η), where $\eta=\lambda/\mu$. The traffic intensity for this process is zero, since nobody ever needs to wait to be served.

Finally we look at an M/M/m queue, with $1<m<\infty$, so $\lambda_n=\lambda$ and $\mu_n=\min(m,n)\mu$. The traffic intensity is $\rho=\lambda/m\mu$. Let $\eta=\lambda/\mu$. From equation (3.88) the ergodicity condition is the finiteness of

$$S=\sum_0^{m-1}\frac{\eta^n}{n!}+\frac{\eta^m}{m!}\sum_0^\infty\rho^n \tag{3.97}$$

or, equivalently, $\rho<1$. The equilibrium distribution is

$$\pi_n=\begin{cases}\dfrac{1}{S}\dfrac{\eta^n}{n!} & 0\le n\le m\\[2ex] \dfrac{1}{S}\dfrac{\eta^m}{m!}\rho^{n-m} & n>m.\end{cases} \tag{3.98}$$

Thus $\boldsymbol{\pi}$ is a combination of the M/M/∞ solution on $\{0,\ldots,m\}$, where there is no waiting, and the M/M/1 solution on $\{m+1,m=2,\ldots\}$ where some customers wait. We see that

$$\mathbf{P}^\pi(\text{all servers busy})=\pi_m+\pi_{m+1}+\cdots=\frac{1}{S}\frac{\eta^m}{m!}\frac{1}{1-\rho} \tag{3.99}$$

and

$$\mathbf{E}^\pi X(t)=\sum k\pi_k=\frac{1}{S}\left[\sum_{n=1}^{m-1}\frac{\eta^n}{(n-1)!}+\frac{\eta^m}{m!}\left\{\frac{\rho}{(1-\rho)^2}+\frac{m}{1-\rho}\right\}\right]. \tag{3.100}$$

As we saw in Chapter 2, every ergodic birth and death chain, and consequently every ergodic birth and death process, is time reversible. Suppose we have a stationary version of a Markovian queue $X(t), t\in\mathbf{R}$. Let $\tilde{X}(t)=X(-t-0)\equiv\lim_{s\uparrow -t}X(s)$. A departure from X at time t corresponds to an arrival to $\tilde{X}$ at time $-t$. If $\lambda_n=\lambda$, so we have Poisson arrivals, then the instants $-t_i$ of arrivals to $\tilde{X}$, corresponding to departures from X at times t_i, form a Poisson process by reversibility. This Poisson process has the same intensity λ, and the departures up to time t are independent of $\tilde{X}(-t)$, which, with probability one, is equal to $X(t)$ since X has probability 0 of jumping at time t. We have shown the following result.

Proposition 3.5 The departure times of an ergodic M/M/• queue form a Poisson process with the same rate as the arrival process. The departure process at time t is independent of past departures up to time t.

Remark This result is astonishing. The output process does not depend on the service rate or the number of servers, just on the input rate. Of course, in equilibrium what comes in must go out, but intuition still balks at the result at first! □

Consider a series of K Markovian queues, where arrivals to the first are Poisson of rate λ, and the servers in each queue act independently of the servers in the other queue, with service rate $\mu_n^{(k)}$ in the kth queue. If the input queue is ergodic, and we assume that the output from the kth queue is the input to the $(k+1)$th queue, then the whole system is ergodic, with independent queue lengths each governed by its respective stationary distribution. This is immediate from repeated use of Proposition 3.5.

Application (Communications center) Hatzikostandis and Howe (1967) provide some data on manual message-handling operators in a military communications terminal. We restrict attention to messages of type A, which arrive at a receipt center (staffed by trained personnel) where they are logged and immediately sent to the processing center (staffed by relatively untrained personnel). After input to the processing center routing is determined by message content and multiple copies are made. The messages are then logged out and placed in a pigeon-hole box until messengers pick them up for distribution to the users.

The data in Table 3.1 are on priority 1 messages, which are immediately dealt with.

Table 3.1 Quarter-hourly arrival counts

No. of messages	0	1	2	3
Receipt center	39	33	19	5
Processing center	41	31	17	7

They consist of 24 hours' worth of 15-minute counts of arrivals to the receipt center and the processing center. If we can consider the transport time between the centers approximately constant, then the arrivals at the processing center have the same distribution as the departures from the receipt center. If the receipt center queue can be modeled as M/M/•, the arrivals and departures should both be Poisson with mean $\lambda/4$, where λ is the hourly intensity of priority 1 message arrivals. Both the input and the output intensities are estimated to be $4\times86/96=3.6$ messages per hour. The variance of the input process is 0.8, and that of the output process is 0.9, both essentially the same as the mean 0.9, indicating that a Poisson distribution is a reasonable description. □

While queues other than M/M/• are non-Markovian, it is sometimes the case that one can find a closely related process which is Markovian, thereby simplifying much of the analysis. We give an example below, and leave another as Exercise 7.

Example (The M/G/1 queue and its imbedded branching process) Consider an M/G/1 queue, starting with one individual arriving at time 0 and being served immediately. During the service time of this individual, a number of new customers arrive. We consider these customers the offspring of the original individual. Doing this for each customer, we obtain, by looking at the process only at the instances of departure, a Bienaymé-Galton-Watson process with offspring distribution

$$p_j = \int_0^\infty \frac{(\lambda t)^j}{j!} e^{-\lambda t} dB(t) \tag{3.101}$$

where B is the cdf of the service time. If the service time has mean μ the offspring mean is $\lambda\mu$, and we have a subcritical process, provided that $\rho - \lambda\mu < 1$. Once the queue becomes empty, we get an independent realization of the process starting from the next arrival.

Letting Y_n denote the queue length just after the departure of customer $n-1$, and Z_n the number of offspring of the nth individual (i.e., the number of customers arriving during the nth service time), we can write

$$Y_{n+1} = Y_n - 1 + Z_n + 1(Y_n = 0). \tag{3.102}$$

Hence Y_n has transition matrix

$$\mathbb{P} = \begin{pmatrix} p_0 & p_1 & p_2 & p_3 & \cdots \\ p_0 & p_1 & p_2 & p_3 & \cdots \\ 0 & p_0 & p_1 & p_2 & \cdots \\ 0 & 0 & p_0 & p_1 & \cdots \\ 0 & 0 & 0 & p_0 & \cdots \\ \cdots & \cdots & \cdots & \cdots & \cdots \end{pmatrix} \tag{3.103}$$

which is irreducible, since all the p_i are positive. The stationary distribution satisfies the system of equations

$$\begin{cases} \pi_0 = \pi_0 p_0 + \pi_1 p_0 \\ \pi_1 = \pi_0 p_1 + \pi_1 p_1 + \pi_2 p_0 \\ \pi_2 = \pi_0 p_1 + \pi_1 p_2 + \pi_3 p_1 + \pi_4 p_0 \\ \cdots \end{cases} \tag{3.104}$$

Writing $q_r = p_{r+1} + p_{r+2} + \cdots$, adding the first n equations and solving for $\pi_{n+1} p_0$ we obtain

$$\begin{cases} \pi_1 p_0 = \pi_0 q_0 \\ \pi_2 p_0 = \pi_0 q_1 + \pi_1 q_1 \\ \pi_3 p_0 = \pi_0 q_2 + \pi_1 q_2 + \pi_2 q_1 \\ \cdots \end{cases} \tag{3.105}$$

Equilibrium analysis of (3.102) yields that the equilibrium mean $\mathbf{E}Y$ satisfies

$$\mathbf{E}Y = \mathbf{E}Y - 1 + \mathbf{E}Z + \mathbf{P}(Y=0) \tag{3.106}$$

so, since $\mathbf{E}Z=\rho$, we have $\mathbf{P}(Y=0) = \pi_0 = 1-\rho$, from which the system (3.105) can be solved recursively. □

3.6. An improved model for cosmic radiation

At the end of Section 3.1 we were discussing shortcomings of the Furry model of cosmic radiation showers. Following Arley (1943), we will try to adapt the model to the physics of the situation. The first problem is that particles should disappear as their energy becomes depleted. The pure birth model therefore needs to be replaced by a birth and death model. Roughly speaking, the rate at which an individual dies should be proportional to the age of the family. This yields a time-dependent birth and death process with birth rate $\lambda_n(t)=\lambda n$ and death rate $\mu_n(t)=\mu nt$. The model is still overly simplified, since the probability of being absorbed (as well as the probability of giving birth to new particles) really depends on the energy of the particular particle at that time. This energy, which depends on the antecedents of each particle, will fluctuate around the average energy, while this model sets it equal to the average energy at that time. However, we will see that it nonetheless adds some amount of realism to the description.

Let us first compute an equation for the mean value $m(t)$ of the process. This mean value is $\sum kP_{1k}(t)$. From the forward equation we see that

$$\frac{dP_{1k}(t)}{dt} = \lambda(k-1)P_{1,k-1}(t)-(\lambda+\mu t)kP_{1k}(t) + \mu t(k+1)P_{1,k+1}(t) \quad (3.107)$$

from which we derive the following equation for the mean:

$$m'(t) = (\lambda-\mu t)m(t) \quad (3.108)$$

which has solution ($m(0)=1$)

$$m(t) = \exp(\lambda t-\frac{\mu}{2}t^2). \quad (3.109)$$

As shown in Figure 3.5, this curve has a maximum of $\exp(\lambda^2/2\mu)$ at $t=\lambda/\mu$. The probability generating function of $X(t)$ satisfies the partial differential equation

$$\frac{\partial G(s;t)}{\partial t} = (\lambda s^2-(\lambda+\mu t)s+\mu t)\frac{\partial G(s;t)}{\partial s}. \quad (3.110)$$

The solution satisfying $G(s;0)=s$ is

$$G(s;t) = (-A)^{-1}(t-A(s)) \quad (3.111)$$

where

$$A(s) = \exp(-\mu(s-s^2/2))\int_0^s \lambda u(1-u)\exp(\mu(u-u^2/2))du \quad (3.112)$$

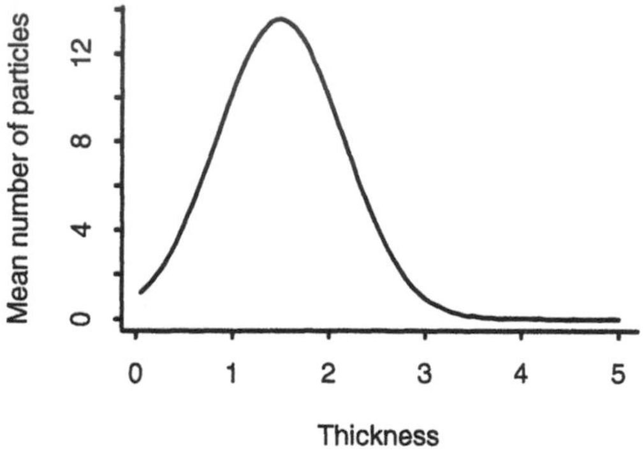

Figure 3.5. The mean curve for the modified model of cosmic radiation.

which can be expressed in terms of incomplete gamma functions.

The final step towards a realistic approximate model for cosmic radiation showers is to take into account the different kinds of particles (electrons and photons) that occur in alternating generations. We will assume that the process is symmetric with respect to the two kinds of particles, apart from the initial condition. In other words, we will assume that the birth and death process described above adequately describes both the production of electrons from photons and the production of photons from electrons. More precisely, we assume that particles have an intensity rate of λ to be absorbed and give birth to two particles of the other type, and that all particles have an intensity rate μt of being absorbed without giving birth to a new particle. Arley (1943, section 5.2) discusses the reasonableness of these assumptions. Let $(X(t),Y(t))$ be the two-dimensional process of electrons and photons, with $X(0)=1$ and $Y(0)=0$. The equations for mean values, letting $n(t)=\mathbf{E}X(t)$ and $m(t)=\mathbf{E}Y(t)$, are

$$n(t+dt) = n(t) + m(t)2\lambda dt - n(t)\lambda dt - n(t)\mu t dt + o(dt) \quad (3.113)$$

and

$$m(t+dt) = m(t) + n(t)2\lambda dt - m(t)\lambda dt - m(t)\mu t dt + o(dt). \quad (3.114)$$

Turning these equations into differential equations we see that

$$n'(t) = -(\lambda+\mu t)n(t) + 2\lambda m(t) \quad (3.115)$$

and

$$m'(t) = -(\lambda+\mu t)m(t) + 2\lambda n(t). \quad (3.116)$$

Solving the first equation for m and substituting that in the second equation yields

$$n''(t) + 2(\lambda+\mu t)n'(t) + ((\lambda+\mu t)^2+\mu-4\lambda^2)n(t) = 0. \quad (3.117)$$

The solutions are

$$n(t) = \tfrac{1}{2}\exp(\lambda t - \mu t^2/2) + \tfrac{1}{2}\exp(-3\lambda t - \mu t^2/2) \tag{3.118}$$

and

$$m(t) = \tfrac{1}{2}\exp(\lambda t - \mu t^2/2) - \tfrac{1}{2}\exp(-3\lambda t - \mu t^2/2). \tag{3.119}$$

When λt is large we have that $n(t) \approx m(t)$, each approximately being equal to half the mean value in the one-dimensional model. The effects of the initial conditions (which are the only asymmetries in the model) disappear exponentially fast.

Arley approximates the marginal time-dependent probability distributions by Pólya distributions, and obtains very good fits to experimental data. However, the tendency in physics has later been to attempt to exploit the branching structure of electron showers. An interesting description of such models is given in Harris (1963, Chapter VII).

3.7. Statistical inference for continuous time Markov chains

We restrict attention to uniform semigroups, so that, in particular, $\sum_j Q_{ij} = 0$. In order to compute the likelihood from complete observation of a continuous time Markov chain $(X(t), t \le T)$ we first note that all we need to know are the number of jumps $N(T)$, the successive jump times τ_i, and the jump chain. In other words, we need to know

$$V_T = \{N(T), X(0), \tau_1, X(\tau_1+), \tau_2, \ldots, X(\tau_{N(T)}+)\}. \tag{3.120}$$

In statistical terminology, V_T is a **sufficient statistic**. The likelihood of Q at $v = (n, x_0, t_1, \ldots, x_n)$ is

$$\begin{aligned} &p^0(x_0)q(x_0)\exp(-q(x_0)t_1)\prod_{j=1}^{n-1}\frac{Q(x_{j-1},x_j)}{q(x_{j-1})}q(x_j)\exp(-q(x_j)(t_{j+1}-t_j)) \\ &\quad \times \frac{Q(x_{n-1},x_n)}{q(x_{n-1})}\exp(-q(x_n)(T-t_n)) \text{ if } n \ge 1 \\ &p^0(x_0)\exp(-q(x_0)T) \text{ if } n=0 \end{aligned} \tag{3.121}$$

Here q_k is written $q(k)$ and Q_{ij} is $Q(i,j)$ for typographic reasons. The derivation of (3.121) is immediate from Theorem 3.3: there are independent exponential holding times, and jumps according to the jump chain matrix. We can write the likelihood, given that $X(0) = x$, as

$$L(Q; V_T) = \left\{\prod_{i=1}^{K_T}\prod_{j \ne i} Q(i,j)^{N_T(i,j)}\right\}\exp\left(-\sum_1^{K_T} A_T(i)q_i\right) \tag{3.122}$$

where the number K_T of observed states up to time T is finite, since (with probability one) only a finite number of states are observed in finite time, $N_t(i,j)$ is the number of transitions from i to j up to time t (so $N_T(i,i) = 0$), and $A_t(i)$ is the total time spent in state i up to time t. It follows that $(A_T(i), N_T(i,j), i,j \in D)$ is a sufficient statistic (it is, in fact, minimal sufficient).

Theorem 3.4 Suppose that X is ergodic at the true parameter Q_0, and that the set $D=\{(i,j): i\neq j, Q_0(i,j)>0\}$ is known. Then with probability one

$$\hat{Q}(i,j;T) = \frac{N_T(i,j)}{A_T(i)} \tag{3.123}$$

are unique solutions to the likelihood equations, and for T large enough they provide a maximum with probability one.

Proof Write the logarithm of the likelihood

$$\begin{aligned}\log L(Q) &= \sum_i 1(X(0)=i)\log p^0(i) - \sum_i A_T(i) q_i + \sum_{i\neq j} N_T(i,j)\log Q_{ij} \\ &= c - \sum_{i\neq j} A_T(i) Q_{ij} + \sum_{i,j} N_T(i,j)\log Q_{ij}\end{aligned} \tag{3.124}$$

where we used the fact that $\sum_j Q_{ij}=0$. Taking derivatives with respect to Q_{ij} and setting them equal to zero yields the likelihood estimates

$$Q_{ij} = \frac{N_T(i,j)}{A_T(i)}. \tag{3.125}$$

The second derivative has diagonal elements $N_T(i,j)/Q_{ij}^2$, and off-diagonal elements 0, so the solution is a maximum whenever $N_T(i,j)>0$, which will eventually happen with probability one since X is ergodic. □

Corollary Let $N_T(i)=\sum_j N_T(i,j)$ be the number of transitions out of i in $(0,T)$. Then $\hat{q}_i=N_T(i)/A_T(i)$.

Proof By invariance of the mle under reparametrization we have that

$$\hat{q}_i = \sum_{j\neq i}\hat{Q}_{ij} = \sum_{j\neq i}\frac{N_T(i,j)}{A_T(i)} = \frac{N_T(i)}{A_T(i)}. \tag{3.126}$$

□

It is fairly straightforward to derive first and second order properties (means, variances, and covariances) of the components of the sufficient statistic.

Proposition 3.6 Let $p_i(t) = \mathbf{P}(X(t)=i)$. Then for $i\neq j$,

$$\mathbf{E}N_t(i,j) = Q_{ij}\int_0^t p_i(u)\,du. \tag{3.127}$$

$$\mathbf{E}N_t(i,j)N_t(k,m) = 1((i,j)=(k,m))Q_{ij}\int_0^t p_i(u)du \tag{3.128}$$

$$+ Q_{ij}Q_{km}\int_0^t\int_0^v (P_{mi}(v-u)p_k(u)+P_{jk}(v-u)p_i(u))dudv.$$

Proof Divide $[0,t)$ into intervals $[(k-1)h,kh)$ for $k=1,\ldots,m$, and let $n_k(i,j)$ count the number of transitions $i\to j$ in the kth interval. Then $N_t(i,j)=\sum n_k(i,j)$. For h small enough, each $n_k(i,j)$ is 0 or 1 with probability $1-o(h)$. For $r\neq j$, $l\neq i$,

$$\begin{aligned}\mathbf{P}(n_k(i,j)=1,X(kh)=r,X((k-1)h)=l)&\\ &= (Q_{li}h+o(h))(Q_{ij}h+o(h))(Q_{jr}h+o(h))+o(h) \\ &= O(h^3) = o(h).\end{aligned} \tag{3.129}$$

Thus

$$\begin{aligned}\mathbf{P}(n_k(i,j)=1) &= \sum_{l,r}\mathbf{P}(n_k(i,j)=1,X((k-1)h)=l,X(kh)=r) \\ &= \mathbf{P}(n_k(i,j)=1,X((k-1)h)=i,X(kh=j)) + o(h) \\ &= p_i((k-1)h)Q_{ij}h + o(h)\end{aligned} \tag{3.130}$$

and

$$\begin{aligned}\mathbf{E}N_t(i,j) &= \sum_k \mathbf{P}(n_k(i,j)=1) = Q_{ij}\sum_1^m p_i((k-1)h)h + o(h) \\ &\to Q_{ij}\int_0^t p_i(u)du\end{aligned} \tag{3.131}$$

as $h\to 0$. The second equality is similar, only having more terms to keep track of. □

The following result deals with the times spent in different states.

Proposition 3.7

$$\mathbf{E}A_t(i)=\int_0^t p_i(u)du. \tag{3.132}$$

$$\mathbf{E}A_t(i)A_t(j) = \int_0^t\int_0^u (P_{ji}(u-v)p_j(v) + P_{ij}(u-v)p_i(u))dvdu. \tag{3.133}$$

Proof Use $A_t(i) = \int_0^t 1(X(u)=i)du$, so, upon taking expectations, we get $\mathbf{E}A_t(i) = \int_0^t \mathbf{P}(X(u)=i)du$. The second part is similar. □

The joint behavior between transitions and time spent in each state is as follows.

Proposition 3.8

$$\mathbf{E}N_t(i,j)A_t(r) = Q_{ij}\int_0^t\int_0^u (P_{ri}(u-v)p_r(v) + P_{jr}(u-v)p_i(v))dvdu. \quad (3.134)$$

$$\mathbf{E}(N_t(i,j)-Q_{ij}A_t(i))(N_t(r,s)-Q_{rs}A_t(r)) = \delta(i,j\,;r,s)\int_0^t p_i(u)du. \quad (3.135)$$

Proof Use arguments similar to those of the previous two proposition. □

As in the discrete time case, an ergodic theorem holds. Let T_i be the time of the first return to i.

Theorem 3.5 Let $X(t)$ be an ergodic chain. Assume that f satisfies

$$\mathbf{E}^i \left| \int_0^{T_i} f(X(s))ds \right| < \infty. \quad (3.136)$$

Then

$$\lim_{t\to\infty}\frac{1}{t}\int_0^t f(X(s))ds = \frac{\mathbf{E}^i\int_0^{T_i} f(X(s))ds}{\mathbf{E}^i T_i} \text{ with probability one.} \quad (3.137)$$

□

A proof is in Bhattacharaya and Waymire (1990, p. 306). Now write $N_t(i,j)=\int_0^t f_{ij}(X(s-),X(s))ds$ where $f_{ij}(u,v)=1(u=i,v=j,i\neq j)$. We deduce (with some handwaving) the following result.

Proposition 3.9 If X is ergodic then as $t\to\infty$

$$\mathbf{E}N_t(i,j)/t \to \pi_i Q_{ij}, \quad (3.138)$$

$$\mathbf{E}N_t(i,j)N_t(r,s)/t^2 \to \pi_i\pi_r Q_{ij}Q_{rs}, \quad (3.139)$$

$$\mathbf{E}A_t(i)/t \to \pi_i, \quad (3.140)$$

$$\mathbf{E}A_t(i)A_t(j)/t^2 \to \pi_i\pi_j, \quad (3.141)$$

$$\mathbf{E}N_t(i,j)A_t(r)/t^2 \to \pi_i\pi_r Q_{ij} \tag{3.142}$$

and

$$\mathbf{E}(N_t(i,j)-Q_{ij}A_t(i))(N_t(r,s)-Q_{rs}A_t(r))/t \to \delta(ij\,;rs)\pi_i. \tag{3.143}$$

□

It follows, using Chebyshev's inequality, that

$$N_t(i,j)/t \overset{\mathbf{P}}{\to} \pi_i Q_{ij} \tag{3.144}$$

and

$$A_t(i)/t \overset{\mathbf{P}}{\to} \pi_i. \tag{3.145}$$

It is possible to show (Adke and Manjunath, 1984, Theorem 5.4.2) that if

$$\xi_{ij}(t) = \frac{N_t(i,j)-A_t(i)Q_{ij}}{t^{1/2}}, \quad i \neq j \tag{3.146}$$

then the ξ_{ij} are asymptotically independent normal variates with mean 0 and variance $\pi_i Q_{ij}$. Hence the following result, summarizing the asymptotic behavior of the nonparametrics mle's.

Theorem 3.6 (i) $\hat{Q}(i,j\,;T) \overset{\mathbf{P}}{\to} Q_{ij}$ for $i,j \in D$.

(ii) The $t^{1/2}(\hat{Q}(i,j\,;T)-Q_{ij})$, $i \in D$, converge in distribution to independent, mean zero, normal variates with variances Q_{ij}/π_i, respectively. □

Corollary $(A_T(i)\hat{Q}(i,j\,;T))^{1/2}\,(\hat{Q}(i,j\,;T) - Q_{ij}) \overset{d}{\to} \mathrm{N}(0,1)$

Proof Use Slutsky's theorem, Theorem 3.6, and (3.145). □

Application (Zoology) S. A. Altman studied the vital events in a baboon troupe (Cohen, 1969) in the Amboseli nature reserve in Kenya. The data are given in Table 3.2. Ignoring the information about which events are births, deaths, immigrations or emigrations, we compute the nonparametric mle. Since all changes are by one unit, the resulting $\hat{Q}$ is tri-diagonal. Figure 3.6 shows the estimated intensities for population increases and decreases, plotted against population size. Popular parametric models for this type of data (birth and death or BIDE processes) are linear in population size. Thus, we may want to fit a line to the points in the figure. Note, however, that the estimators have different

Table 3.2 Vital events in a baboon troupe

After this many days	at this troop size	this event occurred.
41	40	B
5	41	B
22	42	B
2	43	D
17	42	D
26	41	I
0	42	I
55	43	B
35	44	I
20	45	E
5	44	D
6	43	E
32	42	D
4	41	D
0	40	D
22	39	D
10	38	B
0	39	B
7	40	D
4	39	B
17	40	D
11	39	E
3	38	B
4	39	D
8	38	D
2	37	D
5	36	B
10	37	B

variances. Using the corollary to Theorem 3.6 the asymptotic variances are estimated by $\hat{Q}(i,j;T)/A_T(i)=N_T(i,j)/A_T(i)^2$. A weighted least squares line with weights inversely proportional to the estimated variances is shown in the figure. The slope is estimated to be –0.004, and the intercept to be 0.193. Neither of these estimates is significantly different from zero. Of course, a linear model with negative slope is not acceptable, since eventually the estimated intensity may become negative. Fitting a line through the origin, the estimated slope is 5.9×10^{-4} with a standard error of 1.5×10^{-4}.

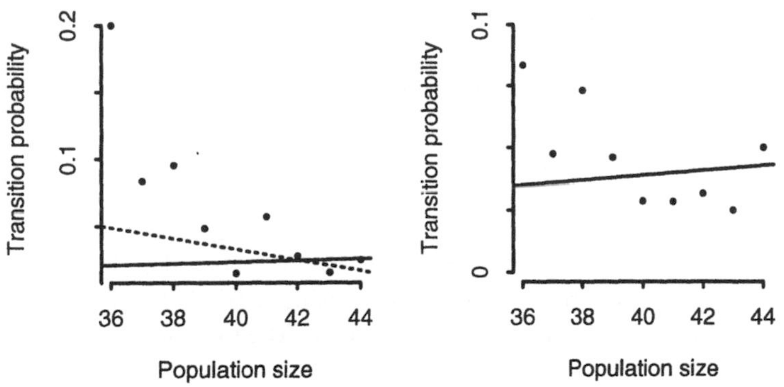

Figure 3.6. Estimated intensity of increase (left) and decrease (right) versus population size.

The estimated population decrease intensities are shown in Figure 3.6. The most common model for such data would be linear with zero intercept. The appropriate weighted least squares line has slope 9.7×10^{-4}, with standard error 2×10^{-4}. The quality of this fit is partly assessed by the (weighted) correlation coefficient of 0.91. These results indicate that a linear birth and death model may be sensible. □

In many applications we are interested in parametric models for Q, and would only use the nonparametric estimators for preliminary data-analytic purposes, such as assessment of linearity. As in the discrete time case one must assume some regularity conditions for Q. Again, we consider only ergodic chains. Assume that the k-dimensional parameter $\boldsymbol{\theta}$ takes values in an open subset of $\mathbf{R}^k$.

Assumptions B

B1: The set $\{(i,j):Q_{ij}(\boldsymbol{\theta})>0\}$ does not depend on $\boldsymbol{\theta}$.

B2: The functions $Q_{ij}(\boldsymbol{\theta})$ are three times continuously differentiable.

B3: The $k\times k$ matrix $\Sigma(\boldsymbol{\theta})$ with elements

$$\Sigma_{lm}(\boldsymbol{\theta}) = \sum_{i,j} \frac{\pi_i(\boldsymbol{\theta})}{Q_{ij}(\boldsymbol{\theta})} \frac{\partial Q_{ij}(\boldsymbol{\theta})}{\partial\theta_l} \frac{\partial Q_{ij}(\boldsymbol{\theta})}{\partial\theta_m} \tag{3.147}$$

is positive definite for all $\boldsymbol{\theta}$, where $\boldsymbol{\pi}(\boldsymbol{\theta})$ is the stationary distribution.

Theorem 3.7 Under Assumptions B the likelihood equations $(\partial/\partial\theta_i)\log L(\boldsymbol{\theta}) = \mathbf{0}$ have, for sufficiently large t, a solution $\hat{\boldsymbol{\theta}}(t)$ with probability one. This solution is consistent and maximizes the likelihood. If, in addition, $Q(\boldsymbol{\theta})$ has all eigenvalues distinct for all $\boldsymbol{\theta}$, then

$$\left[-D^2\log L(\hat{\boldsymbol\theta})\right]^{1/2}(\hat{\boldsymbol\theta}-\boldsymbol\theta)\overset{d}{\to}\mathrm{N}(\mathbf{0},\mathbb{I}) \tag{3.148}$$

where D^2 denotes the matrix of second order partial derivatives.

This result is a simple version of Theorem 7.3 in Billingsley (1961), where a proof can be found.

Application (Zoology, continued) Returning to the Altman baboon troupe data, and still neglecting the information about the character of each event, we may model the population sizes using a BIDE model (the birth and death model indicated earlier is a special case). The likelihood is

$$L(\lambda,\nu,\kappa)\propto\left[\prod_{i=n_{\min}(T)}^{n_{\max}(T)}(\lambda i+\nu)^{n_i(T)}\right]\kappa^{D(T)+E(T)}\exp(-\nu T-(\lambda+\kappa)S(T)) \tag{3.149}$$

where $n_{\min}(T)$ and $n_{\max}(T)$ are the smallest and largest observed population sizes for which a population increase has been observed in the observation time $(0,T)$, $n_i(T)=N_T(i,i+1)$, $S(T)=\int_0^T X(s)ds$, and all terms that do not depend on unknown parameters have been grouped into the constant of proportionality. Numerically we have $T=373$, $n_{\min}(T)=36$, $n_{\max}(T)=44$, $\boldsymbol{n}(T)=(n_{36},\ldots,n_{44})=(1,1,2,2,1,2,2,1,1)$, $S(T)=15407$, and $D(T)+E(T)=15$. Maximizing (3.149) yields $(\hat\lambda,\hat\nu,\hat\kappa)=(1.65\times10^{-9},3.49\times10^{-2},9.74\times10^{-4})$. Note that the terms involving κ can be separated from those containing (λ,ν) whence asymptotically $\hat\kappa$ is independent of $(\hat\lambda,\hat\nu)$. Differentiating $\log L(\lambda,\nu,\kappa)$ twice yields

$$\begin{aligned}
\frac{\partial^2}{\partial\lambda^2}\log L(\lambda,\nu,\kappa)&=-\sum_{i=n_{\min}(T)}^{n_{\max}(T)}\frac{i^2n_i(T)}{(\lambda i+\nu)^2}\\
\frac{\partial^2}{\partial\lambda\partial\nu}\log L(\lambda,\nu,\kappa)&=-\sum_{i=n_{\min}(T)}^{n_{\max}(T)}\frac{in_i(T)}{(\lambda i+\nu)^2}\\
\frac{\partial^2}{\partial\nu^2}\log L(\lambda,\nu,\kappa)&=-\sum_{i=n_{\min}(T)}^{n_{\max}(T)}\frac{n_i(T)}{(\lambda i+\nu)^2}\\
\frac{\partial^2}{\partial\lambda^2}\log L(\lambda,\nu,\kappa)&=-\frac{D(T)+E(T)}{\kappa^2}.
\end{aligned} \tag{3.150}$$

We obtain the asymptotic covariance matrix of the mle by plugging in the mle's into (3.150) and invert the negative of the resulting matrix (with zeros at the cross-derivatives between κ and λ or ν). Here $\mathrm{se}(\hat\lambda)=2.5\times10^{-4}$, $\mathrm{se}(\hat\nu)=1.9\times10^{-3}$, $\mathrm{corr}(\hat\lambda,\hat\nu)=0.19$, and $\mathrm{se}(\hat\kappa)=2.5\times10^{-4}$.

When we ignore the information about the character of each event, the process behaves like an immigration-death (ID) process. In fact, the likelihood ratio statistic for testing the null hypothesis $\lambda=0$ is zero to six decimal places. As seen in Exercise 5, the stationary distribution for an ID process is geometric with mean ν/κ, which we estimated to be 35.8. Note that the entire path of this process lies above its stationary mean, with an observed mean of 41.3. Using a Taylor expansion (Exercise 8), we estimate the variance of $\hat{\nu}/\hat{\kappa}$ to be $(\text{se}(\hat{\nu})/\hat{\kappa})^2 + (\text{se}(\hat{\kappa})\hat{\nu}/\hat{\kappa}^2)^2$, so the standard error of the stationary mean is 9.5. The observed mean is therefore well within range of the usual.

Since in this case we have information about the character of each event, we can use this to write the likelihood

$$L(\lambda,\nu,\mu,\eta) \propto \lambda^{B(T)}\nu^{I(T)}\mu^{D(T)}\eta^{E(T)}\exp(-\nu T-(\lambda+\mu+\eta)S(T)). \tag{3.151}$$

Here $B(T)=10$, $I(T)=E(T)=3$, and $D(T)=12$. The mle's are $\hat{\lambda}=B(T)/S(T)$, $\hat{\nu}=I(T)/T$, $\hat{\mu}=D(T)/S(T)$, and $\hat{\eta}=E(T)/S(T)$. Numerically these are $\hat{\lambda}=6.5\times10^{-4}$, $\hat{\nu}=8.0\times10^{-3}$, $\hat{\mu}=7.8\times10^{-4}$, and $\hat{\eta}=1.9\times10^{-4}$. The main difference from the previous estimates is the value of $\hat{\lambda}$. Since the likelihood factors into four terms, one for each parameter, the estimates are asymptotically independent. Standard errors are again estimated using the second derivative of the log likelihood, which is

$$\begin{aligned}
\frac{\partial^2}{\partial\lambda^2}\log L(\lambda,\nu,\mu,\eta) &= -B(T)/\lambda^2\\
\frac{\partial^2}{\partial\nu^2}\log L(\lambda,\nu,\mu,\eta) &= -I(T)/\nu^2\\
\frac{\partial^2}{\partial\mu^2}\log L(\lambda,\nu,\mu,\eta) &= -D(T)/\mu^2\\
\frac{\partial^2}{\partial\eta^2}\log L(\lambda,\nu,\mu,\eta) &= -E(T)/\eta^2.
\end{aligned} \tag{3.152}$$

Again, we substitute the estimates and invert the negative of the resulting (diagonal) matrix to compute the asymptotic covariance matrix. The standard errors are $\text{se}(\hat{\lambda})=2.1\times10^{-4}$, $\text{se}(\hat{\nu})=4.6\times10^{-3}$, $\text{se}(\hat{\mu})=2.2\times10^{-4}$, and $\text{se}(\hat{\eta})=1.1\times10^{-4}$. One possible explanation to why the population gain parameters ν and λ are so different when we do or do not include detailed information could be that only a small proportion of the population is fertile.

The stationary distribution of a BIDE process is, as we have seen earlier, negative binomial with parameters $r=\nu/\lambda$ and $p=1-\lambda/(\mu+\eta)$, so the mean is $r(1-p)/p$ which we estimate to be 24.8 with a standard error of 14.3. Again, the entire path lies above the stationary mean, this time by 1.2 standard errors. □

Application (Communications center, continued) We return to the queue in the receipt center of the communications terminal, mentioned in section 3.5. Figure 3.7 depicts the queue length at 5-minute intervals for four hours.

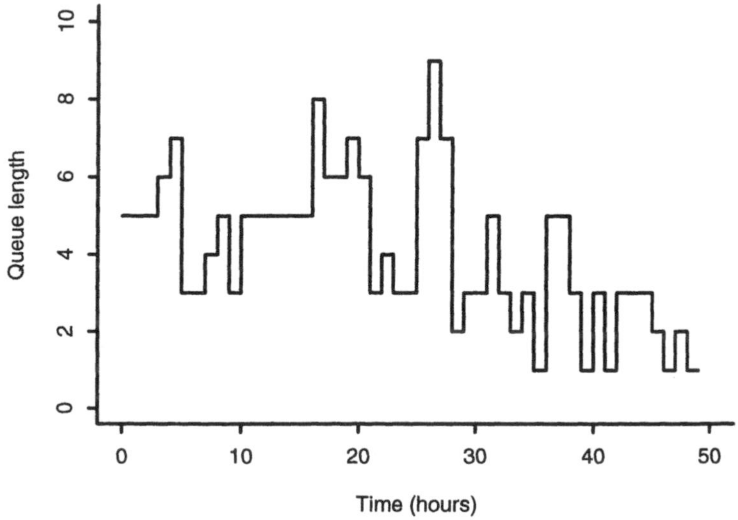

Figure 3.7. Receipt center queue size. Adapted from Hatzikostados and Howe (1967), in R. Cruon (ed.), *Queueing Theory: Recent Developments and Applications*, published by American Elsevier, 1967.

If we had continuous observations we could test the assumptions of an M/M/1 model directly, but since we only observe the discrete skeleton to scale 1/12 of an hour, this is not so straightforward. We will, however, assume an M/M/1 model, and derive the likelihood of the skeleton chain. This is a discrete time Markov chain with transition probabilities $p_{ij}=P_{ij}(1/12)$. The main point of this application is how difficult this estimation problem is, in spite of this being one of the simplest possible models for a queue.

In order to determine $P_{ij}(t)$ we first note that this is a birth and death process with rates λ and μ and no loss possible when the population size is zero. The forward equation is

$$\frac{dP_{ij}(t)}{dt} = \lambda P_{i,j-1}(t) - (\lambda+\mu)P_{ij}(t) + \mu P_{i,j+1}(t), \quad j\geq 1,$$

$$\frac{dP_{i0}(t)}{dt} = -\lambda P_{i0}(t) + \mu P_{i1}(t). \tag{3.153}$$

Using the Laplace transform approach (and a lot of complicated algebra) the solution is (Bailey, 1975, equation 11.65)

$$P_{ij}(t) = \rho^{\frac{1}{2}(j-i)}\{\exp(-(\lambda+\mu)t)I_{i-j}(2\nu t) + \int_0^t \exp(-(\lambda+\mu)s)(\lambda I_{i+j+2}(2\nu s) - 2\nu I_{i+j+1}(2\nu s) + \mu I_{i+j}(2\nu s))ds\} \tag{3.154}$$

where $\rho=\lambda/\mu$ is the traffic intensity, $\nu=(\lambda\mu)^{\frac{1}{2}}$, and $I_n(x)$ is the modified Bessel function of the first kind, defined by

$$I_n(x) = (x/2)^n \sum_{k=0}^{\infty} \frac{(x^2/4)^k}{k!(n+k)!}. \tag{3.155}$$

For small values of x we have $I_n(x) \approx (x/2)^n/n!$ Using this approximation in (3.154) we can evaluate the likelihood explicitly in terms of incomplete gamma functions, or, equivalently, cdf's of gamma distributions. It is astonishing how complex the exact probabilities are for this the simplest of all queueing models! No wonder equilibrium approaches are important in queueing theory (and simulation studies, rather than algebraic derivations, the rule).

Evaluating the likelihood (using the approximation to I_n), Figure 3.8 obtains.

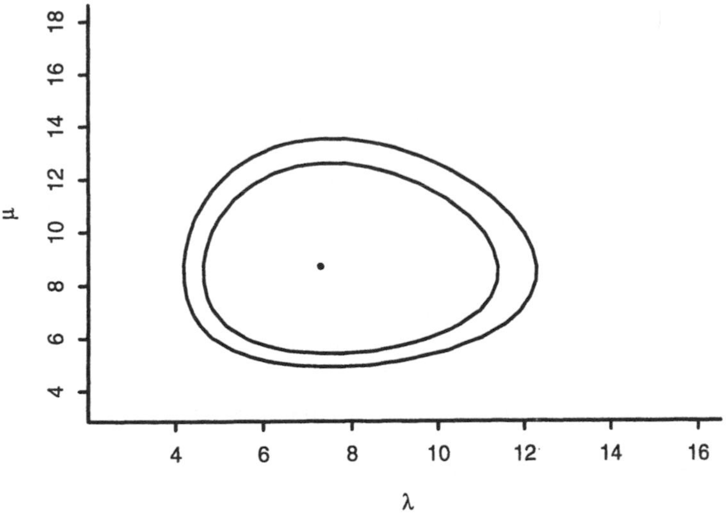

Figure 3.8. Approximate 95% (3 units of log likelihood below the maximum) and 99% (4.6 units below the maximum) contours for the queue log likelihood.

The mle is $\hat{\lambda}=7.5$, $\hat{\mu}=8.5$, for an estimated traffic intensity of $\hat{\rho}=\hat{\lambda}/\hat{\mu}=0.88$. Note that the likelihood surface is fairly flat near the maximum. The size of the 95% likelihood region is considerable.

A much simpler estimate of ρ can be computed from equilibrium considerations. The theory developed in section 3.5 indicates that the stationary distribution is geometric with mean $\rho/(1-\rho)$. Since the observed mean is 4.08, this corresponds to a ρ-value of 0.80, reasonably close to $\hat{\rho}$. This method does not, however, yield an estimate of λ. □

3.8. Modeling neural activity

The central nervous system is a complicated system of cells interacting using electric impulses. Apparently there is substantial non-deterministic behavior, i.e., given the same input current to a nerve cell (**neuron**) it may or may not react in a sense that we shall describe below. In this section we study single neurons, and try to deduce facts about their structure (in a probabilistic sense) from measurements of currents, as was described in the last application of the previous chapter.

Figure 3.9 depicts a neuron. It has three main parts: the cell body or **soma**, containing the cell nucleus; the **dendrites** which are a series of root-like extrusions from the soma, and the **axon**, a long wire-like structure connecting the cell to other neurons. The electrical activity in nerve cells is due to the presence of internal and external fluids in the form of salts containing potassium (K^+), sodium (Na^+), or chloride (Cl^-) which are ionized in solution. The ions move from one side of the cell membrane to the other through channels that are specific to a certain ion (such as a potassium channel), and which can be open or closed, as described in the previous chapter. Different neurons have different permeabilities for the various ions of interest. If the concentration of potassium ions is much higher on the inside of the cell than on the outside, this yields a concentration gradient across the membrane, with excess potassium moving outwards. This diffusion of positive ions disturbs the charge balance and results in an electric field opposing the chemical field. When the fields are balanced, an equilibrium state is obtained. This equilibrium has an electric potential difference across the membrane, with the inside more negative, so the **membrane potential** (inside minus outside potential) is negative. This is called the **rest potential**, and is about –60 to –70 mV. The equilibrium can be disturbed either by applying an electric current across the membrane, or by changing the ionic concentration on the outside of the cell. If the change makes the membrane potential more positive it is called **depolarization**. The key feature of these cells is that when depolarization in a neuron exceeds a time-dependent threshold, the potential shoots up dramatically to about +30 mV, and then decreases exponentially to the rest level. This **firing** is called an **action potential**.

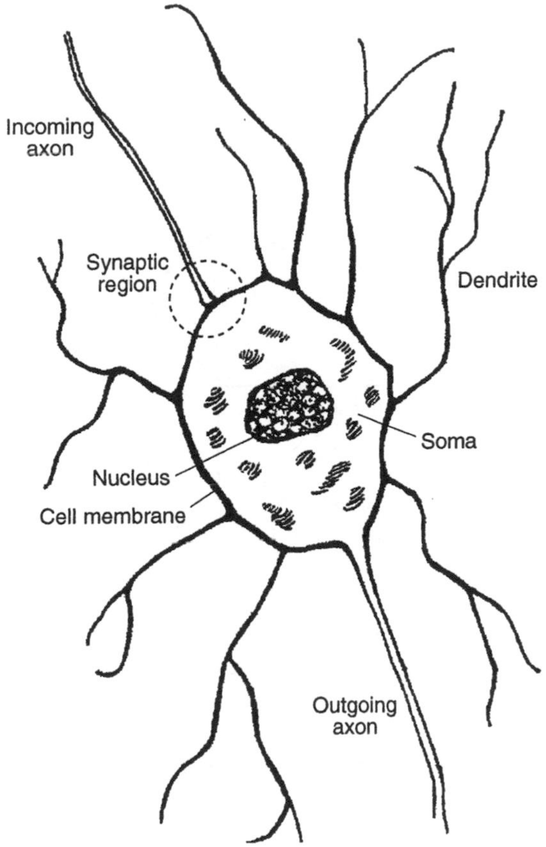

Figure 3.9. A schematic depiction of a neuron.

At the **synapse**, the place where an incoming axon connects with the next cell body, action potential from one neuron releases a chemical, called **neurotransmitter**, from the bouton of the incoming axon (see Figure 3.9). The most common neurotransmitter is acetylcholine (ACh). Binding of acetylcholine to a **receptor**, which protrudes from the protein mass of the receiving neuron, produces a depolarization of the postsynaptic membrane. The soma accumulates all the postsynaptic potentials from the incoming axons. When the integrated effect exceeds the threshold, the neuron fires, as described above. In this section we shall concentrate on the currents in a given synapse, while we return to the action potentials in Chapter 5.

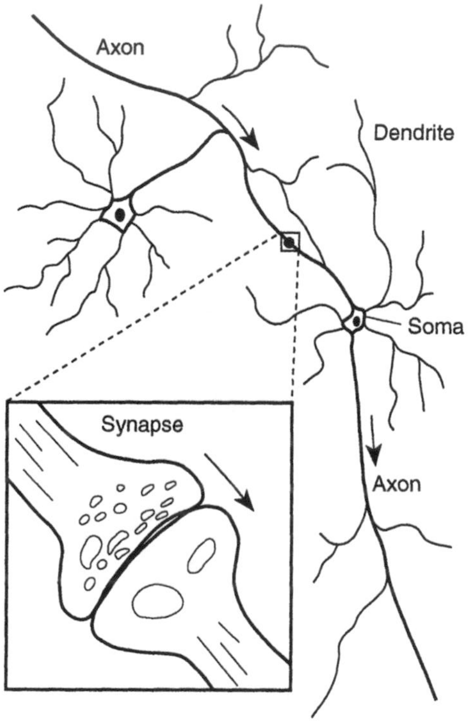

Figure 3.10. A schematic drawing of a network of neurons. The synapse is shown in the detailed blowup.

It is well established that a channel has multiple open and closed states. There are several possible models. The simplest allows for two different closed states. In the resting state, the ligand (ACh) is free. It is known that the ligand must be chemically bound to the receptor for the channel to open. In state two, the ligand is bound to the receptor, but the channel remains closed. The third state is open, with ligand again bound to the receptor. We show this schematically in Figure 3.11, where the arrows denote possible transitions, C stands for closed, O for open, and L for ligand.

$$\mathrm{C+L} \longleftrightarrow \mathrm{CL} \longleftrightarrow \mathrm{OL}$$

Figure 3.11. A three-state model for the ACh channel.

We shall see that this model is too simple. There is evidence of at least two kinds of open states. One possible such model allows for zero, one or two units of ligand bound to the receptor. In order for the channel to open, at least one ligand must be bound. The schematic is given in Figure 3.12.

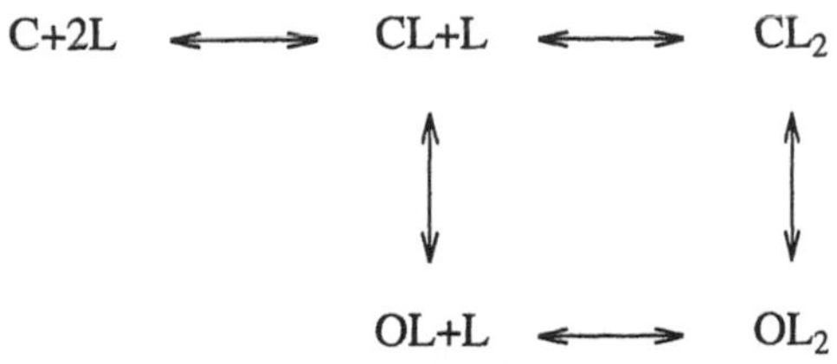

Figure 3.12. A more realistic model for the ACh channel.

We will regard the underlying process as a Markov chain in continuous time, with state space $\{1, 2, 3, 4, 5\}$ corresponding to states $\{C+2L, CL+L, CL_2, OL+L, OL_2\}$, respectively. We are not able to observe this process: all we can tell is whether the channel is open or closed. However, we can write down the intensities based on the chemistry proposed in Figure 3.12. Suppose that we have a concentration c of ligand present. We assume that the binding rate is proportional to the concentration of ligand, so that, for example, $Q_{12}=ck_{12}$. It is useful to group the **Q**-matrix into transitions between closed states (1, 2, 3), between open states (4, 5), and actual opening or closing. We write the intensity matrix

$$\mathbf{Q} = \begin{pmatrix} \mathbf{Q}_{cc} & \mathbf{Q}_{co} \\ \mathbf{Q}_{oc} & \mathbf{Q}_{oo} \end{pmatrix} \tag{3.156}$$

where

$$\mathbf{Q}_{cc} = \begin{pmatrix} -ck_{12} & ck_{12} & 0 \\ k_{21} & -(k_{21}+ck_{23}+k_{24}) & ck_{23} \\ 0 & k_{32} & -(k_{32}+k_{35}) \end{pmatrix} \tag{3.157}$$

covers transitions between closed states, and c is the concentration of ligand present;

$$\mathbf{Q}_{oc} = \begin{pmatrix} 0 & k_{42} & 0 \\ 0 & 0 & k_{53} \end{pmatrix} \tag{3.158}$$

and

$$\mathbf{Q}_{co} = \begin{pmatrix} 0 & 0 \\ k_{24} & 0 \\ 0 & k_{35} \end{pmatrix} \tag{3.159}$$

correspond to transitions between the open and closed states, and

$$\mathbf{Q}_{oo} = \begin{bmatrix} -(k_{42}+ck_{45}) & ck_{45} \\ k_{54} & -(k_{53}+k_{54}) \end{bmatrix} \tag{3.160}$$

describes transitions within the open states.

Since we cannot observe the actual process, but only whether the channel is open or closed, we need to use indirect methods to assess the underlying model. Consider a general model with N_o open and N_c closed states. We first try to determine the distribution of an observable quantity such as T, the **open dwell time**. Let

$$F_{ij}(t) = \mathbf{P}(T \le t, \text{ exit to } j \mid \text{in open state at } t=0). \tag{3.161}$$

Define a new process where the closed states are absorbing:

$$\overline{\mathbf{Q}} = \begin{bmatrix} \mathbf{0} & \mathbf{0} \\ \mathbf{Q}_{oo} & \mathbf{Q}_{oc} \end{bmatrix}. \tag{3.162}$$

Then $F_{ij}(t) = (\overline{P}_{oc}(t))_{ij}$. From the forward equation

$$\overline{P}'(t) = \overline{P}(t)\overline{\mathbf{Q}} = \begin{bmatrix} \mathbf{0} & \mathbf{0} \\ \overline{P}_{oc}(t) & \overline{P}_{oo}(t) \end{bmatrix} \begin{bmatrix} \mathbf{0} & \mathbf{0} \\ \mathbf{Q}_{oc} & \mathbf{Q}_{oo} \end{bmatrix} \tag{3.163}$$

$$= \begin{bmatrix} \mathbf{0} & \mathbf{0} \\ \overline{P}_{oo}(t)\mathbf{Q}_{oo} & \overline{P}_{oo}(t)\mathbf{Q}_{oc} \end{bmatrix}$$

and, since $\overline{P}_{oo}(t) = \exp(\mathbf{Q}_{oo}t)$,

$$f_{ij}(t) = F_{ij}'(t) = (\exp(\mathbf{Q}_{oo}t)\mathbf{Q}_{oc})_{ij}. \tag{3.164}$$

Let $\eta_i = \mathbf{P}$(open aggregate entered via open state i). Then the density of T, called f_o, is given by

$$f_o(t) = \sum_{i \in O} \sum_{j \in C} \eta_i f_{ij}(t) = \boldsymbol{\eta} \exp(\mathbf{Q}_{oo}t)\mathbf{Q}_{oc}\mathbf{1}^T \tag{3.165}$$

where $\boldsymbol{\eta}$ is the vector of η_i and $\mathbf{1}$ is a vector of N_c ones. In order to compute f_o explicitly we need to do some linear algebra. Every symmetric matrix A (so $A^T=A$) can be diagonalized by an orthogonal matrix B (so $B^TB=\mathbb{I}$ and B is invertible with $B^{-1}=B^T$) in the sense that

$$B^{-1}AB = D \tag{3.166}$$

where D is a diagonal matrix with the eigenvalues of A on the diagonal. Equivalently we can write A in the form

$$A = BDB^{-1}. \tag{3.167}$$

An advantage with diagonalization is that matrix powers become very easy to compute:

$$A^n = BDB^{-1} \times BDB^{-1} \times \cdots \times BDB^{-1} = BD^n B^{-1} \tag{3.168}$$

where D^n is diagonal with diagonal elements d_i^n. In particular, matrix exponentials can be computed explicitly:

$$\begin{aligned} \exp(At) &= \sum_{k=0}^{\infty} (At)^k/k! = \sum_{k=0}^{\infty} B(Dt)^k B^{-1}/k! \\ &= B \sum_{k=1}^{\infty} (Dt)^k/k!\, B^{-1} = BE(t)B^{-1} \end{aligned} \tag{3.169}$$

where $E(t)$ is a diagonal matrix with diagonal elements $\exp(d_i t)$. Hence if $\mathbf{Q}_{oo}$ were symmetric, it would be straightforward to compute f_o. Unfortunately, $\mathbf{Q}_{oo}$ is not, in general, symmetric. But we are considering a closed chemical system, so the law of detailed balance holds (cf. the discussion in section 2.4). Thus there is a stationary distribution $\boldsymbol{\pi}$ satisfying

$$\pi_i Q_{ij} = \pi_j Q_{ji} \tag{3.170}$$

for all i and j. Let $\Pi = \text{diag}(\pi_i)$. Then the equilibrium condition (3.170) can be written

$$\Pi \mathbf{Q} = \mathbf{Q}^T \Pi. \tag{3.171}$$

Let $\Pi^{1/2} = \text{diag}(\pi_i^{1/2})$. Pre- and post-multiplying (3.171) by $\Pi^{-1/2}$ we see that

$$\Pi^{1/2} Q \Pi^{-1/2} = \Pi^{-1/2} \mathbf{Q}^T \Pi^{1/2} = (\Pi^{1/2} \mathbf{Q} \Pi^{-1/2})^T \tag{3.172}$$

so $\Pi^{1/2} \mathbf{Q} \Pi^{-1/2}$ is symmetric, and therefore diagonalizable. In particular, writing

$$\Pi = \begin{bmatrix} \Pi_c & \mathbf{0} \\ \mathbf{0} & \Pi_o \end{bmatrix}, \tag{3.173}$$

we see that

$$\begin{bmatrix} \Pi_c^{1/2} & \mathbf{0} \\ \mathbf{0} & \Pi_o^{1/2} \end{bmatrix} \begin{bmatrix} \mathbf{Q}_{cc} & \mathbf{Q}_{co} \\ \mathbf{Q}_{oc} & \mathbf{Q}_{oo} \end{bmatrix} \begin{bmatrix} \Pi_c^{-1/2} & \mathbf{0} \\ \mathbf{0} & \Pi_o^{-1/2} \end{bmatrix} = \begin{bmatrix} \Pi_c^{1/2} \mathbf{Q}_{cc} \Pi_c^{-1/2} & \Pi_c^{1/2} \mathbf{Q}_{co} \Pi_o^{-1/2} \\ \Pi_o^{1/2} \mathbf{Q}_{oc} \Pi_c^{-1/2} & \Pi_o^{1/2} \mathbf{Q}_{oo} \Pi_o^{-1/2} \end{bmatrix}. \tag{3.174}$$

Since $\Pi_o^{1/2} \mathbf{Q}_{oo} \Pi_o^{-1/2}$, in particular, is symmetric, it is diagonalizable by an orthogonal matrix B. Consequently $\mathbf{Q}_{oo}$ is diagonalizable using the matrix $\Pi_o^{-1/2} B$, so we can write

$$\mathbf{Q}_{oo} = \Pi_o^{-1/2} BD (\Pi_o^{-1/2} B)^{-1} = \Pi_o^{-1/2} BDB^{-1} \Pi_o^{1/2}. \tag{3.175}$$

Inserting this into equation (3.169) we see that

$$\exp(\mathbf{Q}_{oo} t) = \Pi_o^{-1/2} BE(t) B^{-1} \Pi_o^{1/2}. \tag{3.176}$$

Finally returning to (3.165) we can write

$$f_o(t) = \boldsymbol{\eta}\, \Pi_o^{-1/2} BE(t) B^{-1} \Pi_o^{1/2} \mathbf{Q}_{oc} \mathbf{1}^T. \tag{3.177}$$

The important thing about this formula is that it is a certain linear combination of the elements of $E(t)$, i.e.,

$$f_o(t) = \sum_{i=1}^{N_o} \alpha_i \exp(-\beta_i t) \tag{3.178}$$

for some α_i, β_i. In other words, the dwell time in the open state is distributed as a mixture of exponential distributions. If we knew N_o (as in the model depicted in Figure 3.12) it would be relatively straightforward to estimate the parameters $\boldsymbol{\alpha}$ and $\boldsymbol{\beta}$ using maximum likelihood. If, on the other hand, we do not know N_o, we can perform the estimation for a sequence of possible values of it. We then see which is the smallest N_o that adequately describes the data, e.g., by computing a χ^2-statistic for goodness of fit. This way of estimating N_o gives some information about the number of open states in the system. Since some of the α_i may be zero, this estimate is actually only a lower bound to the number of open states.

Example (The states of an ACh receptor) Using 163 observed dwell times in open states, obtained from applying a voltage of +100 mV to a planar lipid bilayer containing purified ACh receptor, activated by ACh at the concentration of 50 μM, we get the results in Table 3.3.

Table 3.3 Goodness of fit for different values of N_o

N_o	df	χ^2	P-value	i	α	$-1/\lambda$
1	161	393.8	0	1	.0030	48.7
2	159	201.9	.01	2	.0977	5.83
3	157	138.8	.88	3	.8408	.65

We conclude that there are at least 3 open states. □

The same type of argument shows that the two-dimensional densities are

$$\begin{aligned} f_{oc}(t,s) &= \pi_o \exp(\mathbf{Q}_{oo}t)\mathbf{Q}_{oc}\exp(\mathbf{Q}_{cc}t)\mathbf{Q}_{co}u_o \\ &= \sum_{i=1}^{N_o}\sum_{j=1}^{N_c} \alpha_{ij}\exp(\lambda_i t+\omega_j s) \end{aligned} \tag{3.179}$$

and

$$f_{co}(s,t) = \sum_{i=1}^{N_c}\sum_{j=1}^{N_o} \beta_{ji}\exp(\lambda_i t+\omega_j s). \tag{3.180}$$

We can check the Markov assumption by estimating parameters in the one-dimensional distributions, and then compare the observed and predicted two-dimensional distributions.

In some cases the successive dwell times are independent. Two examples are shown in Figure 3.13. The basic idea is that there is only one closed state that connects directly to any of the open states, so an open dwell time is independent of the preceding closed one (since we know which is the gateway between the two classes of states).

$$C+2L \quad \longleftrightarrow \quad CL + L \quad \longleftrightarrow \quad OL+L \quad \longleftrightarrow \quad OL_2$$

$$C+2L \quad \longleftrightarrow \quad CL + L \quad \longleftrightarrow \quad OL+L$$
$$\searrow \qquad \qquad \nearrow$$
$$OL_2$$

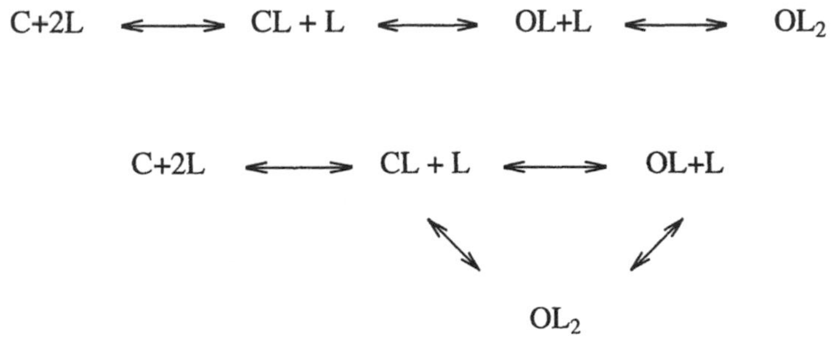

Figure 3.13.   Two models with independent dwell times in open and closed states.

The following definitions will be useful. A **graph** is a collection of **nodes**, with certain pairs of nodes connected by **edges**. The graph is **connected** if there is an edge-path between any two nodes. If we think of the state space as a graph with states as nodes, and edges between nodes for which a transition is possible, a **gateway state** is a state such that if it is removed, the graph is disconnected into two components, one with open states and one with closed states. Then the successive dwell times are independent, since the resulting joint density factors: only one ω_j combines with the λ_i. Thus, looking at the covariance between successive dwell times, we may be able to rule out all models with a gateway state. In fact, a tedious computation shows that if S_i are closed and T_i open dwell times we have that

$$\Gamma_c(k) = \mathbf{Cov}(S_i, S_{i+k}) = \sum_{i=1}^{M-1} u_i \kappa_i^{|k|} \tag{3.181}$$

and

$$\Gamma_o(k) = \mathbf{Cov}(T_i, T_{i+k}) = \sum_{i=1}^{M-1} v_i \sigma_i^{|k|} \tag{3.182}$$

where M is the minimum of the number of open entry states, open exit states, closed entry states, and closed exit states.

Example (Autocorrelation function for an ACh receptor) The autocorrelation function, i.e., the correlations $\Gamma_o(k)/\Gamma_o(0)$ as a function of k, for the open aggregate of the ACh receptor channel, based on 1600 openings with ACh concentration 50 μM and voltage +100 mV, is shown in Figure 3.14.

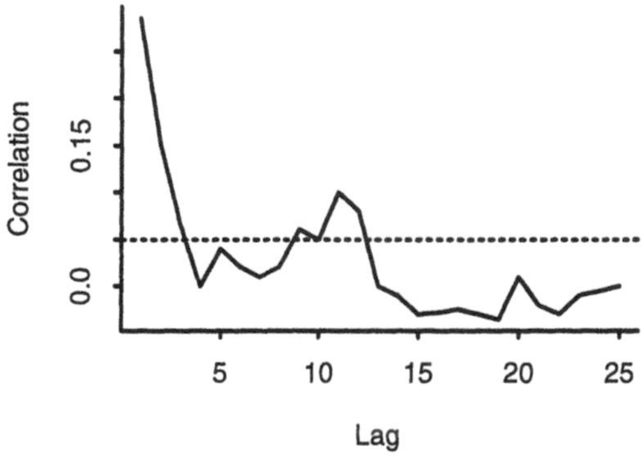

Figure 3.14. Autocorrelation function for the open aggregate of the ACh receptor. Adapted from Labarca et al. (1984); with permission from the Biophysics Society.

The dashed line in the figure corresponds to twice the standard error under the assumption of independence (white noise). The estimate is clearly significantly different from zero, and therefore there must be at least two entry/exit states through which the open and closed aggregates communicate. □

3.9. Blood formation in cats

The bone marrow of humans and other vertebrata contains a relatively small number of very remarkable cells, the **hematopoietic stem cells**. These cells produce by replication and differentiation all the different kinds of blood cells: red cells which transport oxygen throughout the body; white cells which form the immune defense; and platelets which initiate clotting. It has proven very hard to isolate stem cells from the bone marrow. To maintain a reserve for the lifetime of the animal, the stem cell must be able to self-replicate. Also, to supply cells for development into mature blood cells, stem cells must be able to produce differentiated cells. When an uncommitted cell divides it may replicate itself and/or differentiate and become the head of a sequence of cell divisions leading to mature blood cells. We assume that the regulation of the stem cell is similar. As differentiation proceeds, cells become committed to particular developmental pathways. This complex process of specialization of stem cells into mature blood cells is called **hematopoiesis**. Near the beginning of these pathways we find totipotent cells, i.e., cells that can produce all the different types of mature blood cells. Further along the path a partially committed cell (called a **progenitor** cell) may still be able to produce several different types of mature cells. Apparently the capacity to produce different types of cells is lost randomly along the pathway until only one type is produced in successive cell divisions. It is impossible to distinguish morphologically stem cells from progenitor cells.

There are two different theories regarding the kinetics of early hematopoiesis. In both, a large supply of stem cells is postulated. These stem cells represent the only source for mature blood cells, and must last the lifetime of the animal. One theory, which we may call the standard theory, is that the entire supply of stem cells is proliferating (actively dividing), perhaps with a very slow rate of division. Thus, at any time all stem cells contribute to hematopoiesis. The other, Kay's (1965) theory of **clonal succession**, hypothesizes that most stem cells are inactive, and that at any time a small number are proliferating. This theory postulates that the stem cells have a finite lifespan, at the end of which they are replaced by previously dormant stem cells (see Brecher et al. (1986) for a discussion). Evidence of clonal succession may have important implications for research on cancer treatments, bone marrow transplantation, and gene transfer methods.

Much of the work on stem cells, particularly the development of the standard theory of hematopoiesis, has been done using experiments on mice. Since the lifetime of mice is relatively short, it is possible that this evidence may be spurious. Single cells (stem cells) from one mouse can maintain hematopoiesis in another throughout its life time. Retrovirally marked stem cells can contribute to hematopoiesis through 2–3 serial transplants. However, this does not by itself prove that stem cells have very long life times, only that these life times are long compared to the life time of a mouse. For this reason, Abkowitz et al. (1988, 1990, 1993) embarked on a series of experiments using female Safari cats: a cross between the domestic cat (*Felis catus*) and the South American Geoffroy wild cat (*Leopardus geoffroyi*).

Although stem cells cannot be isolated in the laboratory, it is possible to analyze different types of progenitor cells. In female cells only one of the X-chromosomes, chosen at random early in embryogenesis, remains active. Hence each cell can be classified as domestic type or Geoffroy type according to which chromosome remains active. The X-linked enzyme glucose phosphate dehydrogenase (G6PD) forms a neutral genetic marker for each cell: the G6PD in each cell is determined from the active X-chromosome. G6PD types are determinable by electrophoresis. The label is conserved through replication and differentiation, so that the G6PD type is the same for a stem cell and all its progeny (the **clone** of the stem cell). Measurements of the proportion of domestic cells were done among colonies grown from progenitor cells sampled from the marrow of the experimental animals. Typically there were 50–100 such colonies grown, so that the G6PD type of 50–100 progenitors could be determined.

Normal cats provided no evidence for or against the clonal succession hypothesis. The model given below indicated that a relatively large number (more than 40) of stem cells were operating, so that small changes in the number of active domestic stem cells would have little influence on the observed proportions among progenitor cells (Figure 3.15).

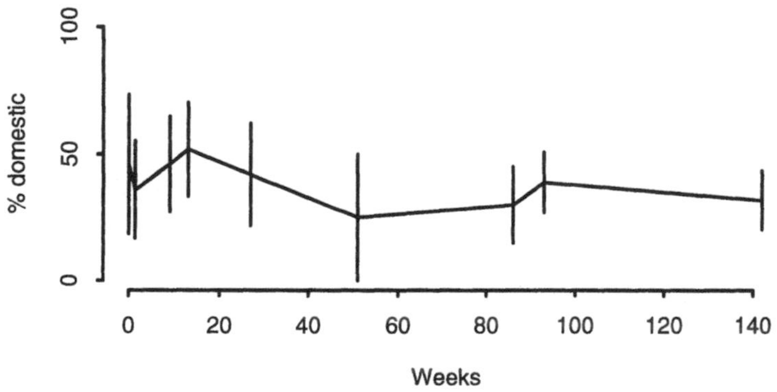

Figure 3.15. Observed proportions of domestic progenitor cells for a normal cat. The vertical lines are pointwise binomial confidence bands. Adapted from Guttorp et al. (1990), published by the Institute of Mathematics and Its Applications.

In order to obtain a more focused view of the system in operation, autologous bone marrow transplantations were done. Some bone marrow cells were extracted, the subject was irradiated to kill all remaining marrow, and only very few of the reserved cells were returned. As before the transplantations, percentages of domestic type committed progenitor cells were recorded every 2–3 weeks following the recovery of normal blood cell counts. Hematopoiesis was operating as before, but G6PD analysis suggested that much fewer cells were responsible for maintaining normal blood production. Figure 3.16 shows the data for one of the treated cats.

The complex processes which control the differentiation and amplification of hematopoietic cells cannot be determined from the G6PD data alone. Any model designed to learn something from these data must necessarily ignore some details, while at the same time be rich enough to account for most of the structure of the cell kinetics of early hematopoiesis. The first simplifying assumption is to consider the cells of interest as falling into one of two compartments: a stem-cell compartment, denoted C_1, and a progenitor-cell compartment, denoted C_2. The essential modeling problems are well illustrated in Figure 3.17 which is an idealized view of the progenitor-cell compartment at two time points. In a real system, many thousands of cells inhabit C_2, although only about 30 or so are shown in this idealization. At time 1, C_2 is composed of three clones; that is the offspring from three stem cells released into C_2 from

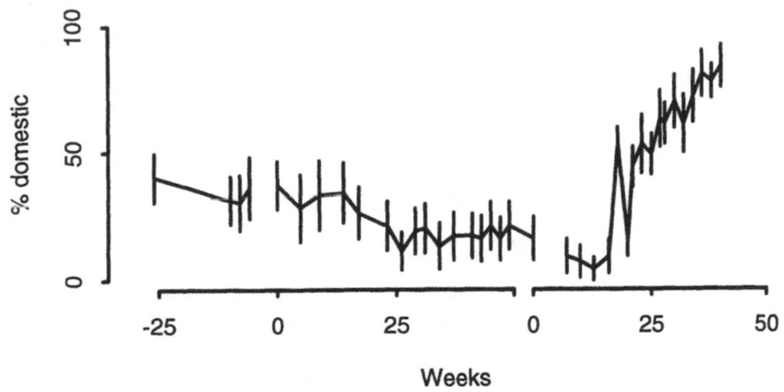

Figure 3.16. Observed proportions of domestic progenitor cells for a cat with two autologous bone marrow transplantations. Negative time is before the first transplant. Time starts over at the time of the second transplant. Adapted from Guttorp et al. (1990), published by the Institute of Mathematics and Its Applications.

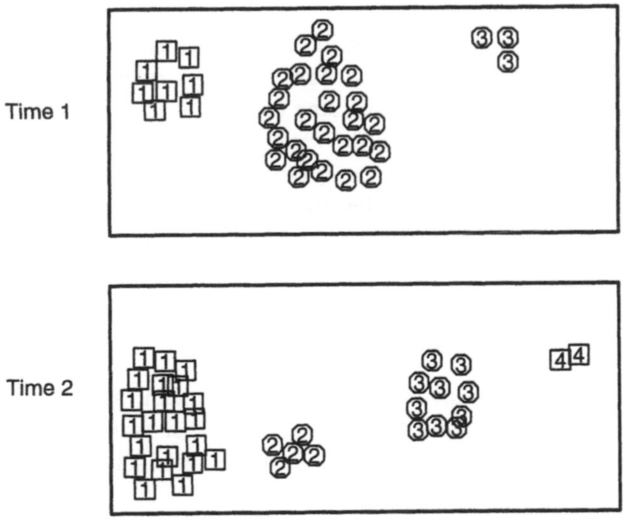

Figure 3.17. An idealization of the progenitor-cell compartment: each number corresponds to a progenitor cell. Cells with the same number are in the same clone. Boxes represent cells having the domestic type G6PD, and circles cells having Geoffroy type G6PD.

C_1. In this schematic, clone 2 dominates the pool, having more cells than clones 1 and 3 combined. For the cells sampled at time 1, the G6PD type (square denoting domestic and circle Geoffroy) is recorded. Later, at time 2, the composition of C_2 has changed—this evolution being driven by several factors:

- expansion of a clone (e.g., clones 1 and 3)
- terminal differentiation (e.g., most of clone 2)
- release of new stem cells (e.g., clone 4)

We refer to the lifetime of a clone as the time from release of its stem-cell ancestor until terminal differentiation of its constituent cells.

One limitation of the data is that the clone of a cell cannot be identified. For example, cells from clone 2 cannot be distinguished from cells of clone 3, given the binary nature of the marker. If clonal succession explains cellular development, then we would expect the clones to be relatively few in number, and to last a relatively short amount of time. If the standard theory is more appropriate, then clones would have a relatively longer lifetime.

If the first compartment, C_1, has a large number of dormant or self-replicating cells, it is reasonable to assume that the proportion of domestic type cells in this compartment is constant over time. We denote this proportion by p_d.

As a dynamic process, the number of clones composing C_2 may fluctuate because of terminal differentiation and new stem-cell release. A balance is expected if the process is stable. We assume (at least as a first approximation) that the total number of clones, denoted N, stays constant over time. A number $X(t)$ of these N clones are of the domestic type at time t. The fraction $X(t)/N$ fluctuates between 0 and 1 because a depleted clone may be replaced by a clone having a different G6PD type. Let $X(t)$ be a continuous-time, finite-state, birth and death process. The transition intensities are

$$\begin{array}{l} x \to x+1 \text{ with intensity } (N-x)\lambda p_d \\ x \to x-1 \text{ with intensity } x\lambda(1-p_d) \end{array} \tag{3.183}$$

This continuous-time process induces a finite Markov chain $(X_1, X_2, \ldots, X_m)$ by restriction to the sampling times $t_1, \ldots, t_m$. Because sampling times are unequal, the transition probabilities for this chain are not stationary. They can be written

$$\mathbf{P}(X_i = k \mid X_{i-1} = j) = \sum_{l=\max(0, j+k-N)}^{\min(j,k)} \binom{j}{l} \binom{N-j}{k-l} \times p_{00}^{N-j-k+l} p_{01}^{k-l} p_{10}^{j-l} p_{11}^{l} \tag{3.184}$$

where, writing $\mu=\exp(-\lambda)$ and $\Delta t_i=t_i-t_{i-1}$, we have

$$\begin{aligned} p_{00} &= \mu^{\Delta t_i}+(1-\mu^{\Delta t_i})(1-p_d), \\ p_{01} &= (1-\mu^{\Delta t_i})p_d, \\ p_{10} &= (1-\mu^{\Delta t_i})(1-p_d), \\ p_{11} &= \mu^{\Delta t_i}+(1-\mu^{\Delta t_i})p_d. \end{aligned} \tag{3.185}$$

The number $X(t)$ of domestic type clones in C_2 influences the number of domestic type cells in this progenitor pool. The proportion p_t of domestic type cells in C_2 differs from $X(t)/N$ because all the clones do not have exactly the same number of cells at every point in time. Intuitively, we expect p_t to equal $X(t)/N$ on average, and in fact for certain models the fluctuations of p_t can be quantified. A first approximation is to assume that these fluctuations are negligible, and under this assumption the observation distribution is

$$(Y_i \mid X(t_i),N) \sim \mathrm{Bin}(n_i, \frac{X(t_i)}{N}) \tag{3.186}$$

where Y_i, n_i, and t_i are as defined above. This model is a continuous time **hidden Markov chain**, very much like the discrete time hidden Markov chain in the previous chapter.

Evaluation of the likelihood function L at a given parameter triple $\theta=(N,\lambda,p)$ is a nontrivial exercise because the model probabilities are specified in terms of an unobservable process. The likelihood function is

$$L(\theta) = \mathbf{P}_\theta(Y_1=y_1,Y_2=y_2,\ldots,Y_m=y_m) = \mathbf{P}_\theta(Y=y). \tag{3.187}$$

where $y=(y_1,y_2,\ldots,y_m)$ are the observed counts. Because the model is specified in two stages, it is natural to rewrite the likelihood

$$\begin{aligned} L(\theta) &= \sum_{x\in S}\mathbf{P}_\theta(Y=y \mid X=x)\mathbf{P}_\theta(X=x) \\ &= \sum_{x\in S}\prod_{i=1}^{m}\left[\mathbf{P}_\theta(Y_i=y_i \mid X_i=x_i)\mathbf{P}_\theta(X_i=x_i \mid X_{i-1}=x_{i-1})\right] \end{aligned} \tag{3.188}$$

where the summation is over the set S of possible X-values at the sampling times

$$S = \{x=(x_1,x_2,\ldots,x_m) : 0\le x_i\le N\}. \tag{3.189}$$

The cardinality of S is $(N+1)^m$, making an algorithm based on equation (3.188) unviable. A recursive algorithm can be developed by writing the likelihood

$$L(\theta) = \mathbf{P}_\theta(Y_1=y_1)\prod_{i=2}^{m}\mathbf{P}_\theta(Y_i=y_i \mid Y_1=y_1,\ldots,Y_{i-1}=y_{i-1}) \tag{3.190}$$

and noting that each factor in this product can be expanded into a sum over the $N+1$ possible levels of the state at that time. Expansion of the first factor uses the stationary marginal binomial distribution of X_1. The latter factors can be expressed as

$$\mathbf{P}_\theta(Y_i{=}y_i \mid Y_1^{i-1}{=}y_1^{i-1}) = \sum_{j=0}^{N} \mathbf{P}_\theta(Y_i{=}y_i \mid X_i{=}j)\mathbf{P}_\theta(X_i{=}j \mid Y_1^{i-1}{=}y_1^{i-1})$$
$$\equiv \sum_{j=0}^{N} u_i(j;\theta)\, v_i(j;\theta). \tag{3.191}$$

Now compute

$$\mathbf{P}_\theta(X_i{=}j \mid Y_1^{i-1}{=}y_1^{i-1}) = \sum_k \mathbf{P}_\theta(X_i{=}j, X_{i-1}{=}k \mid Y_1^{i-1}{=}y_1^{i-1})$$
$$= \mathbf{P}_\theta(X_i{=}j \mid X_{i-1}{=}k)\mathbf{P}(X_{i-1}{=}k \mid y_1^{i-1}{=}y_1^{i-1}) \tag{3.192}$$

using that the hidden state is Markov. Furthermore, using Bayes' theorem,

$$\mathbf{P}_\theta(X_{i-1}{=}k \mid Y_1^{i-1}{=}y_1^{i-1})$$
$$= \frac{\mathbf{P}_\theta(Y_{i-1}{=}y_{i-1} \mid X_{i-1}{=}k)\mathbf{P}_\theta(X_{i-1}{=}k \mid Y_1^{i-2}{=}y_1^{i-2})}{\sum_{l=0}^{N} \mathbf{P}_\theta(Y_{i-1}{=}y_{i-1} \mid X_{i-1}{=}l)\mathbf{P}_\theta(X_{i-1}{=}l \mid Y_1^{i-2}{=}y_1^{i-2})}. \tag{3.193}$$

Consequently,

$$v_i(j;\theta) = \sum_{k=0}^{N} \mathbf{P}_\theta(X_i{=}j \mid X_{i-1}{=}k)\frac{u_{i-1}(k;\theta)v_{i-1}(k;\theta)}{\sum_{l=0}^{N} u_{i-1}(l;\theta)v_{i-1}(l;\theta)}. \tag{3.194}$$

According to the observation distribution (3.186) the u_i are binomial probabilities. Importantly, equation (3.194) shows that the v_i can be computed recursively for increasing i (and known value of θ) using the past u_j's and v_j's. Thus the likelihood surface can be evaluated recursively. This recursion allows computation of the likelihood at a given parameter value. The entire likelihood function can be approximated by evaluating L on a large grid. Figure 3.18 shows the contours of L for the parameters N and $1/\lambda$ from one experimental cat. The third parameter, p_d, is replaced by an estimate from pre-transplant data. The likelihood is multimodal, and has a fairly unusual shape. Since one of the parameters is integer-valued, the usual likelihood theory does not apply. Instead the contours in the figure are associated with approximate coverage probabilities determined from a **bootstrap calibration** described below.

Having a way to compute the likelihood on a grid, the following question arises. How far down from the mle do you go in likelihood units to have an approximate $100(1-\alpha)\%$ confidence set, for some small number α? This is a question of frequency calibration of the likelihood-based confidence set, which we now address. The set of parameters θ whose likelihood is at least $100(1-r)\%$ of the maximum $L(\hat{\theta})$ is called a $100(1-r)\%$ likelihood region for θ.

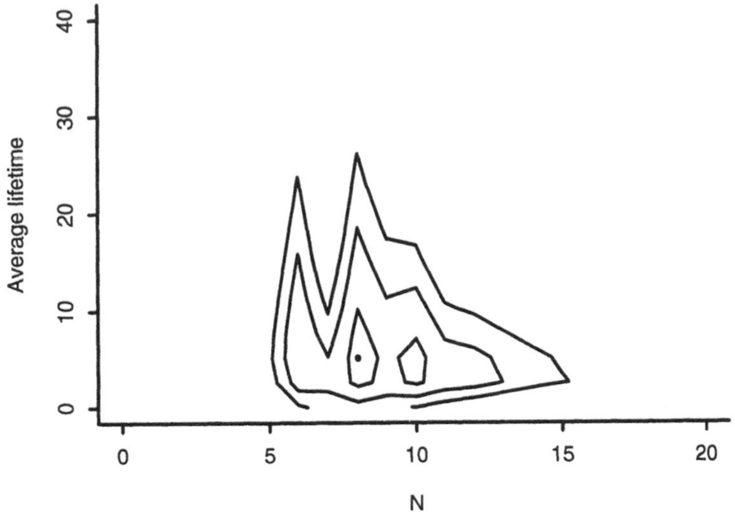

Figure 3.18. Calibrated contours of the joint likelihood of N and $1/\lambda$ for the data from an experimental cat. The contours correspond to 50%, 95%, and 99%, and the maximum is at the dot. Adapted from Guttorp et al. (1990), published by the Institute of Mathematics and Its Applications.

Such a set is

$$\mathrm{LR} = \{\theta : R = \frac{L(\theta)}{L(\hat{\theta})} \geq r\ \}. \tag{3.195}$$

The cutoff r of the likelihood region determines that region's coverage probability, and so it is important to know how to choose r to achieve approximate 95% coverage, say. In general, the coverage also depends on θ_0, and is not knowable for any particular procedure for choosing r. Hence we aim for approximate coverage. Standard theory states that asymptotically $2\log R \sim \chi^2_{\dim(\theta)}$, but, as we have noted, it does not apply here. Without theoretical results on the distribution of R, we perform approximate calibration using the bootstrap.

To apply the bootstrap, it is convenient to modify our notation slightly. Since the likelihood depends on both data and parameters, we use the notation $L(\theta;y)$ for the likelihood determined by an observed time series $y=(y_1, \ldots, y_m)$. Introduce

$$T(y\,;\theta) = 2\log\frac{L(\theta;y)}{L(\hat{\theta}(y);y)} \tag{3.196}$$

as the random variable whose distribution we want to determine using the bootstrap. The bootstrap algorithm starts with a model fit, that is θ is estimated from the data y by the maximum likelihood estimate $\hat{\theta}(y)$. (This involves the recursive updating procedure and a grid search.) The next step is to mimic the sampling process via simulation. On the computer, generate *nboot* time series $z^1, z^2, \ldots, z^{nboot}$; each z^j is a series like y. Generation of each z^j takes two steps. First a hidden state X^* is generated by running continuous time Markov process with parameters determined by the fitted model. Then, restriction to the fixed sampling times gives us the binomial success probabilities of the observation distribution which are used to get z^j. For each bootstrapped series z^j, the log likelihood ratio statistic is computed;

$$T_j = T(z^j; \hat{\theta}(z^j)). \tag{3.197}$$

Note that the entire maximum likelihood computation has to be redone for every bootstrapped series z^j. The empirical distribution of the T_j's converges to the bootstrap distribution of T as *nboot* increases. This bootstrap distribution is used to determine the cutoff r for the likelihood-based confidence set.

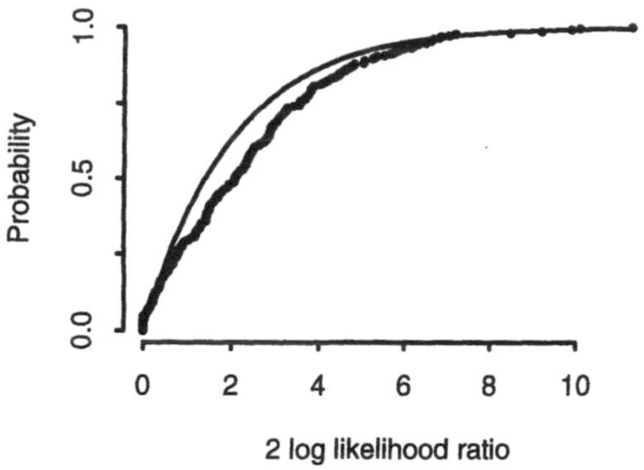

Figure 3.19. Estimated distributions of the log likelihood ratio statistic: The solid curve is the cdf of a χ_2^2 random variable, and the dots determine the empirical distribution function of 240 bootstrapped statistics.

Figure 3.19 compares the cdf of a χ_2^2 random variable to the empirical distribution function of 240 T_j's computed using the bootstrap algorithm described above. The bootstrapped T_j are stochastically larger than a χ_2^2 random variable, which means that confidence sets produced by bootstrap calibration are larger than those produced appealing to standard theory.

The scientific conclusion (Abkowitz et al., 1990) from this analysis (as well as the analysis of two other treated cats) was that clonal succession is the mechanism of blood production, as evidenced by the short stem cell lifetimes of the treated animals.

3.10. Bibliographic remarks

The description of the various models for cosmic radiation follows Arley (1943). Hooper and Scharff (1958) is an introduction to the physics of cosmic radiation. My main sources for general theory have been Freedman (1983) and Bhattacharaya and Waymire (1990). Asmussen (1987) is a good source for queueing material.

The material on inference has Adke and Manjunath (1984) as source, while the application to baboon data was inspired by Keiding (1977). The work by John Rice and David Brillinger on neurophysiology has been of interest to me for two decades by now. Most of the material in section 3.8 is from Labarca et al. (1984) and Fredkin et al. (1985).

Janis Abkowitz got me interested in hematopoiesis, and section 3.9 builds largely on joint work with her and Michael Newton. A general review of what is known about stem cells is Golde (1991).

3.11. Exercises

Theoretical exercises

1. Prove Lemma 1.

2. For the Pólya birth process, verify equation (3.14).

3. Prove Lemma 3.3.

4. Let $X(t)$ be a continuous time Markov chain with stationary transition probabilities. Determine the stationary distribution of the jump chain, and compare it to the stationary distribution of $X(t)$.

5. Find the stationary distribution of a linear immigration-death process.

6. Derive the stationary distribution of an M/M/1 queue.

7. Consider a G/M/1 queue, and let Y_n be the number of individuals in the queue at the time of the nth arrival. Let D_n be the number of departures between the nth and $(n+1)$th arrivals.

(a) Show that $Y_{n+1}=Y_n+1-D_n$.

(b) Show that Y_n is a Markov chain, and determine its transition matrix.

8. For a linear immigration-death process, find the asymptotic distribution of $\hat{\nu}/\hat{\kappa}$.

Hint: Do a bivariate Taylor expansion of the function $f(x,y)=x/y$, and use the asymptotic distribution of $(\hat{\nu},\hat{\kappa})$.

9. Consider a queue with $\lambda_n=\lambda$ and $\mu_n=\mu$, but in which the queue can become negative. An example is a taxi queue, where taxis waiting with no passengers has a positive queue length, while passengers waiting with no taxis corresponds to a negative queue length. Both passengers and taxis arrive according to a Poisson process. Loading time is negligible. Derive the stationary distribution of queue length.

10. Show that the discrete skeleton of a linear BID-process is a Bienaymé-Galton-Watson process with immigration.

11. Show that the jump chain of a linear birth-death process is a random walk.

12. Elephants wander around in herds of different sizes. It can happen that when two herds meet, they amalgamate to form a single herd. It is of course also possible that a herd may break in two. Holgate's (1967) model for this is the following. Let $X(t)$ be the number of herds at time t, and assume that the total number K of elephants is constant. If there are j herds, the probability that two merge within the next h time units is $(j-1)\mu h+o(h)$, and the probability that one herd will split off is $(K-j)\lambda h+o(h)$. Thus $Q_{i,i+1}=(K-i)\lambda$, $Q_{i,i-1}=(i-1)\mu$, and $Q_{ii}=-(K-i)\lambda-(i-1)\mu$. We are interested in estimating the unknown parameter $\boldsymbol{\theta}=(\lambda,\mu)$.

(a) Show that the log likelihood is

$$l_T(\lambda,\mu) = \sum_1^{K-1} N_T(i,i+1)\log((K-i)\lambda) + \sum_2^{K} N_T(i,i-1)\log((i-1)\mu)$$
$$- \sum_1^{K} A_T(i)((K-i)\lambda + (i-1)\mu)$$

where $A_T(i)=\int_0^T 1(X(t)=i)dt$.

(b) Let U_t be the number of jumps up before time t, and D_t the number of jumps down. Show that

$$\hat{\mu}_T = \frac{U_T}{\sum_1^{K-1}(K-i)A_T(i)} \quad \text{and} \quad \hat{\lambda}_T = \frac{D_T}{\sum_2^{K}(i-1)A_t(i)}.$$

(c) Find the asymptotic covariance matrix for the mle.

13. Let $X(t)$ be a Poisson process with rate λ.

(a) Show that, given $X(t)=n$ the jump times $\tau_1,\ldots,\tau_n$ have the same distribution as an ordered sample of size n from the uniform distribution on $(0,t)$.

(b) Now assume that the rate depends on time, so $\lambda=\lambda(t)$. Show that the jump times now are distributed as an ordered sampled of size n from a distribution with cdf $F(x)=\Lambda(x)/\Lambda(t)$ where $\Lambda(x)=\int_0^x \lambda(u)du$.

(c) Let $X(t)$ be as in part (b). Show that $Y(t)=X(\Lambda^{-1}(t))$ is a unit rate Poisson process, so a nonhomogeneous Poisson process is obtained from a homogeneous one by letting the speed with which time goes by vary.

14. Using the results from the previous question, suppose we observe a Poisson process and wish to test the hypothesis that the rate is constant against the hypothesis that it is monotonically increasing with time. A test statistic suggested in the reliability literature (e.g., Barlow et al., 1972, Chapter 6) is the **total time on test** statistic, given by

$$V_n = \sum_{i=1}^{n} \tau_i/\tau_n.$$

Show that the conditional distribution of V_n, given that $\tau_n=t$, is that of the sum of $n-1$ iid uniform random variables, and deduce that V_n under the null hypothesis has approximately a normal distribution. Would you reject for small or large values of V_n?

15. Consider a queue with input $X(t)$ following a nonhomogeneous Poisson process with rate $\lambda(t)$, an infinite number of servers, and service time distribution with density $f(t)$.

(a) Show that the output $Y(t)$ from this queue is a nonhomogeneous Poisson process with rate $\gamma(t)=\int_0^t \lambda(t-u)f(u)du$.

(b) Now suppose that $\lambda(t)\equiv 1$. Show that the mean of the output process is $m(t)=\mathbf{E}Y(t)=\int_0^t F(x)dx$, where F is the cdf of the service time distribution.

(c) Let $S_i=m^{-1}(\tau_i)$. Show that the events S_i are a Poisson process with rate

$$\gamma^T(t) = \frac{\gamma(m^{-1}(t))}{F(m^{-1}(t))}.$$

16. In the context of Exercise 15, suppose that we wish to test the hypothesis that $\lambda(t)=\lambda$ against the hypothesis that $\lambda(t)$ is an increasing function of t, based on observations of the output process for a known service time distribution f. Show that this testing problem is equivalent to testing $\gamma^T(t)=\lambda$ against $\gamma^T(t)$ being an increasing function. In other words, the test based on the output times of one process is equivalent to a test based on the input times of a related process, such as that in Exercise 14.

17. Let $X(t)$ be a Poisson process with rate

$$\lambda(t) = \begin{cases} 1+\theta t^{-1/2} & t\geq 1 \\ 0 & 0\leq t\leq 1 \end{cases}.$$

This is called a **weak trend** (Brown, 1972).

(a) Derive the likelihood equation for estimating θ based on observing $X(t), t\leq T$, and comment on its solution.

(b) A simpler estimate is the moment type estimator given by the solution to the equation $X(T)=\mathbf{E}_\theta X(T)$. Show that it is unbiased but inconsistent.

18. A system contains j energy quanta. It can lose one (if $j\geq 0$) with intensity μ, or gain one with intensity λ.

(a) Show that the equilibrium distribution is a **Gibbs distribution** of the form $\pi_j \propto \exp(-\beta j)$.

(b) Now suppose that at energy level j there are several states between which the process moves micro-reversibly, so that $Q(x,x')=Q(x',x)$ if x and x' have the same energy, and $\lambda Q(x,x')=\mu Q(x',x)$ if the energy $E(x')=E(x)+1$. Show that this system also has equilibrium distribution of the Gibbs form

$$\pi(x) \propto \exp(-\beta E(x)).$$

19. Consider a telephone exchange with k lines. When all lines are busy, any incoming telephone call is lost. Assume that calls arrive according to a homogeneous Poisson process, and that their lengths are exponentially distributed.

(a) Find the stationary distribution of the chain.

(b) Find the probability that an incoming call is lost (Erlang's loss formula).

20. Let

$$\mathbf{Q}_1 = \begin{pmatrix} -1 & 1 & 0 \\ 0 & -1 & 1 \\ 1 & 0 & -1 \end{pmatrix} \quad \text{and} \quad \mathbf{Q}_2 = \begin{pmatrix} -1 & ½ & ½ \\ ½ & -1 & ½ \\ ½ & ½ & -1 \end{pmatrix}$$

be the intensity matrices of two Markov processes. Show that their discrete skeletons to scale $4\pi/\sqrt{3}$ are identical.

21. Consider the autocatalytic reaction

$$A + X \underset{k'}{\overset{k}{\leftrightarrow}} 2X$$

where the catalyst A is abundant, so that we can assume that it has constant concentration c_A. The number of X-molecules can be viewed as a birth-death process, where the birth rate is $\lambda_n = kc_A n$ and the death rate is $\mu_n = k'n(n-1)$. Find the stationary distribution for this reaction.

22. Consider a Markov chain with nonstationary intensity of the form $q_n(t)$. Let $\lambda(t)$ be a continuous increasing function of time, and consider changing the time scale using $\lambda(t)$, so that we observe $Y(t)=X(\lambda(t))$.

(a) Show that Y has intensity $q_n(\lambda(t))\lambda'(t)$.

(b) Consider a Pólya process observed using the time change $\lambda(t)=-\log P_0(t)$. Show that the resulting process is a (nonhomogeneous) Poisson process.

23. Let $X(t)$ be a linear birth and death process starting from 1 individual.

(a) Show that $\mathbf{P}(X(t)=i)=(1-\alpha)(1-\beta)\beta^{i-1}$, $i\geq 1$, where

$$\alpha=\mu\frac{1-\exp((\mu-\lambda)t)}{\lambda-\mu\exp((\mu-\lambda t)} \quad \text{and} \quad \beta=\lambda\frac{1-\exp((\mu-\lambda)t)}{\lambda-\mu\exp((\mu-\lambda t)}.$$

(b) Deduce that if $\mu<\lambda$ the probability of extinction is μ/λ.

(c) Let T_n be the first time to hit n. Show that

$$\mathbf{P}(T_n \leq t) \approx \frac{(1-\alpha)\beta^{n-1}}{1-\mu/\lambda}.$$

24. A simple model for the folding of protein molecules into its native structure is given by a chain of $N+1$ amino acids, with N bonds between them. The connecting bond between two neighboring amino acids can either be characterized as correct (native) or incorrect (non-native). These bonds change over time. An incorrect bond becomes correct at rate k_0, while a correct bond becomes incorrect at rate k_1. Let $X(T)$ denote the number of correct bonds at time t. We are interested in the mean time τ_i until $X(t)=0$, starting from i.

(a) Show that $X(t)$ is a finite birth and death process, and write down its intensity matrix.

(b) Show that $\boldsymbol{\tau}=(0,\tau_1,\ldots,\tau_N)$ satisfies $\boldsymbol{\tau}\mathbf{Q}=-\mathbf{1}$.

(c) Verify that

$$\tau_i = \frac{1}{Nk_0}\sum_{n=0}^{i-1}\binom{N-1}{n}^{-1}\sum_{m=n+1}^{N}\binom{N}{m}K^{m-n}$$

where $K=k_0/k_1$.

(d) Statistical thermodynamics suggests that $K=\nu\exp(-U/kT)$ where $\nu+1$ is the number of possible bonds and U is an energy penalty for making an incorrect bond. Here k is Boltzmann's constant and T the absolute temperature. Show that for large N and k_0 not too small we have, approximately, that

$$\tau_i \approx \frac{1}{Nk_0}(1+\nu\exp(-U/kT))^N$$

regardless of the initial state i.

Remark: The case $U=0$, or a completely random bond replacement strategy, yields an enormously long mean folding time, while a search strategy biased towards correct bonds yields much more realistic folding times (Zwanzig et al., 1992).

Computing exercises

C1. Given an exponential random number generator, and a function to simulate a discrete time Markov chain, show how to generate a continuous time Markov chain.

C2. How would you solve the stationary distribution equation $\mathbf{Q}\boldsymbol{\pi}=\mathbf{0}$?

C3. Given a **Q**-matrix of your choice, how big a sample do you need to estimate it accurately? How does the sample size depend on the dimensionality of **Q**?

C4. Generate a linear BIDE-process with parameters given by the fit to the baboon data from page 157 and look at successive stretches of 373 days. Assess the probability that any such stretch lies entirely above its stationary mean.

C5. (a) Show that equation (3.100) is monotone as a function of m.

(b) The design of a service station is a compromise between the cost of decreasing the queue length (by adding servers) and the cost of additional idle time for the servers. Study the relationship between these costs.

Data exercises

D1. The data in Table 3.4 are counts of the number of alpha-particles emitted from a piece of polonium in successive 7.5-second intervals (Rutherford and Geiger, 1910).

Table 3.4 Alpha-particle counts

3	7	4	4	2	3	2	0	5	2
5	4	3	5	4	2	5	4	1	3
3	1	5	2	8	2	2	2	3	4
2	6	7	4	2	6	4	5	10	4

These data are part of a larger experiment, from which the average number of particles per minute was found to be 31.0. Investigate the hypothesis that the data come from a homogeneous Poisson process with this rate.

D2. A controversial issue in the study of multiple sclerosis (MS) is whether or not the disease is contagious. Kurtzke and Hyllested (1986) noted the apparently sudden occurrence of MS in the Faroe Islands shortly after the arrival of British troops in 1941. As there had been no cases earlier in the century, they argued that this was an indication of an infectious agent at work. Supposing that MS cases occur following a Poisson process, one would expect that the intensity $\lambda(t)$ would be increasing in time if the disease is infectious, while if the British troops introduced some non-infectious agent precipitating MS, the intensity would be constant.

The data for disease onset in years from 1941 are given in Table 3.5 (actually, the data were only given to nearest year, and have been uniformly redistributed between $j-\frac{1}{2}$ and $j+\frac{1}{2}$; cf. Joseph et al., 1990).

Table 3.5 MS onset for Faroe Island population

2.43	2.62	2.94	3.28	3.92	4.21	4.31	4.66
4.99	5.40	5.60	6.24	6.79	7.01	7.79	8.50
11.19	11.59	12.95	13.85	14.82	15.54	16.90	18.10
18.13	19.32	20.01	23.68	26.79	27.89	29.02	31.92

Assume that the incubation time for the disease (corresponding to the service time in the queueing description) has a Weibull distribution with density

$$f(x) = cx^{c-1}\exp(-x^c)$$

and $c = 1/3$, corresponding to a mean incubation time of 6 years. Using the results in Exercises 14–16, assess the contagion theory for MS based on these data.

D3. Using the baboon data from page 157, use the jump chain of births and deaths/emigrations (ignoring the few immigrations) to test the hypothesis that these form a random walk (cf. Exercise 11).

D4. The data in data set 4 are successive counts of the number of gold particles observed by Westgren (1916, Experiment C) in a fixed small volume of water every 1.39 seconds. Physical theory suggests that these counts should be described by a linear immigration-death process.

(a) Determine a confidence set for the rates in this process.

(b) Test the goodness of fit of the model using the autocorrelation approach of the application on page 100.

D5. Table 3.6 contains data from Sartwell (1950) on the incubation period of streptococcial infection.

Table 3.6 Streptococcial incubation time in 12-hr periods

Incubation period	0–1	1–2	2–3	3–4	4–5	5–6
Number	0	1	7	11	11	7
Incubation period	6–7	7–8	8–9	9–10	10–11	11–12
Number	5	4	2	2	0	1

Assuming, following Morgan and Watts (1980), that the infections started with one bacterium, use the theory developed in Exercise 23 to estimate μ, λ, and the incubation period n (we assume that the incubation period is the time needed to reach a threshold n of bacteria in the organism). Assess the precision of your estimates. The form of the parameters α and β suggest that $\lambda-\mu$ is an important parameter. How well can it be estimated from these data?

D6. The whooping crane is a very rare migratory bird with breeding grounds in Canada's Northwest Territories and wintering grounds in Texas. Miller et al. (1974) give annual counts from 1938–1972 of whooping cranes arriving in Texas in the fall. Using a linear BID model to fit these data, try to distinguish between the hypothesis of a constant breeding population and that of a breeding population proportional to the population size. How would you estimate the extinction probability at the beginning of the observation period? At the end?

Table 3.7 Texas whooping crane counts

14	22	26	15	19	21	18	17	25	31
30	34	31	25	21	24	21	28	24	26
32	33	36	38	32	33	42	44	43	48
50	56	57	56	51					

CHAPTER 4

Markov random fields

When "time" is a spatial index, the random process is called a random field. We generalize the Markov assumption to deal with this situation, and discuss some classes of processes that originated in statistical physics, and have proved quite useful in a variety of applications. The Ising model of ferromagnetism is applied to an agricultural problem. A simple autoregressive model is useful for improving some blurred astronomical pictures, while a hidden Markov random field is applied to the statistical analysis of pedigrees in population genetics.

4.1. The Ising model of ferromagnetism

The magnetization of a permanent magnet diminishes in strength as the magnet is heated. Above a certain critical temperature the magnet stops being a magnet. At the other extreme, for very low temperatures non-magnets may exhibit spontaneous magnetization. At the atomic level, each atom is by itself a small magnet. A material is magnetized when all (or most) of the atoms align magnetically. The interaction between these tiny magnets explains spontaneous magnetization.

Consider N sites (or atoms), ordered linearly, and associate with each site a magnetic dipole (spin) which can be either positive or negative (Figure 4.1).

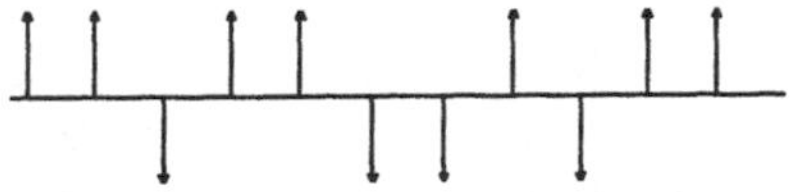

Figure 4.1. Schematic picture of a linear magnet. Adapted from *Markov Random Fields and their Applications,* by Ross Kindermann and J. Laurie Snell, Contemporary Mathematics, Volume 1, by permission of the American Mathematical Society.

The state space is $\{-1,1\}^N$. In order to describe magnetic behavior, Ising (1925) in his PhD dissertation introduced a probability measure in the following fashion. For each configuration $\mathbf{x}=(x_1, \ldots, x_N)$ of N signs, associate an **energy**

$$U(\mathbf{x}) = -J\sum_{i\sim j} x_i x_j - mH\sum_i x_i \tag{4.1}$$

where $i\sim j$ means that sites i and j are neighbors. This assumes that only neighboring sites affect any given site. The constant J describes the material. If $J>0$ we have the **attractive** case: configurations with a lot of sign changes have larger energy. Since nature strives towards the lowest possible energy, an attractive material would tend to be magnetized quite easily. The opposite case is called **repulsive**, and magnetization of such materials is hard. The second term of (4.1) represents the influence of an external magnetic field of (signed) intensity H. The factor m is a property of the material. In the attractive case, energy is minimized when all sites line up (first term) in the direction of the external field (second term). We can write $U(\mathbf{x})=\sum U_i(\mathbf{x})$, where

$$U_i(\mathbf{x}) = -J\sum_{|j-i|=1} x_i x_j - mHx_i \tag{4.2}$$

describes the energy contribution from site i.

To go from energy to probability we use thermodynamic considerations. Gibbs[1] showed that for a system in thermal equilibrium, the probability of finding the system in a state (or configuration) with energy E is proportional to $\exp(-E/kT)$ where T is the absolute temperature, and k is Boltzmann's constant (see Exercise 3.18). In our case, with energy $U(\mathbf{x})$, we obtain the **Gibbs measure**

$$\mathbf{P}(\mathbf{x}) \propto e^{-U(\mathbf{x})/kT}. \tag{4.3}$$

The constant of proportionality is called the **partition function**, and written $Z=\sum_{\mathbf{x}}\exp(-U(\mathbf{x})/kT)$.

Ising was mainly interested in the case of $H=0$, hoping to be able to explain spontaneous magnetization (low energy) at very low temperatures. Let $n_+(\mathbf{x})=\#\{x_i=+1\}$, and let $n_-(\mathbf{x})=N-n_+(\mathbf{x})$. Then the **total magnetization** is $M(\mathbf{x})=n_+(\mathbf{x})-n_-(\mathbf{x})$. If there is spontaneous magnetization, and we start in a random configuration, we ought to have a bimodal density for M since then M would tend to be either large positive or large negative. Suppose that $J=H=0$, so that $U=0$. Then all states have equal probability, i.e., independent assignment of signs to each site. By the law of large numbers, $M_N/N\to 0$ with probability one. If we allow interaction, so $J\neq 0$, but have no external field, so $H=0$, the Gibbs measure can be written

[1]Gibbs, Josiah Willard (1839–1903) American physicist. One of the founders of statistical thermodynamics.

$$\mathbf{P}(\mathbf{x}) = \frac{1}{Z}\exp(\frac{J}{kT}(n_e(\mathbf{x}) - n_o(\mathbf{x}))) \tag{4.4}$$

where $n_e(\mathbf{x})$ are all neighbors with equal magnetic spin (even parity), and $n_o(\mathbf{x})$ are those with opposite spin (odd parity). Thus

$$\mathbf{P}(\mathbf{x}) = \frac{1}{Z}\exp(\frac{J}{kT}(N(N-1) - 2n_o(\mathbf{x}))) \propto \exp(-bn_o(\mathbf{x})). \tag{4.5}$$

Suppose now that we generate spins from the left according to a Markov chain with

$$\mathbb{P} = \begin{bmatrix} p & 1-p \\ 1-p & p \end{bmatrix}. \tag{4.6}$$

Starting with a random sign, the distribution of spins is

$$\mathbf{P}(\mathbf{x}) = ½ p^{n_e(\mathbf{x})}(1-p)^{n_o(\mathbf{x})} = c(\frac{p}{1-p})^{-n_o(\mathbf{x})} \tag{4.7}$$

which, writing $b = \log(p/1-p)$, is of the same form as Ising's Gibbs measure. Hence the Ising model in one dimension is a reinvention of the two-state Markov chain. This also explains why Ising did not pursue his idea any further: since the law of large numbers and the central limit theorem both hold for one-dimensional Markov chains, the resulting magnetization distribution has to be unimodal. Also,

$$\lim_{N\to\infty} \mathbf{P}(x_N=1 \mid x_0=1) = \lim_{N\to\infty} \mathbf{P}(x_N=1 \mid x_0=-1) = \pi_1 = ½ \tag{4.8}$$

since the chain (4.6) is doubly stochastic. Thus the value of the spin at a particular site has little effect on the values of spins far away. But this is not how magnetism at low temperatures work: the effect of changing a spin seems to propagate all over the material. It turns out that the Ising model actually can describe this kind of behavior, but in order to see it one must look at the two-dimensional lattice instead of the one-dimensional one.

4.2. Markov random fields

In the previous section we saw that there is a close relation between Markov chains and Gibbs measures in one dimension. To develop the theory for more than one dimension, we need a version of the Markov property that does not depend on the direction of time. Recall from Exercise 2.1 that the Markov property in one dimension is equivalent to the property that

$$\mathbf{P}(X_k=x \mid X_{k-1}=x_{-1}, \ldots, X_{k-n}=x_{-n}, X_{k+1}=x_1, \ldots, X_{k+m}=x_m) \tag{4.9}$$
$$= \mathbf{P}(X_k=x \mid X_{k-1}=x_{-1}, X_{k+1}=x_1).$$

This definition is symmetric in time. It simply says that the conditional distribution, given everything but the present, just depends on the nearest neighbors. It is this version of the Markov property that can be generalized to any situation, be it a lattice or a graph, where one can talk about neighbors.

Consider a finite state space S, and a connected graph $\mathcal{G}$ with nodes that are labeled by the index set $I=\{1,\ldots,m\}$, $m\leq\infty$. A system $\Delta=\{\delta_i, i\in I\}$ of subsets of $\mathcal{G}$ is a **neighborhood system** if for all $i\in I$

(i) $i\notin\delta_i$

(ii) $j\in\delta_i$ implies that $i\in\delta_j$

On a connected graph we construct a neighborhood system by calling all nodes connected with edges neighbors.

Example (Integer lattices in the plane) There are different ways of introducing a lattice on $\mathbf{Z}^2$. The most common way is the regular **square lattice** shown on the left in Figure 4.2.

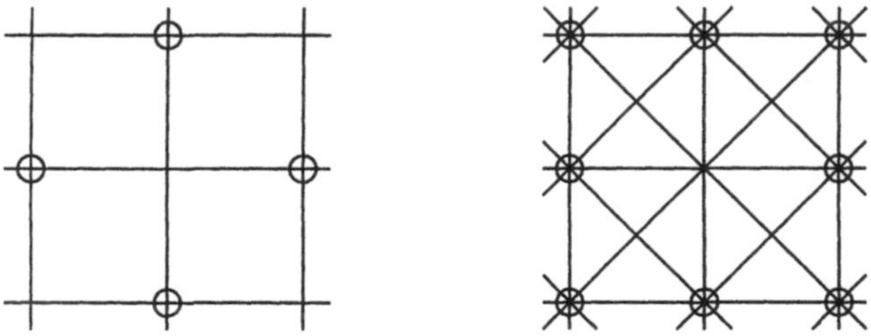

Figure 4.2. The regular integer lattice (left) and the extended integer lattice (right). The circles denote the neighbors of the center node.

Here the neighbors of a node are the nodes directly to the north, east, south, and west of it. The **extended lattice** adds diagonal edges, and thus the neighbors also include the nearest nodes to the northeast, northwest, southeast, and southwest. □

For simplicity we first consider a finite graph. Let $\mathbf{x}_{-i}$ be a configuration (specifying the values at all nodes) on all of $\mathcal{G}$ except the node indexed by i. Using $\mathbf{X}_A$ to denote $\{X_i; i\in A\}$, we say that $\mathbf{X}$ is a **Markov random field** on $\mathcal{G}$ with respect to the neighborhood system Δ if $\mathbf{X}$ is an $S^{\mathcal{G}}$-valued random vector satisfying

$$\mathbf{P}(X_i=x_i \mid \mathbf{X}_{-i}=\mathbf{x}_{-i}) = \mathbf{P}(X_i=x_i \mid \mathbf{X}_{\delta_i}=\mathbf{x}_{\delta_i}) \equiv p_i(x_i \mid x_{\delta_i}) \tag{4.10}$$

where p_i, the **local characteristics** at node i, can be specific to the node, and δ_i is the neighborhood of i. We say that the local characteristics are **stationary** if $p_i\equiv p$. We know from the Brooks expansion (1.20) that there are consistency conditions that conditional probabilities must satisfy, at least if we are dealing with probabilities satisfying the positivity condition. In particular, for each reordering of the nodes of the graph we get a different factorization of $\mathbf{P}(\mathbf{z})/\mathbf{P}(\mathbf{x})$. These factorizations must all be equal.

Remark To produce a consistent definition for infinite graphs one must be careful. We would really like to apply (4.10) directly, but the condition on the left-hand side may be an event of probability zero. Let $\mathbf{P}$ be a positive measure corresponding to an S-valued process $\mathbf{X}$ on a countable graph $\mathcal{G}$. We call $\mathbf{X}$ (or $\mathbf{P}$) a Markov random field on $\mathcal{G}$ with respect to a neighborhood system Δ if (4.10) holds for every finite subset Λ containing i and δ_i, i.e., if $\mathbf{x}$ is a configuration on Λ we have

$$\mathbf{P}(X_i=x_i \mid \mathbf{x}_{-i}) = \mathbf{P}(X_i=x_i \mid \mathbf{x}_{\delta_i}). \tag{4.11}$$

□

Example (Markov chain) We saw in Exercise 2.1 that the defining property of a Markov random field is equivalent to the Markov property on the integers. Hence all Markov chains with stationary transition probabilities are Markov random fields, and vice versa. The nodes are instances of (discrete) time. The neighbors of a node k are $k-1$ and $k+1$. If we denote the local characteristics by $f_{yz}(x)=\mathbf{P}(X_k=x \mid X_{k-1}=y, X_{k+1}=z)$, then in terms of the transition matrix $\mathbb{P}$, we have

$$f_{yz}(x) = \frac{P_{yx}P_{xz}}{(P^2)_{yz}}. \tag{4.12}$$

□

A set C is a **clique** if every pair of nodes in C are neighbors.

Example (The two-dimensional lattice) For the regular lattice on $\mathbf{Z}^2$ any clique is the empty set, or a node, or a pair of nodes. For the extended lattice we need to add triangular and square configurations of neighbors. □

A **potential** V is a way of assigning a number $V_A(\mathbf{x})$ to every subconfiguration $\mathbf{x}_A$ of a configuration $\mathbf{x}$. The energy corresponding to a potential V is

$$U(\mathbf{x}) = \sum_A V_A(\mathbf{x}) \tag{4.13}$$

where the sum is over all subsets A of the nodes of the graph. As before, the Gibbs measure induced by U is defined by

$$\mathbf{P}(\mathbf{x}) = \frac{e^{-U(\mathbf{x})}}{Z} \tag{4.14}$$

where Z is the partition function $\sum_{\mathbf{x}} \exp(-U(\mathbf{x}))$. V is a **nearest neighbor potential** if $V_A(\mathbf{x})=0$ whenever A is not a clique.

Example (A model for political affinity) Weidlich (1971) suggested a model for the political leaning of individuals, in which pair-wise interactions affected the affinity of an individual. For example, an individual whose neighbors are all conservative would have a tendency towards conservatism, as compared to an individual whose neighbors were all liberal. An external field would, in this interpretation, correspond to the current state of government, and a model with minimum energy corresponds to maximum boredom, in that all individuals agree with each other and with the government. □

Theorem 4.1 On a finite graph, any nearest neighbor potential induces a Markov random field.

Proof Consider a finite graph $\mathcal{G}$. Then, with C denoting the set of cliques,

$$p_i(x_i \mid \mathbf{x}_{-i}) = \frac{\mathbf{P}(\mathbf{x})}{\sum_{\mathbf{x}'} \mathbf{P}(\mathbf{x}')} = \frac{\exp(-\sum_{C \in C} V_C(\mathbf{x}))}{\sum_{\mathbf{x}'} \exp(-\sum_{C \in C} V_C(\mathbf{x}'))} \tag{4.15}$$

where $\mathbf{x}'$ is any configuration which agrees with $\mathbf{x}$ everywhere except possibly at i. For any clique C which does not contain i, $V_C(\mathbf{x})=V_C(\mathbf{x}')$, so the terms corresponding to non-neighbors of i cancel, and we have a Markov random field. □

Remark The theorem holds for an infinite graph as well, provided that the V_C are such that

$$\sum_{\mathbf{x}} \exp(-\sum_{C \in C} V_C(\mathbf{x})) < \infty. \tag{4.16}$$

□

Example (The two-dimensional lattice, continued) Define a **pair potential** as a potential with $V(i,j)=V(j,i)$. In $\mathbf{Z}^2$, a nearest neighbor pair potential is determined by three parameters: $V(i,i)=V(0,0)=v_0$, $V(0,e_N)=v_N$ and $V(0,e_E)=v_E$, where e_N is a unit step up (northward) and e_E is a unit step to the right (eastward). This is because $V(i,j)=V(0,i-j)$ and $V(0,-j)=V(j,0)=V(0,j)$. If x is a configuration over a finite set Λ, then, writing $\mathbf{x}(i)$ for x_i

$$\mathbf{P}(X_i=1 \mid \mathbf{X}_{-i}=\mathbf{x}_{-i}) \tag{4.17}$$

$$= \frac{1}{1+\exp(v_0+v_N(\mathbf{x}(i+e_N)+\mathbf{x}(i-e_N))+v_E(\mathbf{x}(i+e_E)+\mathbf{x}(i-e_E)))}.$$

It is convenient to rewrite (4.17) by using the **logit** transformation

$$\text{logit}\,\mathbf{P}(X_i=1 \mid \mathbf{X}_{-i}=\mathbf{x}_{-i}) = \log\left[\frac{\mathbf{P}(X_i=1 \mid \mathbf{X}_{-i}=\mathbf{x}_{-i})}{1-\mathbf{P}(X_i=1 \mid \mathbf{X}_{-i}=\mathbf{x}_{-i})}\right] \tag{4.18}$$

$$= -(\nu_0+\nu_N(\mathbf{x}(i+e_N)+\mathbf{x}(i-e_N))+\nu_E(\mathbf{x}(i+e_E)+\mathbf{x}(i-e_E))).$$

This type of model is called **autologistic** (Besag, 1972). Usually the autologistic model is defined on $S=\{0,1\}$. The Ising model, with $S=\{-1,1\}$ is given by a a pair potential with $\nu_N=\nu_E$. Such models are called **locally isotropic**. The Ising model for two dimensions was introduced by Peierls (1936). □

Application (Plant disease) When studying the propagation of plant disease, a common method is to assess all the plants on a field or in a greenhouse for signs of the disease at fixed intervals of time. One can then produce a dynamic map, showing the spread of the disease throughout the study area. Cochran (1936) describes some data collected by Bald at the Waite Institute in Australia on the spread of spotted wilt of tomatoes. This is a virus disease, carried by a species of thrips. The field had 16 plots, arranged in a 4×4 Latin square design, each with 6 rows containing 15 plants each, for a total of 1440 plants.

Figure 4.3. Presence of spotted wilt on a tomato field.

The tomatoes were planted on November 26, 1929, and the counts displayed in Figure 4.3, where each square corresponds to a diseased plant, were made on December 18. Further counts were made on December 31, and at three times in January of the following year. We are interested in modeling this using the Ising model (or, equivalently, the isotropic first-order autologistic process of Besag, 1972). For now we will just consider those plants that were diseased at the first count, leaving the next set of observations, and their modeling, to the Exercises.

□

4.3. Phase transitions in Markov random fields

We have seen that nearest neighbor potentials induce Markov random fields. In fact, whenever $\mathbf{P}(\mathbf{x})>0$ for all $\mathbf{x}$, the Markov random fields are exactly those Gibbs measures (Hammersley and Clifford, 1971).

Theorem 4.2 (Hammersley–Clifford) Assume that $P(\mathbf{x})>0$ for all configurations $\mathbf{x}$. Then $\mathbf{P}$ is a Markov random field on a finite graph $\mathcal{G}$ with respect to a neighborhood system Δ if and only if it is a Gibbs measure corresponding to a nearest neighbor potential.

Proof We already did the 'if' part above, so it remains to do the 'only if' part. For simplicity we do it for $S=\{0,1\}$, although the argument can be modified to work for a general finite state space (see Besag, 1974). Write $\mathbf{x}\leq\mathbf{y}$ if $x_i\leq y_i$, $i=1,\ldots,m$, and let $|\mathbf{x}|=\#\{x_i=1\}$. We first need a lemma.

Lemma 4.1 (Möbius inversion formula) Let U and V be functions from $\{0,1\}^{\mathcal{G}}\to\mathbf{R}$. Then

$$U(\mathbf{x})=\sum_{\mathbf{y}\leq\mathbf{x}}V(\mathbf{y}) \tag{4.19}$$

if and only if

$$V(\mathbf{x})=\sum_{\mathbf{y}\leq\mathbf{x}}(-1)^{|\mathbf{x}-\mathbf{y}|}U(\mathbf{y}). \tag{4.20}$$

Remark An arbitrary U can be written in the form (4.19) by choosing V as in (4.20). This representation is unique. For example, if $m=2$ we get

$$\begin{cases} V(0,0)=U(0,0)\\ V(0,1)=-U(0,0)+U(0,1)\\ V(1,0)=-U(0,1)+U(1,0)\\ V(1,1)=U(0,0)-U(0,1)-U(1,0)+U(1,1)\end{cases} \tag{4.21}$$

so

$$V(0,0)+V(0,1)=U(0,0)-U(0,0)+U(0,1)=U(0,1). \tag{4.22}$$

□

Proof of Lemma Assume (4.20). Then for a given $\mathbf{x}$,

$$\sum_{\mathbf{y}\leq\mathbf{x}}V(\mathbf{y})=\sum_{\mathbf{y}\leq\mathbf{x}}\sum_{\mathbf{z}\leq\mathbf{y}}(-1)^{|\mathbf{y}-\mathbf{z}|}U(\mathbf{z})=\sum_{\mathbf{z}\leq\mathbf{x}}U(\mathbf{z})c(\mathbf{z}) \tag{4.23}$$

where

$$c(\mathbf{z})=\sum_{\mathbf{z}\leq\mathbf{y}\leq\mathbf{x}}(-1)^{|\mathbf{y}-\mathbf{z}|}=\sum_{\mathbf{u}\leq\mathbf{x}-\mathbf{z}}(-1)^{|\mathbf{u}|}=\begin{cases}1 & \text{if } \mathbf{z}=\mathbf{x}\\ 0 & \text{otherwise,}\end{cases} \tag{4.24}$$

since each non-empty finite set has as many subsets of even as of odd cardinality. Here $\mathbf{u} \le \mathbf{x}-\mathbf{z}$ means that the ones in $\mathbf{u}$ occur only in spots where $\mathbf{x}$ and not $\mathbf{z}$ are one. □

Proof of Theorem, cont. Given P, let $U(\mathbf{x})=\log(\mathbf{P}(\mathbf{x})/\mathbf{P}(\mathbf{0}))$ where $\mathbf{0}$ is the pattern with all zeros. Let $V(\mathbf{x})$ be given by (4.20). We need to show that if $\mathbf{P}$ is Markov, then $V(\mathbf{x})=0$ unless $\mathbf{x}=1(C)$ where C is a clique (i.e., the configuration which is 1 on C and 0 elsewhere). Suppose that $\mathbf{x}$ is not of this form, so there are two points i and j, not neighbors, with $x_i=x_j=1$. Let $A=\{k: x_k=1\}$. Every subset of A is of the form D, $D\cup\{i\}$, $D\cup\{j\}$, or $D\cup\{i,j\}$, where $D\subset A\setminus\{i,j\}$. Let $\mathbf{z}=1(D)$. Collecting terms with the same sign, we get

$$V(\mathbf{x}) = \sum_{\mathbf{z}\le\mathbf{x}-1(i)-1(j)} (-1)^{|\mathbf{x}-\mathbf{z}|} \log\left[\frac{\mathbf{P}(\mathbf{z})\mathbf{P}(\mathbf{z}+1(i)+1(j))}{\mathbf{P}(\mathbf{z}+1(i))\mathbf{P}(\mathbf{z}+1(j))}\right]. \tag{4.25}$$

We need to show that the argument of the logarithm is 1, or equivalently that

$$\frac{\mathbf{P}(\mathbf{z})}{\mathbf{P}(\mathbf{z})+\mathbf{P}(\mathbf{z}+1(i))} = \frac{\mathbf{P}(\mathbf{z}+1(j))}{\mathbf{P}(\mathbf{z}+1(j))+\mathbf{P}(\mathbf{z}+1(i)+1(j))}. \tag{4.26}$$

Notice that $\mathbf{P}(\mathbf{z}) + \mathbf{P}(\mathbf{z}+1(i)) = \mathbf{P}(\mathbf{X}_{-i}=\mathbf{z}_{-i})$, since the left-hand side represents the sum over the two possible values of X_i. Hence the left-hand side of (4.26) is $\mathbf{P}(X_i=0 \mid \mathbf{X}_{-i}=\mathbf{z}_{-i})$, while the right-hand side is $\mathbf{P}(X_i=0 \mid \mathbf{X}_{-i}=\mathbf{z}_{-i}+1(j))$. But by the Markov property, since j is not a neighbor of i, these two probabilities are equal. □

Here is an important question: given a nearest neighbor potential, is there only one random field having this potential? On the line we already know the answer: there is only one Markov chain with a given transition matrix. For a finite graph, an argument similar to that in the proof above shows that the same holds there, provided the positivity condition is satisfied, at least. But in $\mathbf{Z}^2$ the answer can be different! This was first shown by Onsager (1944), who computed an asymptotic approximation to the partition function, resulting in a function not being analytic at a critical temperature (this is how physicists determine so-called phase transitions). We will use an argument suggested by Peierls (1936) and made precise by Griffiths (1964). The first step is to show the effect of the boundary configuration on an interior point.

Consider a finite lattice where all the boundary nodes are fixed to be +. Using the Ising model we have

$$\begin{aligned}\mathbf{P}(\mathbf{x}) &= \frac{1}{Z}\exp\left(\frac{J}{kT}\sum_{i\sim j}x_ix_j+\frac{mH}{kT}\sum_i x_i\right)\\ &= \frac{1}{Z}\exp(-bn_o(\mathbf{x})+hM(\mathbf{x}))\end{aligned} \tag{4.27}$$

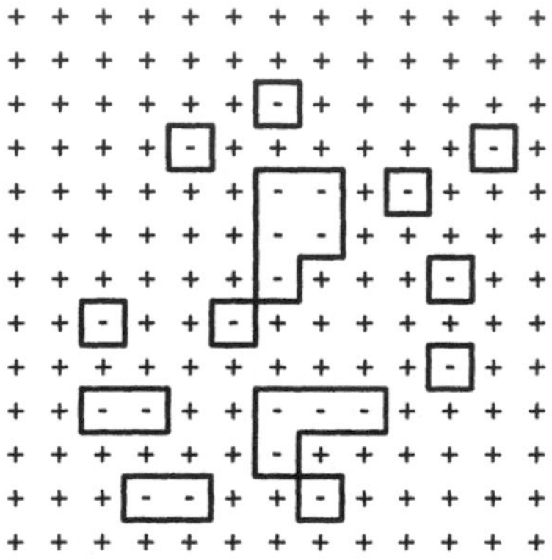

Figure 4.4. A finite lattice with positive boundary.

where $b=2J/kT$ and $h=mH/kT$. If $h=0$, the energy only depends on the b $2J/KT$, strength of Ising model nearest neighbor interactions number n_o of odd parity neighbors, i.e., neighbors with opposite signs. Notice that this is just the length of the boundaries (heavy lines) in the figure. Let x_0 denote the spin at a fixed interior node which we call 0.

Theorem 4.3 For an Ising model on a finite 2-dimensional regular lattice Λ, the boundary conditions affect the distribution of x_0.

Proof With a + boundary, if $x_0=-$ there must be a closed boundary containing 0. Call this a **circuit**, and denote its length by L. All nodes inside a circuit have negative spin. Let $\bar{\mathbf{x}}$ be an arbitrary configuration with the same circuit, and let $\mathbf{x}'$ agree with $\bar{\mathbf{x}}$ except that the negative spins inside the circuit are switched to positive. Then there are L fewer odd bonds: $n_o(\mathbf{x}')=n_o(\bar{\mathbf{x}})-L$. Hence

$$\mathbf{P}_+^\Lambda(\text{given circuit around } 0) = \frac{\sum_{\bar{\mathbf{x}}} e^{-bn_o(\bar{\mathbf{x}})}}{\sum_{\mathbf{x}} e^{-bn_o(\mathbf{x})}} \leq \frac{\sum_{\bar{\mathbf{x}}} e^{-bn_o(\bar{\mathbf{x}})}}{\sum_{\mathbf{x}'} e^{-bn_o(\mathbf{x}')}} \tag{4.28}$$

$$= \frac{\sum_{\bar{\mathbf{x}}} e^{-bn_o(\bar{\mathbf{x}})}}{\sum_{\bar{\mathbf{x}}} e^{-bn_o(\bar{\mathbf{x}})+bL}} = e^{-bL}$$

where the notation $\mathbf{P}_+^\Lambda$ is intended to be a reminder of the boundary condition as well as the finite lattice. We now have that

$$\mathbf{P}_+^\Lambda(x_0=-1) \le \sum_{L=4,6,\ldots} r(L)e^{-bL} \tag{4.29}$$

where $r(L)$ is the number of configurations $\bar{x}$ with circuits of length L. Now consider a circuit of length L. It must have each node a distance no more than $L/2$ from 0. There are at most L^2 ways of choosing the starting node for this circuit. Given a starting node, there are at most 4 ways of choosing the next node, and then at most 3 to choose each succeeding node (since we cannot double back). Thus there are at most $4L^2 3^L$ circuits of length L around 0, so

$$\mathbf{P}_+^\Lambda(x_0=-1) \le \sum_{L=4,6,\ldots} 4L^2(3e^{-b})^L < \frac{1}{3} \tag{4.30}$$

if b is large enough. Notice that this is true even as we let the size of the lattice grow to infinity. If we had started with a negative boundary, we would by symmetry get that $\mathbf{P}_-^\Lambda(x_0=1)<1/3$, so $\mathbf{P}_-^\Lambda(x_0=-1)>2/3$. Thus, the effect of the boundary is felt at the center, provided b is large enough. Since the estimate (4.30) does not depend on the size of the set Λ (as long as it contains an orbit of length 4), the effect of the boundary on the center is there regardless of how far away the boundary is. A more careful version of this argument can be used to show the existence of a **critical temperature** T_c such that for $T<T_c$ (corresponding to $b>b_c$) the boundary effect is felt, and otherwise it is lost.

It should be intuitively reasonable, at least, that if we have a strictly increasing sequence Λ_n of finite sets with + boundary, the probability of a positive spin at the origin (in the attractive case) ought to be decreasing with n, and does in fact converge to a limiting value $\mathbf{P}_+(X_0=1)=\lim_n \mathbf{P}_+^{\Lambda_n}(X_0=1)$. The argument can be generalized to more than one point, allowing us to specify $\mathbf{P}_+(\mathbf{X}_A=\mathbf{x}_A)$ for all finite sets A. As in the previous section, this specification determines a probability on the infinite lattice. We can produce the same kind of limit for a $-$ boundary, yielding a probability $\mathbf{P}_-$. If b is large enough, we know that $\mathbf{P}_-(X_0=1) \ne \mathbf{P}_+(X_0=1)$. Hence we have two different probabilities, determined by the same local characteristics. This phenomenon is called **phase transition** in physics. It can be shown that if $h\ne0$ and $b>0$, or if $h=0$, $0<b\le b_c$ there is no phase transition (i.e., $\mathbf{P}_-=\mathbf{P}_+$). But in the case we studied here, with $h=0$ and $b>b_c$, one can show that any Gibbs measure on the two-dimensional lattice is a mixture of $\mathbf{P}_+$ and $\mathbf{P}_-$. Returning now to Ising's original work, this last statement means that when the temperature is low enough, and there is no external field, the magnetization will be a mixture between $\mathbf{P}_+$ (which tends to have large chunks of +) and $\mathbf{P}_-$ (with large chunks of $-$). In other words, the magnetism of the material at low temperatures will have a bimodal distribution, corresponding (qualitatively, at least) to reality.

It is instructive to look at how an Ising model without external field changes as a function of b. Figure 4.5 shows a sequence of simulated 32×32 Ising fields, for different values of b. For ease of comparison, the more common color is always black.

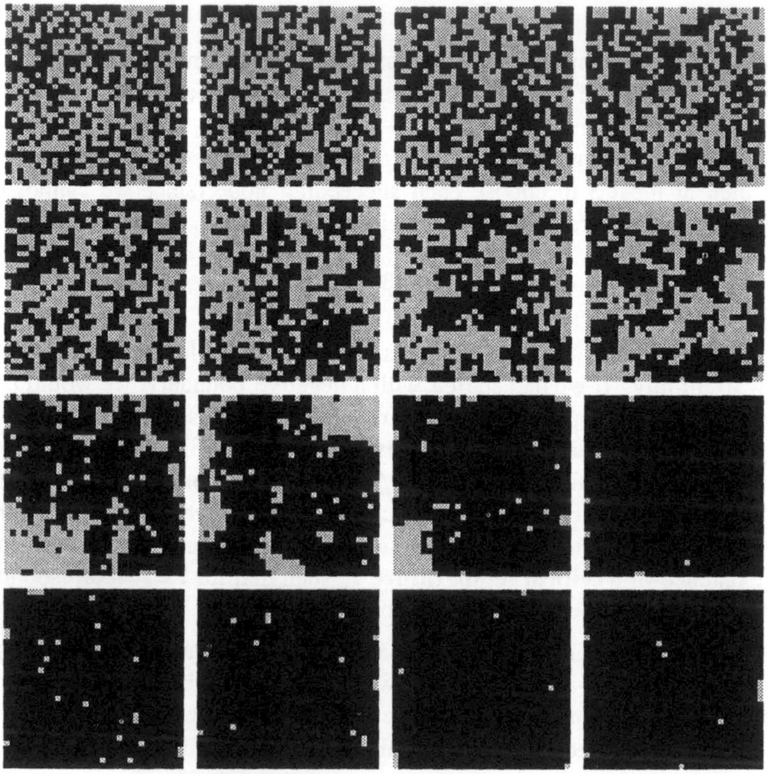

Figure 4.5. A sequence of finite Ising fields. The parameter b is 0 in the upper left-hand figure, and increases by steps of 0.1 going across the rows, so that the lower right-hand figure has $b = 1.5$.

4.4. Likelihood analysis of the Ising model

The statistical analysis of Markov random field models used to be impracticable. To see why, consider the partition function for a nearest neighbor potential: it is the sum over all configurations of some numbers. But even in the Ising model there are 2^N configurations for a lattice subset having N elements. For the application at the end of section 2, N is 1276 (ignoring the boundary) and it is not possible to compute the partition function directly. Statistical physicists have spent much time and effort to develop ways of approximating partition functions, but only recently have statisticians started to develop manageable approaches to inference for Markov random fields (beginning with Besag, 1972). In this section we use the Markov chain Monte Carlo approach to compute the likelihood function for the Ising model.

The likelihood of an observed pattern $\mathbf{x}$ with a fixed boundary which we, by analogy to the treatment of initial values of a Markov chain, will argue conditionally upon, is

$$L(\boldsymbol{\theta}) = \frac{1}{Z(\boldsymbol{\theta})}\exp(t_1\theta_1+t_2\theta_2) \tag{4.31}$$

where $t_1=\sum x_i$ is the excess of positive over negative observations, and $t_2=\sum_i\sum_{j\sim i}x_ix_j$ is the excess of even over odd parity neighbors. Here t_1 does not contain any boundary values, while the boundary values enter in the calculation for t_2 for the sites adjacent to the boundary. These are the sufficient statistics (cf. Appendix A) for the Ising model.

There is a neat trick for computing certain likelihoods (in particular for so-called exponential families of distributions; see Bickel and Doksum, section 2.3), called **exponential tilting**. It was first used to improve asymptotic expansions (Reid, 1988, is a nice review). Notice that the log likelihood ratio $\lambda = \log(L(\boldsymbol{\theta})/L(\boldsymbol{\phi}))$ can be written

$$\lambda = t_1(\theta_1-\phi_1) + t_2(\theta_2-\phi_2) - \log(Z(\boldsymbol{\theta})/Z(\boldsymbol{\phi})). \tag{4.32}$$

Now compute

$$\frac{Z(\boldsymbol{\theta})}{Z(\boldsymbol{\phi})} = \mathbf{E}_{\phi}\exp(T_1(\theta_1-\phi_1)+T_2(\theta_2-\phi_2)). \tag{4.33}$$

Consider a sequence $\mathbf{x}^{(n)}, n\leq N$, of observations of the Ising model with parameter $\boldsymbol{\phi}$, with corresponding natural sufficient statistics $t_i^{(n)}$, $i=1,2$. An obvious estimate of the quantity in (4.33) is

$$\frac{1}{N}\sum_{n=1}^{N}\exp(t_1^{(n)}(\theta_1-\phi_1)+t_2^{(n)}(\theta_2-\phi_2). \tag{4.34}$$

If the sequence $\mathbf{x}^{(n)}$ is chosen appropriately, the average in (4.34) converges with probability one to its expected value (4.33). Hence we can estimate the likelihood ratio λ by substituting the logarithm of (4.34) into (4.32). Notice that we only need to simulate for one particular value of $\boldsymbol{\phi}$, and can compute the likelihood (approximately) for all values simultaneously. Computationally this may still be a very large problem, with numerical difficulties abounding.

To maximize (4.32) with respect to $\boldsymbol{\theta}$ we compute the likelihood equations

$$\begin{cases} \dfrac{\partial\lambda}{\partial\theta_1} = t_1 - \dfrac{\dfrac{\partial Z(\boldsymbol{\theta})}{\partial\theta_1}/Z(\boldsymbol{\phi})}{Z(\boldsymbol{\theta})/Z(\boldsymbol{\phi})} \\ \dfrac{\partial\lambda}{\partial\theta_2} = t_2 - \dfrac{\dfrac{\partial Z(\boldsymbol{\theta})}{\partial\theta_2}/Z(\boldsymbol{\phi})}{Z(\boldsymbol{\theta})/Z(\boldsymbol{\phi})} \end{cases} \tag{4.35}$$

We estimate $Z(\boldsymbol{\theta})/Z(\boldsymbol{\phi})$ by (4.34), and $(\partial Z(\boldsymbol{\theta})/\partial\theta_i)/Z(\boldsymbol{\phi})$ by

$$\frac{1}{N}\sum_{n=1}^{N} t_i^{(n)}\exp(t_1^{(n)}(\theta_1-\phi_1)+t_2^{(n)}(\theta_2-\phi_2)). \tag{4.36}$$

The asymptotic covariance matrix is given by $\left[-D^2\log(Z(\boldsymbol{\theta})/Z(\boldsymbol{\phi}))\right]^{-1}$, which can also be estimated in a similar fashion.

In order to simulate the paths $\mathbf{x}^{(n)}$ we employ the Metropolis algorithm with a slight twist. To implement it, we start with all sites having the same value. Next we go through each site in turn and attempt to change its sign. Thus the Q-matrix is

$$Q = \begin{bmatrix} 0 & 1 \\ 1 & 0 \end{bmatrix}. \tag{4.37}$$

The odds ratio can be written

$$r = \exp(\phi_1\Delta t_1+\phi_2\Delta t_2) \tag{4.38}$$

where Δt_i is the change in sufficient statistics from the original configuration to the configuration with the sign of the value at site i changed. Clearly, this only depends on the neighbors of site i. After updating all sites, Geyer (1991) suggests trying to change all signs of the current pattern in order to avoid the problem with being stuck in one of the two modes of a bimodal distribution. We saw in the previous section that this can happen to the Ising model when ϕ_1 is small and ϕ_2 is large. For other parameter values this step has little effect, but it is not very time-consuming.

The Ising model is aperiodic and irreducible, so Markov chain Monte Carlo works, in the sense that an ergodic theorem holds. The swapping scheme discussed above is intended to make sure that the algorithm moves through as much as possible of the state space.

Application (Plant disease, continued) The sufficient statistics for the data in Figure 4.3 are t_1=–834 and t_2=2266. Simulating from an Ising process with the same boundary and with ϕ_1=0 and ϕ_2=0.5 yielded the values of t_1 and t_2 depicted in Figure 4.6. We simulated 100,000 steps, keeping every 100 for a total of 1,000 values. Optimizing the likelihood yields the mle $\hat{\boldsymbol{\theta}}$=(0,0.52). It is clear from the simulated values that we are in the part of the parameter space where phase transition occurs, so that t_1 is sometimes positive and sometimes negative. □

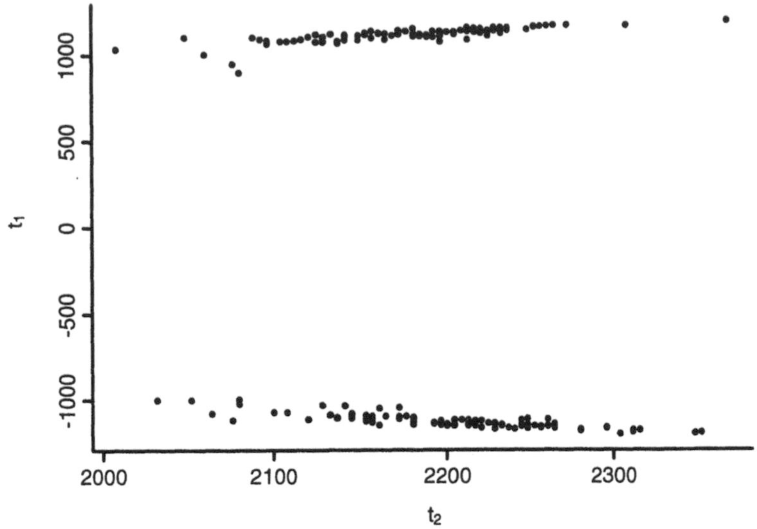

Figure 4.6. Simulated values of t_1 and t_2 for $\phi_1=0$ and $\phi_2=0.5$.

4.5. Reconstruction of astronomical images

In classical astronomy the scientist took photographic pictures of the sky, as seen through an optical telescope. The pictures were used to classify objects, for example, galaxies into spiral, elliptical, etc.; to detect new objects; or to catalog the visible universe. Nowadays astronomers use digital detectors, based on semiconductors called charge-coupled devices (CCD). A digital image is an $n \times k$-array of counts, where n and k are on the order of hundreds or thousands.

There are several sources of noise in these pictures. First, the observed image has been blurred by the turbulent movement of the atmosphere during the period of observation, which is often several hours. In addition, there is a faint background glow due to scattered light from the moon and stars, and also resulting from charged particles from space colliding with the gases in the upper atmosphere. There is electric noise in the CCD's. There are gross errors, such as lines of readings at either minimum or maximum intensity, and, finally, photons have counting errors associated with them. We are trying to measure the flux **S** at earth of a given sector of sky, and are observing a blurred, noisy image **Z**. The task, therefore, is to reconstruct **S** on the basis of **Z**. To do so we shall, much as in the neurophysiological application in section 2.12, write down a simple spatial model for smooth images which, together with the fairly well understood noise and blurring mechanisms, will enable us to reconstruct **S** by maximizing the posterior distribution of **S** given **Z**.

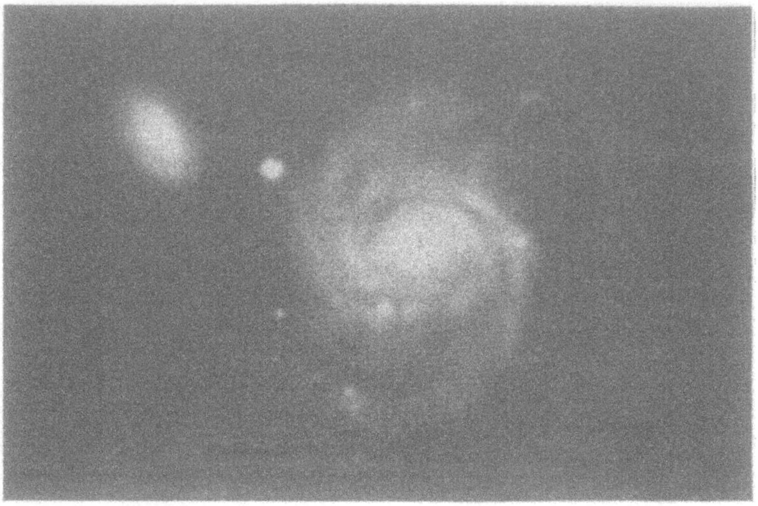

Figure 4.7. A telescope image of two galaxies. Reproduced with permission from Molina and Ripley (1989), published by Carfax Publishing Company.

The blurring can be described by a **point-spread function** (psf) h, which describes how a point image (such as a star) gets blurred during observation. This has been studied extensively, and following Molina and Ripley (1989) we use

$$h(r) = \frac{\beta}{\pi R^2}\left\{1+\left[\frac{r}{R}\right]^2\right\}^{-\beta} \tag{4.39}$$

where r is the distance from the source (i.e., the picture element or **pixel** in the true picture **S**) to the receiving pixel (in **Z**), while R and β are parameters. For the image we are going to study, β=3 and R=3.5 were estimated by a least square fit to images of point sources. This means that a point source is spread out over a few hundred pixels (see Figure 4.8). Let $H=(h(d_{ij}))$ be the **blurring matrix**, where d_{ij} is the distance between pixel i and pixel j, in some ordering. Then the blurred picture has values

$$\boldsymbol{\mu} = \mathbf{S}H. \tag{4.40}$$

The number Z_i' of photons received at pixel i, in the absence of electronic noise, would have a Poisson distribution with mean and variance $\boldsymbol{\mu}=\mathbf{S}H$. Not all photons are actually recorded, but rather a non-random fraction b of them. This is a characteristic of the device, and can be estimated from the data or taken to be known from previous experience. Hence the number of actual arrivals at pixel i has mean and variance μ_i/b, while the corrected value $Z_i''=Z_i'b$ has mean μ_i and variance $b^2(\mu_i/b)=b\mu_i$. Adding independent

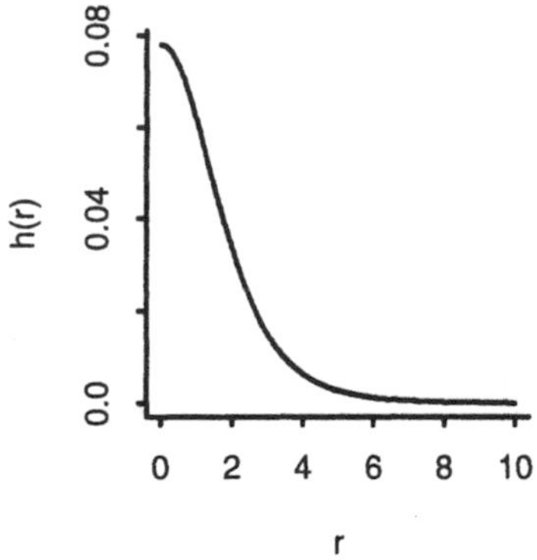

Figure 4.8. The point spread function from Molina and Ripley (1989). Adapted with permission from Carfax Publishing Company.

electronic noise, assumed to have mean 0 and variance a independently for each pixel, suggests that $\mathbf{E}Z_i=(\mathbf{S}H)_i=\mu_i$ and $\mathbf{Var}Z_i=a+(\mathbf{S}H)_i/b=\sigma^2(\mu_i)$.

In order to estimate $\sigma^2(\mu)$, consider

$$Z_i^N = Z_i - \tfrac{1}{4}\sum_{j\sim i} Z_j. \tag{4.41}$$

In a featureless region this should be a mean zero random variable with variance $(5/4)\sigma^2(\mu_i)$. Thus, we could divide μ_i into small intervals and estimate $(5/4)\sigma^2(\mu_i)$ from the variance of the Z_i^N corresponding to those intervals. Of course, we do not really know μ_i, but can estimate it by Z_i (its mean is typically a few hundred, while its variance is equal to its mean, so the relative error in replacing μ_i by Z_i is small). However, the gross errors in the image can be damaging to standard variance estimates. The solution is to use a **robust** estimate of variance, which downweights or ignores the gross errors, such as (Huber, 1981)

$$\hat{\sigma}^2(\mu_i) = (0.6745\,\text{median}\,|\,Z_i^N\,|\,)^2/1.25. \tag{4.42}$$

The constant 0.6745 yields the right value in the absence of gross errors for a Gaussian distribution, which we expect to be a reasonable assumption here (we basically have Poisson counts with a large mean). Figure 4.9 shows a plot of $\hat{\sigma}^2(\mu)$ against μ based on an image of size 1056×656, which contains a central core of size 329×256 pixels, shown in Figure 4.8. Here $\hat{\sigma}^2(\mu)\approx 200+0.7\mu$.

What we want to do next is to compute the conditional distribution of **S**, given data **Z**:

$$\mathbf{P}(\mathbf{S}\,|\,\mathbf{Z}) = \text{const}\times\mathbf{P}(\mathbf{Z}\,|\,\mathbf{S})\,\mathbf{P}(\mathbf{S}). \tag{4.43}$$

To do so, we need an expression for **P(S)**, the unconditional distribution of **S**, i.e., a probabilistic expression of what we expect astronomical images to look

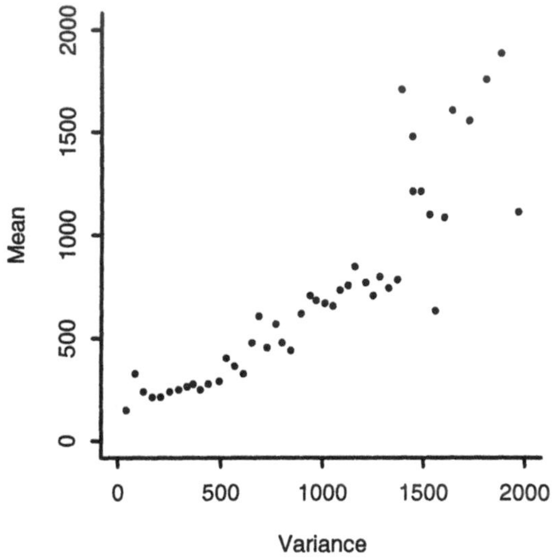

Figure 4.9. Robust variance estimate plotted against mean value, based on a 1056×656 pixels image containing the image of interest. Adapted with permission from Molina and Ripley (1989), published by Carfax Publishing Company.

like. For an image with no stars, but galaxies at smoothly varying luminosity, we would expect $S_i \geq 0$ (with quite a few zeros), and **S** also being spatially smooth. Stars, on the other hand, would of course be spatially non-smooth.

We want to use a continuous state model for the luminosity, while staying within the framework of the Markov random field. A simple approach is to assume that

$$(X_i \mid \mathbf{X}_{-i}) \sim \mathrm{N}(\theta_i, \tau^2) \tag{4.44}$$

where

$$\theta_i = \mathbf{E}(X_i \mid \mathbf{X}_{-i}) = \sum_{j=1}^{n} c_{ij} x_j \tag{4.45}$$

is a weighted average of the values at other pixels. We should have $c_{ii}=0$, and c_{ij} positive only when $i \sim j$. The first order **conditional autoregressive model** (Whittle, 1954), has $c_{ij}=\phi 1(i \sim j)$. Let $C=(c_{ij})$.

Proposition 4.1 The conditional Gaussian autoregressive model given by (4.44) satisfies

$$\mathbf{X} \sim \mathrm{N}(\mathbf{0}, \tau^2(\mathbb{I}-C)^{-1}) \tag{4.46}$$

whenever $\mathbb{I}-C$ is invertible.

Proof By the Brook representation (1.20) we can write

$$\log\frac{f(\mathbf{x})}{f(\mathbf{0})} = -\frac{1}{2\tau^2}\sum_{i=1}^{n}\left[\{x_i-\sum_{j=1}^{i-1}c_{ij}x_j\}^2-\{\sum_{j=1}^{i-1}c_{ij}x_j\}^2\right] \tag{4.47}$$

$$=-\frac{1}{2\tau^2}\mathbf{x}(\mathbb{I}-C)\mathbf{x}^T.$$

□

Remark A sufficient condition for $\mathbb{I}-C$ to be invertible is that $\phi<\frac{1}{4}$. It is easy to allow the X_i to have means μ_i. We just apply the definitions to $X_i-\mu_i$. Likewise, one can allow the conditional variances to depend on i. This changes the variance in the proposition to $\text{diag}(\tau_i^2)(\mathbb{I}-C)^{-1}$. □

We are now ready to estimate the true luminosity $\mathbf{S}$. Assuming, as suggested above, that the conditional distribution of the Z_i given $\mathbf{S}$ are independent $N(\mu_i,\sigma^2(\mu_i))$, and that $\mathbf{S}$ is assumed a priori to follow a conditional Gaussian autoregressive model with ϕ just below $\frac{1}{4}$, we choose $\mathbf{S}$ to maximize

$$-\tfrac{1}{2}\left\{\frac{1}{\tau^2}\mathbf{S}(\mathbb{I}-C)\mathbf{S}^T+\sum_{i=1}^{n}(Z_i-\mu_i)^2/\sigma^2(\mu_i)\right\}. \tag{4.48}$$

Recall that $\mu_i=(\mathbf{S}H)_i$, and that $\sigma^2(\mu_i)=a+b\mu_i$. Write $W=\text{diag}(\sigma^2(\mu_i))$. By differentiating (4.48) we obtain the likelihood equation

$$(\mathbb{I}-C)\mathbf{S}^T=\tau^2H(\mathbf{Z}-\mathbf{S}H)W^{-1}-\tau\sum_{j=1}^{n}\frac{(Z_j-(\mathbf{S}H)_j)^2}{W_{jj}}\frac{\partial W_{jj}/\partial S_i}{W_{jj}} \tag{4.49}$$

$$=\tau^2H\{(\mathbf{Z}-\mathbf{S}H)W^{-1}-b((\mathbf{Z}-\mathbf{S}H)W^{-1})^2\}.$$

We need to take into account the constraint that $S\geq 0$ as well. If we can ignore the second term on the right-hand side, the equation would simplify to

$$\mathbf{S}=\max(0,\ C\mathbf{S}+\tau^2H(\mathbf{Z}-\mathbf{S}H)W^{-1}), \tag{4.50}$$

which can be solved iteratively by starting at some value $\mathbf{S}_0$ and using that in the right-hand side of (4.50) to get $\mathbf{S}_1$, etc. Since $(\mathbf{Z}-\mathbf{S}H)_i$ is of the order of $W_{ii}^{1/2}$, the first term in the braces on the right-hand side of (4.49) is of order $W_{ii}^{-1/2}$, while the second is of order W_{ii}^{-1}, which will be negligible except when W_{ii} is very small. The reconstruction is shown in Figure 4.10. It appears reasonable except for bright spots, such as that between the two galaxies. Such spots get a dark halo, and lose all internal resolution, basically because the prior assumes

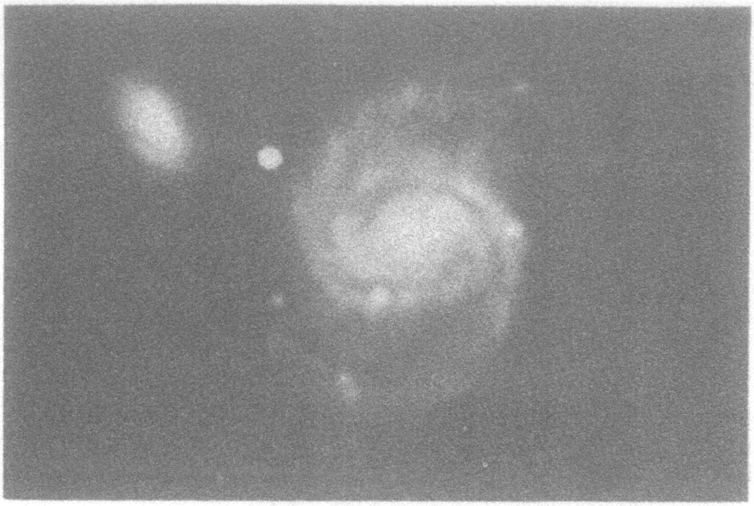

Figure 4.10. CAR prior reconstruction of the galaxy image. Reproduced with permission from Molina and Ripley (1989), published by Carfax Publishing Company.

that the image has the same smoothness both at low and at high observed values. In order to avoid this problem, we need to transform the image.

The distribution of luminosity I (energy radiated per second) over an astronomical object such as a galaxy has been modeled by

$$I(r) = I(0)\exp(-(ar)^{1/4}), \tag{4.51}$$

suggesting that perhaps one should view pictures on a logarithmic scale. This is standard practice in astronomy (and done here). Applying this transformation to $\mathbf{S}$, we let $Y_i = \log(S_i + 100)$. Assuming a conditional autoregressive prior for $\mathbf{Y}$ instead of for $\mathbf{S}$ we have $\mathbf{Y} \sim \mathrm{N}(\mathbf{0}, \tau(\mathbb{I}-C))$, and need to maximize

$$-\frac{1}{2\tau^2}\mathbf{Y}(\mathbb{I}-C)\mathbf{Y}^T + \sum_{i=1}^{n}(Z_i-\mu_i)^2/\sigma^2(\mu_i). \tag{4.52}$$

Using the same approximation as above (ignoring the dependence of the noise variance on $\mathbf{S}$, or equivalently on $\mathbf{Y}$), we iteratively solve

$$\mathbf{Y} = \max(\log(100)\mathbb{I}, CY + \tau\,\mathrm{diag}(e^{Y})HW^{-1}(Z-(e^{Y}-p)H)). \tag{4.53}$$

It remains to try to determine τ. If we could observe $\mathbf{Y}$ we would estimate τ by $\hat{\tau} = \mathbf{Y}(\mathbb{I}-C)\mathbf{Y}^T/n$. However, we only observe $\mathbf{Z}$. Very roughly we have $(\log Z_i) \approx YH$, suggesting that we replace $\mathbf{Y}$ in $\hat{\tau}$ by $H^{-1}(\log Z_i)$. A more sophisticated approach, using Fourier analysis, is presented by Molina and Ripley (1989), but they indicate that we only need to get τ right to within an order of magnitude.

Figure 4.11. Reconstruction with a CAR prior on the logarithmic scale. Reproduced with permission from Molina and Ripley (1989), published by Carfax Publishing Company.

The resulting reconstruction shows more detail, particularly in the smaller galaxy which now looks spiral rather than elliptical. The troublesome small source between the two galaxies is now neatly resolved without a halo.

The approach outlined here was used in some analyses of images obtained from the Hubble Space Telescope before the correction of the optics in 1993. Due to a mirror with incorrect radius (by a very small amount), the images obtained by the telescope were very blurred. No simple point spread function, like that suggested above, was sufficient for analysis of these images. Rather, the psf would cover the entire image (although, being very structured, not losing all that much detail), and the correction due to blurring was considerable. These considerations are discussed in Di Gésu et al. (1992), pp. 75–114 and King et al. (1991). A most interesting discussion of the whole Hubble project, with lots of examples of images and psf's, is the book by Chaisson (1994).

4.6. Image analysis and pedigrees

A **pedigree** (or genealogy) is a group of individuals with a complete specification of all relationships between them. Every individual has a unique label, and the pedigree is often specified as a list of these individual labels followed by the labels of both parents. Such pedigrees are **in triplet form**. All individuals have either both or neither parent specified. Those individuals with no parent specified are called **founders** of the pedigree. Individuals are **spouses** if they have mutual offspring in the pedigree, and every such spouse pair is a **marriage**.

Figure 4.12 shows a depiction of very simple pedigree. Squares denote male individuals, circles are females, and diamonds individuals of unknown gender. The small dots denote marriages, and are connected to individuals: parents on top and children on the bottom.

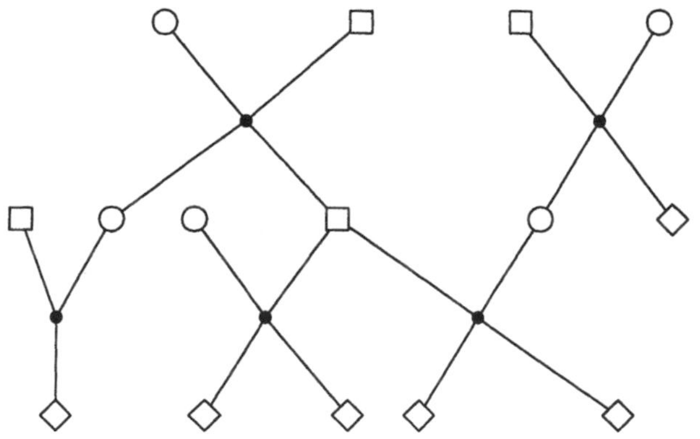

Figure 4.12. A simple pedigree.

This particular pedigree has six founders (the four parents in the top generation, and the two spouses not related to these parents in the middle generation).

We can consider a pedigree to be a finite connected graph with certain properties. A finite **directed graph** has nodes representing entities, and **arcs** (ordered sets of nodes) representing relationships between entities. Thus, an arc (g_1,g_2) has a direction, $g_1 \rightarrow g_2$. Arcs give rise to unordered pairs of nodes, or edges (arcs without direction). A sequence of arcs (g_1,g_2), $(g_2,g_3),\ldots,(g_{n-1},g_n)$ is a **path.** If $g_1=g_n$ the path is a **circuit**. A sequence of edges is a **chain** if each edge has one node in common with the preceding and one node in common with the succeeding edge. If it starts and ends with the same node it is a **cycle**. A graph is **connected** if a chain exists between each pair of nodes, and a **tree** is a connected graph without cycles or circuits.

A pedigree, then, is a directed graph with two kinds of nodes (individuals and marriages), and two kinds of directed arcs, marriage arcs and descendant arcs. Edges can only exists between individual and marriage nodes, so they must alternate along any simple chain in a pedigree. If there is a path from g_1 to g_2 then g_1 is an **ancestor** of g_2. Since an individual cannot be his own ancestor, pedigrees have no circuits. We call a pedigree **looped** if it contains any cycles. The pedigree in Figure 4.12 is unlooped. There are four different loop types:

1. Marriage rings: A is married to B who is married to C who is married to D who is married to A.
2. Inbreeding loops: biologically related individuals marry, so there are two paths to a marriage node from a common individual (parent marrying offspring) or marriage node (first cousins marry).
3. Exchange loops: one individual in each marriage has an ancestor in common with one individual in each preceding and succeeding marriage, having no ancestor in common with his spouse (to rule out type 2).
4. Multiple marriage loops: two members of a set of marriages are related, provided that they are not themselves spouses (which would be type 2) and have no spouse in common (type 3).

Application (The polar eskimo pedigree) The polar eskimos are the northernmost known native population in the world. They live in the Thule district of northern Greenland (75°–79°30'N, 68°–74°W). Until their first contact with white men in 1818 they thought themselves to be the only people in the world. The polar eskimo pedigree (Gilberg et al., 1978; augmented by A. W. F. Edwards) has 1614 individuals, 225 of which are founders, stretching from a birth in 1805 to one in 1975. The pedigree has eleven distinct components, the largest one shown in Figure 4.13 having 1581 individuals. Out of the 1614 individuals recorded, 721 had children. 21 individuals had three marriages, and 8 had four. There is one case where six men and seven women (including 3 brother–sister pairs) are involved in 11 marriages. Consequently, this pedigree is very complicated. □

The key observable in a pedigree is the **phenotype** corresponding to a locus on a chromosome. The phenotype is the actual expression or feature, such as blue eyes, which is determined by the **genotype** of the locus, or the precise description of which alleles have been inherited from the parents. There are three main types of questions relating to pedigrees. One can study the accuracy of the genealogical structure, e.g., for disputed paternity cases. This is done by comparing likelihoods for hypothesized genealogies. The genetic model is known, with the pedigree unknown. Given a pedigree, one may be interested in estimating the genetic model, i.e., quantities such as crossover rate, mutation rate, etc. This can be done by maximum likelihood, and assumes the pedigree known and the genetic model unknown (although known structurally). Finally, it may be of interest to compute the predictive distribution of a genotype (or phenotype) of future individuals, given current information. This is the topic of genetic counseling, but also applies to animal breeding for genetic variability, appropriate for endangered species. In this case both the pedigree and the genetic model are known.

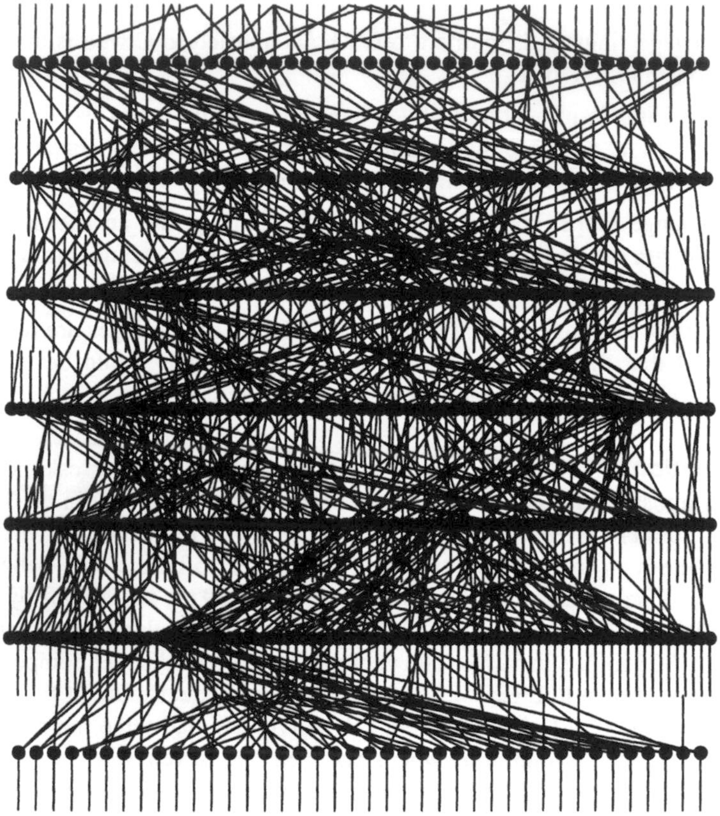

Figure 4.13. The largest component of the polar eskimo pedigree. Adapted from Sheehan (1992), published by the Institute of Mathematics and Its Applications.

Let $\mathbf{X}=(X_1, \ldots, X_m)$ be genotypes for individuals on a pedigree $\mathcal{G}$. By convention we will number individuals in birth order, so that parents always have a smaller index than their offspring. We typically cannot observe the genotypes, but at most the phenotypes $\mathbf{Y}=(Y_1, \ldots, Y_m)$. The stochastic model has three basic ingredients: the data, consisting of phenotypes for a subset D of individuals in the pedigree; the pedigree $\mathcal{G}$ itself; and a genetic model. The latter has in turn three parts. The **penetrance** probabilities are

$$\mathbf{P}(Y_i=y \mid X_i=x) = W(y \mid x). \tag{4.54}$$

These are **simple** if they are 0 or 1. Next, genetic theory tells us that genes are passed from parents to offspring. Consequently, we can compute (from any genetic model) the **transition probabilities**

$$\mathbf{P}(X_i=x \mid X_{Mi}=x_M, X_{Fi}=x_F) = T(x \mid x_M, x_F) \tag{4.55}$$

where Mi denotes the mother of i, and Fi the father. Finally we need the **genotype distribution on founders**, namely

$$\mathbf{P}(\text{random founder } i \text{ has genotype } x) = \mathbf{P}(X_i=x) = R_i(x). \tag{4.56}$$

We can think of $R_i(x)$ as a prior distribution on the genotypes of founder i. Often $R_i(x) \equiv R(x)$ is taken to be the same for all founders.

If we further assume that genotypes of offspring are conditionally independent, given the genotypes of the parents, we can write down a formula for the likelihood. Let F denote the founders. The observed phenotypes $\mathbf{Y}_D$ have probability function

$$\mathbf{P}(\mathbf{Y}_D=\mathbf{y}_D) = \sum_{\mathbf{x}} \left\{ \prod_{i \in F} R_i(x_i) \prod_{j \notin F} T(x_j \mid x_{Mj}, x_{Fj}) \prod_{k \in D} W(y_k \mid x_k) \right\}. \tag{4.57}$$

We are, in effect, setting the penetrance probabilities equal to one for unobserved individuals. For large pedigrees the exact computation of likelihoods is extremely difficult. The method of **peeling** was developed by Elston and Stewart (1971), and for more complex pedigrees by Cannings et al. (1978). It works by successively reducing the pedigree to independent pieces, and is very effective on loop-free pedigrees, but is computationally infeasible for really complicated cases.

Using the same type of calculation as that leading to (4.57) we can compute the conditional probability that a particular individual i takes on genotype x_i, given the genotypes $\mathbf{x}_{\mathcal{G} \setminus i}$ of the remaining pedigree:

$$\begin{aligned} \mathbf{P}(X_i=x_i \mid \mathbf{x}_{\mathcal{G} \setminus i}) &= \frac{\mathbf{P}(\mathbf{X}=\mathbf{x})}{\sum_{x=1}^{k} \mathbf{P}(X_i=x, \mathbf{x}_{\mathcal{G} \setminus i})} \\ &= \frac{\prod_{j \in F} R_j(x_j) \prod_{l \notin F} T(x_l \mid x_{Ml}, x_{Fl})}{\sum_{x=1}^{k} \prod_{j \in F} R_j(x_j) \prod_{l \notin F} T(x_l \mid x_{Ml}, x_{Fl}) \big|_{x_i=x}}. \end{aligned} \tag{4.58}$$

Since the denominator is just the sum over all possible genotypes for i, all terms that do not involve i cancel. The remaining terms has entries corresponding to i's parents Mi and Fi, to i's spouses S_i, and to the children C_i^s for each spouse s in S_i. Hence

$$\mathbf{P}(X_i{=}x_i \mid \mathbf{x}_{\mathcal{G}\setminus i}) = \begin{cases} \dfrac{R_i(x_i)\prod\limits_{s\in S_i}\{\prod\limits_{c\in C_i^s} T(x_c \mid x_i, x_s)\}}{\sum\limits_{x=1}^{k} R_i(x)\prod\limits_{s\in S_i}\{\prod\limits_{c\in C_i^s} T(x_c \mid x, x_s)\}} & \text{if } i\in F \\[2ex] \dfrac{T(x_i \mid x_{Mi}, x_{Fi})\prod\limits_{s\in S_i}\{\prod\limits_{c\in C_i^s} T(x_c \mid x_i, x_s)\}}{\sum\limits_{x=1}^{k} T(x \mid x_{Mi}, x_{Fi})\prod\limits_{s\in S_i}\{\prod\limits_{c\in C_i^s} T(x_c \mid x, x_s)\}} & \text{if } i\notin F, \end{cases} \tag{4.59}$$

so $\mathbf{P}(X_i{=}x_i \mid \mathbf{x}_{\mathcal{G}\setminus i}) = \mathbf{P}(X_i{=}x_i \mid \mathbf{x}_{\delta_i})$, where the neighborhood δ_i consists of the parents, spouses, and children of i (Figure 4.14).

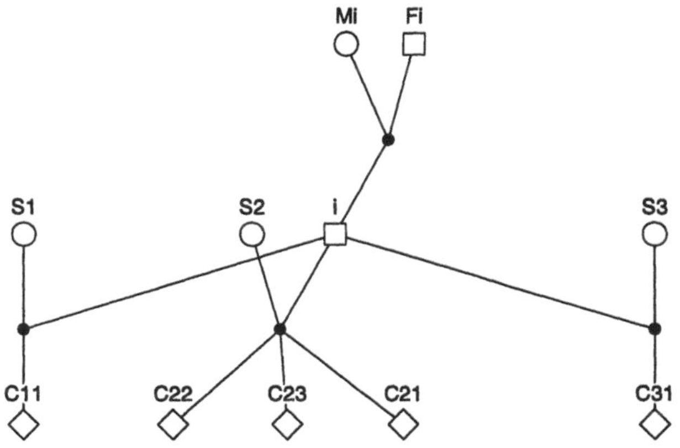

Figure 4.14. A neighborhood for a pedigree. Adapted from Sheehan (1992), published by the Institute of Mathematics and Its Applications.

Hence the genotypes $\mathbf{X}$ form a Markov random field on the pedigree $\mathcal{G}$ with respect to this neighborhood system. The structure of the resulting model for pedigree data is familiar: there is an unobserved state $\mathbf{X}$, which is a Markov random field, and an observed perturbation $\mathbf{Y}$ of the state. Hence it is a hidden Markov random field, much like the image model in the previous section. An important difference, however, is that the state space for $\mathbf{X}$ no longer is of the form $S^{\mathcal{G}}$, as we shall see later.

For the statistical analysis of pedigree data, rather than conditioning on the boundary (i.e., the founders) we put down a probability model for them, so the resulting analysis is different from what we have done before. But the Gibbs sampler can be implemented. Instead of visiting each individual in turn, it proves useful to choose an individual to update at random. Choose individual i with probability c_i. Then replace the current value for i by one drawn at random from the conditional density

$$\mathbf{P}(X_i=x \mid \mathbf{y}, \hat{\mathbf{x}}_{G\setminus i}) \propto W(y_i \mid x)\,\mathbf{P}(X_i=x \mid \hat{\mathbf{x}}_{\delta_i}) \tag{4.60}$$

where $\hat{\mathbf{x}}$ is the current configuration. Thus, every new step of the chain differs from the previous one in at most one individual. The non-zero transition probabilities for the Gibbs sampler occur only when $\mathbf{x}$ and $\mathbf{z}$ differ in at most one element. If they differ only at element i, we have

$$P_{\mathbf{x},\mathbf{z}} = c_i\mathbf{P}(X_i=x_i \mid \mathbf{X}_{\delta_i}=\mathbf{z}_{\delta_i}, y_i), \tag{4.61}$$

while if $\mathbf{x}=\mathbf{z}$

$$P_{\mathbf{x},\mathbf{z}} = \sum_{i=1}^{n} c_i\mathbf{P}(X_i=x_i \mid \mathbf{X}_{\delta_i}=\mathbf{z}_{\delta_i}, y_i) \tag{4.62}$$

As we have seen before, the Gibbs sampler applies when the chain is irreducible and aperiodic (the aperiodicity is Exercise 1). In particular, because of the structure of the Gibbs sampler, one needs to be able to get from one configuration to another via a sequence of switches at a single site at the time. For the case of pedigree data, this may not be the case. Some configurations are simply not allowed, although the algorithm may be trying to move towards them. To start with, we are only interested in genotypic configurations that are consistent with the data, in that $\mathbf{P}(\mathbf{X}=\mathbf{x} \mid \mathbf{Y}=\mathbf{y})>0$. Let Ω be the set of consistent configurations. The following example illustrates the type of problems we encounter.

Example (ABO blood system) The ABO blood classification system has three alleles, A[1], B and O. The A-gene is codominant with B. The O allele is dominated by the other two (Table 4.1).

Table 4.1 The ABO system

Phenotype	Genotype
A	*AA AO*
B	*BB BO*
AB	*AB*
O	*OO*

Suppose now that we observe an *AB* individual. Then each of the parents may be *AA*, but both cannot. Hence it is not sufficient to look at the possible genotypes for individuals separately, but one needs to take into account neighboring values that constrain the possibilities. □

[1]There really are two different *A*-genes, one of which dominates the other, but for simplicity we shall ignore that.

Consider now a trait with two alleles, A and a. For some individuals we can determine their genotypes exactly from the data. Call that set G. Given a consistent configuration $\mathbf{x}$, define a configuration $\mathbf{z}$ by

$$z_i = \begin{cases} x_i & \text{if } i \in G \\ Aa & \text{if } i \notin G. \end{cases} \tag{4.63}$$

In other words, we assume that all incompletely known individuals are heterozygotes. Then $\mathbf{z}$ may or may not be consistent. However, it turns out that under a simple condition on the penetrance probabilities, not only is $\mathbf{z}$ consistent, but it also communicates with any other consistent state, which is precisely what we need to make the Gibbs sampler work.

Theorem 4.4 If the penetrance probabilities W satisfy

$$W(p \mid AA) > 0 \text{ and } W(p \mid aa) > 0 \text{ implies that } W(p \mid Aa) > 0 \tag{4.64}$$

for all phenotypes p, then

(i) any consistent configuration communicates with $\mathbf{z}$, and

(ii) $\mathbf{z}$ is consistent.

Proof Let $\mathbf{X}^{(k)}$ be the kth iteration of the Gibbs sampler defined by (4.60). We first note that if we have two consistent configurations $\mathbf{x}$ and $\mathbf{w}$ which differ only in the genotype of i, then $\mathbf{x}$ and $\mathbf{w}$ communicate, since

$$\begin{aligned} \mathbf{P}(\mathbf{X}^{(k+1)}=\mathbf{w} \mid \mathbf{X}^{(k)}=\mathbf{x}) &= c_i \mathbf{P}(X_i=w_i \mid \mathbf{X}_{\delta_i}=\mathbf{x}_{\delta_i}, y_i) \\ &= c_i \mathbf{P}(X_i=w_i \mid \mathbf{X}_{\delta_i}=\mathbf{w}_{\delta_i}, y_i) > 0. \end{aligned} \tag{4.65}$$

Define a sequence of configurations

$$\begin{aligned} \mathbf{x}^{(0)} &= (x_1, x_2, \ldots, x_m) = \mathbf{x} \\ \mathbf{x}^{(1)} &= (z_1, x_2, \ldots, x_m) \\ \mathbf{x}^{(2)} &= (z_1, z_2, \ldots, x_m) \\ &\cdots \\ \mathbf{x}^{(m)} &= (z_1, z_2, \ldots, z_m) = \mathbf{z} \end{aligned} \tag{4.66}$$

We need to show that if $\mathbf{P}(\mathbf{X}=\mathbf{x}^{(k)} \mid \mathbf{y})>0$ then $\mathbf{P}(\mathbf{X}=\mathbf{x}^{(k+1)} \mid \mathbf{y})>0$. If $k+1 \in G$, so that the genotype of $k+1$ is certain from the data, we have $\mathbf{P}(\mathbf{X}=\mathbf{x}^{(k+1)} \mid \mathbf{y})=\mathbf{P}(\mathbf{X}=\mathbf{x}^{(k)} \mid \mathbf{y})>0$. If $k+1 \notin G$ but $k+1 \in F$ we can compute

$$\begin{aligned} \frac{\mathbf{P}(\mathbf{X}=\mathbf{x}^{(k+1)} \mid \mathbf{y})}{\mathbf{P}(\mathbf{X}=\mathbf{x}^{(k)} \mid \mathbf{y})} &= \frac{R(Aa)}{R(x_{k+1})} \frac{W(y_{k+1} \mid Aa)}{W(y_{k+1} \mid x_{k+1})} \\ &\times \prod_{i \in S_{k+1}} \prod_{j \in C^i_{k+1}} \frac{T(x_j \mid \tilde{x}_i, Aa)}{T(x_j \mid \tilde{x}_i, x_{k+1})} \end{aligned} \tag{4.67}$$

where $\tilde{x}_i$ is w_i if i is before $k+1$ and x_i if i is after $k+1$. Since $\mathbf{P}(\mathbf{X}=\mathbf{x}^{(k)} \mid \mathbf{y})>0$

we must have $T(x_j \mid \tilde{x}_i, x_{k+1})>0$ for all i and j, so $T(x_j \mid \tilde{x}_i, Aa)>0$. By assumption either $W(y_{k+1} \mid Aa)>0$, and we are done with this case, or one of $W(y_{k+1} \mid AA)$ or $W(y_{k+1} \mid aa)$ is zero, in which case $k+1 \in G$, which is a contradiction. The case when $k+1$ is not a founder is similar, having just a different first term on the right-hand side of (4.67). Thus $\mathbf{z} \in \Omega$ and $\mathbf{x} \rightarrow \mathbf{z}$, which is what we wanted to show. □

Example (A counterexample) To see that the condition on the penetrance probabilities is necessary, consider the simple example in Table 4.2.

Table 4.2 A genetic model

	p_1	p_2
AA	½	½
Aa	0	1
aa	½	½

Suppose we observe an individual with phenotype p_1, whose parents are also observed to have p_1. Then $\Omega=\{(AA,AA,AA),(aa,aa,aa)\}$. Here (4.64) is not satisfied, and we cannot move from one consistent configuration to the other, changing only one genotype at the time, without visiting an inconsistent configuration. □

One way out of this situation is to change the penetrance probabilities to make them all positive, by defining

$$W^*(p \mid g) = \begin{cases} W(p \mid g) & \text{if } W(p \mid g)>0 \\ \gamma>0 & \text{if } W(p \mid g)=0. \end{cases} \tag{4.68}$$

The we can define $\mathbf{P}^*(\mathbf{X}=\mathbf{x} \mid \mathbf{Y}=\mathbf{y})$ in the same fashion as we defined $\mathbf{P}$ earlier, but replacing W by W^*. This may not correspond to any genetic model, but will be seen to be a computational device that allows us to move around the state space freely. First note that

$$\mathbf{P}^*(\mathbf{X}=\mathbf{x} \mid \mathbf{y}) = C\mathbf{P}(\mathbf{X}=\mathbf{x} \mid \mathbf{y}) \tag{4.69}$$

where C does not depend on $\mathbf{x} \in \Omega$. Hence

$$\mathbf{P}^*(\mathbf{X}=\mathbf{x} \mid \mathbf{x} \in \Omega, \mathbf{y}) = \frac{\mathbf{P}^*(\mathbf{X}=\mathbf{x}, \mathbf{x} \in \Omega \mid \mathbf{y})}{\sum_{\mathbf{z} \in \Omega} \mathbf{P}^*(\mathbf{X}=\mathbf{z}, \mathbf{z} \in \Omega \mid \mathbf{y})} = \mathbf{P}(\mathbf{X}=\mathbf{x} \mid \mathbf{y}). \tag{4.70}$$

Thus, by rejecting any inconsistent configurations (with respect to the real genetic model) we can sample from $\mathbf{P}(\mathbf{X}=\mathbf{x} \mid \mathbf{y})$. The proportion of rejected samples from $\mathbf{P}^*(\mathbf{X}=\mathbf{x} \mid \mathbf{y})$ is $r=1-C$. Now all states communicate, and the Gibbs sampler therefore works.

Application (ABO blood classification for polar eskimos) Among many characteristics that have been studied for the polar eskimos is the ABO blood classification system, as described in the example above. Not all the individuals in the pedigree have ABO-classifications; in fact, only 290 do. Five of those tested positive for B, indicating that, perhaps, that gene was introduced by a fairly recent founder. Removing all individuals that do not carry information about the B-gene leaves two pedigree components. The smaller one has 20 individuals, 9 of which are founders, with two individuals observed to carry the gene, while the larger has 543 individuals, 105 of which are founders, and three carrying the gene. To simplify matters we reduce the data to a diallelic system, using B and B^c (meaning not B).

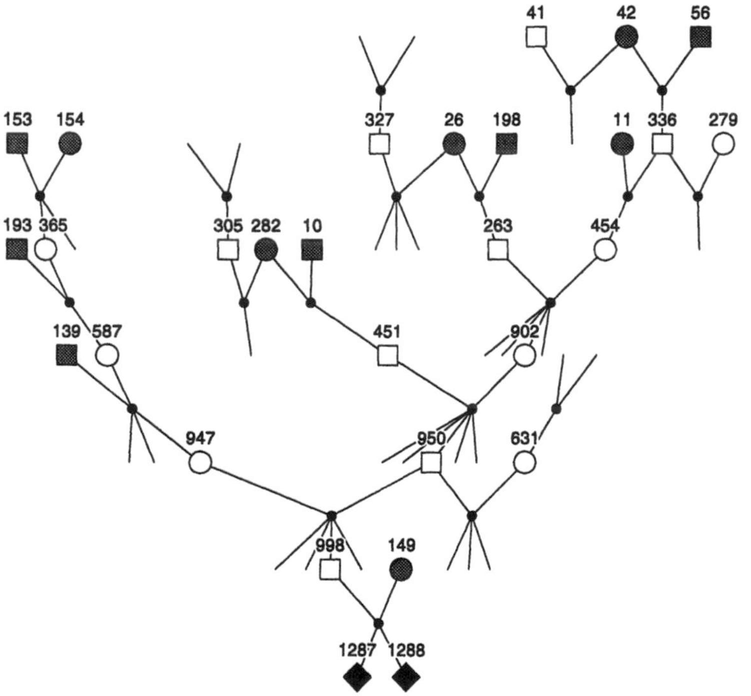

Figure 4.15. The ABO blood type pedigree. Siblings, parents, and spouses who are not of interest are not marked, and their existence is indicated by open-ended lines. Adapted from Sheehan (1992), published by the Institute of Mathematics and Its Applications.

Figure 4.15 shows the relevant part of the pedigree. As it happens, the mother 149 of the two carriers 1287 and 1288 was observed to carry the B-gene, while the father 998 was known to be *OO*. Hence both 1287 and 1288 must be BO, and the ancestry is clear. In order to see how the Gibbs sampler does in this fairly complex pedigree, we shall remove the data on 149 and 998. Sheehan

(1990, 1992) computed 100 sets of gene frequencies for 11 founders, each based on 20,000 cycles of the Gibbs sampler. Table 4.3 shows the results for two different choices of distribution among the founders. The first choice is $R(\mathrm{B})=\frac{1}{2}$. This eliminates the dependence on R in the likelihood formula (4.57), corresponding to what is sometimes called an **ignorance prior** in Bayesian statistics. The other choice, acknowledging that the B-gene appears to bc rare, is to let $R(\mathrm{B})=0.02$. The choice of R is seen to have substantial influence on the results. Using the ignorance type prior yields results that are at variance with the actual situation, while the rare B-gene prior does a good job at explaining the real source of the B-genes.

4.7. Bibliographic remarks

The book by Kindermann and Snell (1980) is a delightful introduction to Markov random fields on graphs. Cressie (1991) and Ripley (1981) are general reference works for spatial statistics.

The theoretical aspects of Markov random fields have been the subject of much work in probability theory as well as in statistical mechanics. I found the books by Preston (1976) and Ruelle (1969) particularly informative. The statistical aspects were developed by Besag starting in the early 70s. His 1974 Royal Statistical Society discussion paper contains a wealth of information (as well as some interesting historical tidbits). Besag's 1986 discussion paper on image analysis is a good place to learn about hidden Markov random fields. Another good source is the paper by Besag et al. (1995). A variety of applications (including some of the papers mentioned here) is in Mardia and Kanji (1993).

The section on astronomy is principally based on Molina and Ripley (1989). Additional material is found in Molina et al. (1992) and in Ripley (1991). An interesting discussion of some statistical problems important to astronomers is in Feigelson and Babu (1992).

Nuala Sheehan (1990, cf. also 1992) wrote her PhD dissertation on the application of Markov chain Monte Carlo methods to the computation of probabilities in pedigrees. Some of the technical details were provided by Sheehan and Thomas (1993). Further work in this direction is in Geyer and Thompson (1992), and Lin et al. (1994).

4.8. Exercises

Theoretical exercises

1. Show that the Gibbs sampler given in (4.61–62) is aperiodic.

2. Consider iid Gaussian random variables ε_i, and define a spatial process Z_i on a finite lattice by

$$Z_i - \theta_1(Z_{N(i)} + Z_{S(i)}) - \theta_2(Z_{E(i)} + Z_{W(i)}) = \varepsilon_i$$

Table 4.3 Mean and sd of B-gene founder frequencies
Flat prior (**P**(B)=0.5)

Genotype	B-ancestors					
	10	11	26	42	56	149
BB	0.025	0.073	0.004	0.072	0.139	0.795
sd	0.002	0.004	0.002	0.004	0.005	0.003
BO	0.333	0.333	0.144	0.290	0.333	0.201
sd	0.003	0.003	0.004	0.004	0.003	0.003
OO	0.642	0.593	0.853	0.639	0.528	0.003
sd	0.004	0.005	0.005	0.005	0.006	0.001
	153	154	193	198	282	
BB	0.163	0.161	0.102	0.074	0.011	
sd	0.006	0.007	0.004	0.005	0.002	
BO	0.315	0.315	0.333	0.333	0.265	
sd	0.004	0.004	0.003	0.003	0.004	
OO	0.522	0.523	0.565	0.592	0.724	
sd	0.007	0.008	0.006	0.006	0.004	

Sharp prior (**P**(B)=0.02)

Genotype	10	11	26	42	56	149
BB	0.000	0.000	0.000	0.000	0.000	0.039
sd	0.000	0.000	0.000	0.000	0.000	0.001
BO	0.021	0.023	0.007	0.020	0.026	0.949
sd	0.001	0.002	0.002	0.002	0.002	0.008
OO	0.979	0.977	0.993	0.980	0.974	0.012
sd	0.001	0.002	0.002	0.002	0.002	0.008
Genotype	153	154	193	198	282	
BB	0.000	0.000	0.000	0.000	0.000	
sd	0.000	0.000	0.000	0.000	0.000	
BO	0.031	0.030	0.030	0.023	0.016	
sd	0.004	0.004	0.003	0.003	0.004	
OO	0.969	0.970	0.970	0.977	0.984	
sd	0.004	0.004	0.006	0.001	0.002	

where $N(i)$ is the northern neighbor of i, etc. This is called a **simultaneously specified Gaussian autoregression** (Whittle, 1954).

(a) Show that for suitable b_{ij}

$$Z_i = \mu_i + \sum b_{ij}(Z_j - \mu_j) + \varepsilon_i.$$

(b) Show that $\mathbf{Z} \sim \mathrm{N}(\boldsymbol{\mu},\sigma^2(\mathbb{I}-B)^{-1}(\mathbb{I}-B^T)^{-1})$ where $B=(b_{ij})$.

(c) Show that this model is equivalent to a CAR process.

3. In agricultural field trials, contiguous experimental units (usually rectangles) are treated in different ways, according to some experimental design. In classical design theory, variations in fertility across the field was dealt with by randomization. This, in essence, incorporates spatial variability in the overall error variability. It is sometimes advantageous to put down a spatial model for the fertility of the field, taking into account the smoothness one would expect in the fertility gradient (Besag, 1991). Thus we may write

$$\mathbf{Z} = \boldsymbol{\alpha}\mathbf{X} + \boldsymbol{\varepsilon}$$

where $\mathbf{X}$ is a design matrix of 0's and 1's, indicating which treatments have been assigned to what experimental units, $\boldsymbol{\alpha}$ is a vector of treatment effects (or contrasts), and $\boldsymbol{\varepsilon}$ is the underlying spatial error. Suppose that $\boldsymbol{\varepsilon}$ follows a conditionally autoregressive Gaussian model with known dependence matrix $\mathbf{C}$. Show that the mle $\hat{\boldsymbol{\alpha}}$ satisfies

$$\hat{\alpha}\mathbf{X}(\mathbb{I}-\mathbf{C})\mathbf{X}^T = \mathbf{Z}(\mathbb{I}-\mathbf{C})\mathbf{X}^T.$$

4. Consider a binary tree. The root r is the zeroth level, and has two branches whose endpoints together constitute the first level, etc. A configuration ω on a tree with n levels is an assignment of + or – to each branch. The nth level is the boundary. Define a potential as in the Ising model by $U(\omega)=-bn_o(\omega)+hM(\omega)$, where n_o is the number of odd bonds (including the boundary), and M is the sum over all interior points. This is called a **Cayley tree** (Spitzer, 1975). Divide the partition function into $Z_-^{(n)}$ and $Z_+^{(n)}$, where $Z_-^{(n)}$ is the sum over all configurations with – at the root. Let $u_n=Z_-^{(n)}/Z_+^{(n)}$. We want to see the influence of a negative boundary, say, on the probability of – at the root.

(a) By looking at the four possible configurations on levels 0 and 1, show that

$$Z_-^{(n)} = e^{-h}(Z_-^{(n-1)} + e^{-b}Z_+^{(n-1)})^2.$$

(b) Deduce that

$$u_n = \frac{(u_{n-1}+e^{-b})^2}{e^{2h}(e^{-b}u_{n-1}+1)^2},$$

and that $u_1=\exp(-2(b+h))$.

(c) Derive an equation for the limit of u_n (assuming it exists).

(d) Show that u_n converges when $b>0$, that there is one solution to the equation in (c) when $0<b<\log 4.5$, while for some values of h there can be three solutions when $b>\log 4.5$. Deduce that the probability of – at the root can depend on infinitely removed boundary conditions.

5. Let Z_i be a random field on a graph $\mathcal{G}$. Define the **pseudolikelihood** function

$$PL(\boldsymbol{\theta}) = \prod P(Z_i \mid Z_j, j{\neq}i; \boldsymbol{\theta}).$$

(a) Show how to estimate θ by maximizing the pseudolikelihood for the Ising model.

(b) Show that the maximum pseudolikelihood estimate for a Markov chain on the integers is different from the maximum likelihood estimate.

(c) Show that the pseudolikelihood is equivalent to the likelihood when data are independent. Can you think of any other situation when they would be the same?

6. This exercise is intended to give you a feel for the complexity of the calculations that occur in image analysis. The following are artificial data from a Markov random field:

1	1
1	0

Denoting the observed configuration by Y, assume that it was derived from a true configuration X by independent flipping of each pixel with probability p (i.e., if the pixel was 1 it stays 1 with probability $1-p$, and changes to 0 with probability p). Assuming that $p<½$, find the maximum likelihood estimate of X. *Hint:* Find the mle of p for each possible X.

(b) Assume now that X was derived from a potential of the form

$$V_i(\mathbf{x}) = \theta \sum_{j \sim i} 1(x_j{\neq}x_i),$$

where the sum is over all nearest neighbor pairs of pixels. There is no external force, just neighbor interaction. Using Bayes' formula, find the conditional distribution of X given Y.

(c) Using the mle for p from part (a), find the possible maxima of the distribution in part (b) as a function of θ. Discuss the interpretation of the resulting potential for the different cases. For what value of X does the overall maximum of the conditional distribution occur?

7. Derive the nearest neighbor pair potential model for a $\{0,1\}$-model on the extended integer lattice.

8. In emission tomography, a compound containing a radioactive isotope is introduced into the body and forms an unknown emitter density $\lambda(\mathbf{x})$ under the body's metabolism. Emissions occur according to a Poisson process of rate $\lambda(\mathbf{x})$. In the common case of a positron emitter, the emitted positron finds a nearby electron and annihilates with it, yielding a pair of photons flying in opposite directions. The photons are attenuated by annihilation and energy absorption into the matter through which their trajectories pass. Assume that we have a linear array of detectors $(\sigma_1, \ldots, \sigma_L)$, equispaced and positioned at angle θ relative to the x_1-axis (see Figure 4.16).

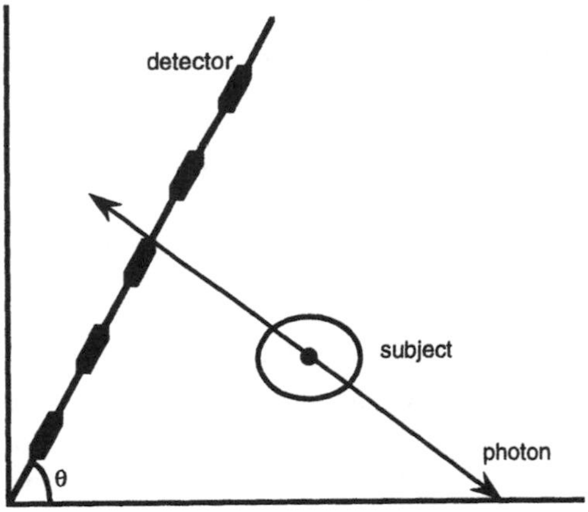

Figure 4.16. Set-up of a positron emission tomography array.

The experimental design consists of positioning the detector array at equispaced angles $\theta_1, \ldots, \theta_N$ for T time units at each angle. We observe the random variables Y_{jk} counting the number of arrivals at detector σ_j and angle θ_k. Let $p(\mathbf{x};\theta,\sigma)$ be the probability that a photon emitted at $\mathbf{x}$ in the direction θ of a detector at σ reaches the detector. This is a known function.

(a) Assume that the orientation of photon trajectories are uniformly distributed on $[0,2\pi)$. Show that the resulting counts are independent Poisson random variables with parameters

$$\mathbf{E}Y_{jk} = T \int_{\theta_k-\frac{1}{2}\Delta\theta}^{\theta_k+\frac{1}{2}\Delta\theta} \int_{\sigma_j-\frac{1}{2}\Delta\sigma}^{\sigma_j+\frac{1}{2}\Delta\sigma} \int_{\mathbf{x}\in *L(\theta,\sigma)} \lambda(\mathbf{x})p(\mathbf{x};\theta,\sigma)d\mathbf{x}d\sigma d\theta \equiv A_T(\lambda)$$

where (σ,θ) is the line at angle θ through the point σ of the detector array.

(b) In order to get practical results, we discretize the domain Ω into pixels, parametrized by discrete points s in a square lattice. Then $\boldsymbol{\lambda}$, the vector with elements consisting of $\lambda(\mathbf{x})$ integrated over each square in the lattice, represents a piecewise constant approximation of the isotope concentration in the continuous domain. For the discretized form we have $\mathbf{EY}=a_T\boldsymbol{\lambda}$, where a_T is a matrix of enormous order $(LN\times\#(S))$. Show that the the log likelihood function can be written

$$l(\boldsymbol{\lambda}) = \sum_{j,k}(-\log Y_{jk}! + Y_{jk}\log(a_T\boldsymbol{\lambda})_{jk} - (a_T\boldsymbol{\lambda})_{jk}).$$

9. In discrete image analysis problems where the posterior distribution needs to be maximized over a rather large number of possible pictures, the computing problem can be quite involved. The method of **simulated annealing**, borrowed from the metallurgic process of annealing an alloy by cooling it so slowly that the atoms can be arranged in a low-energy configuration, was introduced in optimization theory by Pincus (1968). Let P be a probability on a finite (but typically large) set S. We want to find the $\hat{s} \in S$ with the largest $P(s)$. Define

$$P_T(s) = \frac{P(s)^{1/T}}{\sum_{t \in S} P(s)^{1/T}}.$$

(a) Show that as $T \to 0$, P_T concentrates increasingly near $\hat{s}$.

(b) One way to implement simulated annealing is to use a Metropolis type algorithm in which the transition matrix changes over time, so that the stationary distribution becomes P_T for decreasing T. Argue that the resulting Markov chain should approach $\hat{s}$, in the sense of concentrating on values with very high values of P.

10. Suppose you are interested in the value of a pollutant on the points of a 16×16 grid. You have measurements $Z_1, \ldots, Z_k$ at a small number k of locations $(i_1,j_1), \ldots, (i_k,j_k)$. These measurements are normally distributed with mean μ, variance σ^2, and correlation

$$\mathbf{Corr}(Z_l,Z_m) = 1 - \exp(-\theta d_{l,m})$$

where $d^2_{l,m} = (i_l - i_m)^2 + (j_m - j_l)^2$.

(a) Consider a location (i_0,j_0) for which measurements have not been made. How would you predict the value Z_0 at that location? *Hint:* A reasonable predictor is $\mathbf{E}(Z_0 \mid Z_1, \ldots, Z_k)$. This is a normal regression problem, although the data are dependent. In geostatistics this is called a **kriging** problem.

(b) Suppose that you are interested in the total amount of deposition over the entire grid. For some of these locations you have data, but for most you do not. How would you predict that random variable?

Computing exercises

C1. Generate an Ising model on a finite lattice.

C2. Implement the Gibbs sampler for likelihood analysis in the Ising model. Assess the run time as a function of the size of the lattice, comparing a model with phase transition to one without.

C3. Compare the run time in Exercise C2 to that of the pseudolikelihood estimation method of Exercise 3. How does the pseudolikelihood estimate compare to the maximum likelihood estimate in the model with and without phase transition?

C4. An Ising model on a finite rectangular lattice is said to have a **free boundary** when the boundary points are viewed as having fewer neighbors than the interior points. It has a **periodic boundary** when it is wrapped onto a torus, so that the northern boundary has the southern boundary as neighbors, and the eastern boundary has the western boundary as neighbors, and vice versa. For one realization each of an 8×8 and a 32×32 Ising model, with parameters of your choice, compare the maximum likelihood estimates conditionally upon the boundary, with free boundary, and with periodic boundary.

C5. For a Bayesian analysis of the tomography problem of Exercise 8, assume that the values of $\boldsymbol{\lambda}$ are confined to pixel intensities from 0 to 15. Define an energy by

$$U(\boldsymbol{\lambda}) = \sum_{<s,t>} \beta\psi(\lambda_s-\lambda_t) + \sum_{[s,t]} \frac{\beta}{\sqrt{2}}\psi(\lambda_s-\lambda_t)$$

where $<s,t>$ denotes nearest horizontal or vertical neighbors, and $[s,t]$ denotes nearest diagonal neighbors.

(a) Choosing

$$\psi(t) = \frac{-1}{1+(t/\delta)^2},$$

generate a sample from this Gibbs distribution, with δ=0.7, and β=5.

(b) Generate a sample of Poisson random variables, as given in Exercise 6, for an array of your choice.

(c) Compare a maximum likelihood analysis and a Bayesian reconstruction, using the prior in (a), of the simulated data in (b).

Data exercises

D1. The data used in the application of section 4.2 are available as data set 5. Maximize the pseudolikelihood of Exercise 5, and compare it to the likelihood fit obtained in the text. Comment on the results.

D2. Data set 6 contains the second measurement (additional diseased plants) for Cochran's tomato spotted wilt data. A reasonable model of contagion would depend on the number of previously diseased neighbors of a given plant. Develop and fit such a model.

D3. Ghent (1963) counted balsam-fir seedlings in five feet square quadrats. The data are given in Table 4.4. Bartlett (1967) suggested to reclassify the counts into −1, 0, and 1 for low, medium, and high counts, respectively, and to assess the dependence of the reclassified counts on the sum of the neighbor to the left and below (conditional upon the boundary).

(a) Assess this dependence using a chi-square test for independence in r×c-tables.

Table 4.4 Balsam-fir seedling counts

0	1	2	3	4	3	4	2	2	1
0	2	0	2	4	2	3	3	4	2
1	1	1	1	4	1	5	2	2	3
4	1	2	5	2	0	3	2	1	1
3	1	4	3	1	0	0	2	7	0
4	2	0	0	2	0	3	2	3	2
2	2	2	0	3	4	7	4	3	3
2	3	1	2	3	8	5	5	1	2
1	1	2	1	4	4	5	3	2	3
3	1	6	1	3	5	4	7	4	3

(b) If independence can be rejected, fit a nearest-neighbor Gibbs model to the data, and assess the standard errors of your estimates.

D4. The mountain pine beetle (*Dendroctonus ponderosa Hopkins*) attacks lodgepole stands, and can destroy a stand in a few years. The data in data set 7 were collected in central Oregon for an unmanaged stand of lodgepole pines attacked by mountain pine beetle in 1984. They contain x- and y-coordinates, a measure of vigor (in g stemwood per m^2 of crown leaf area), the diameter at breast height (in inches), a measure of leaf area (in m^2), age (in years at breast height), and the response variable (0 for not attacked, 1 for attacked). Preisler (1993) suggests an autologistic model where the covariates enter linearly in the potential. Here the distance to nearest attacked tree may be particularly important. Develop a model for the probability that a given lodgepole pine suffers a pine beetle attack as a function of appropriate covariates. Estimate the parameters, and assess the standard errors of your estimates.

D5. The hairy-nosed wombat (*Lasiorhinus latifrons*) of South Australia is a marsupial with a broad, squat body, short neck, and large flat head with a hairy muzzle. The wombat is nocturnal. During the day they sleep in burrows. Often a group of wombats will inhabit burrows together, forming a warren. The data in data set 8 are counts of the number of warrens in 450×450 m^2 quadrats in a grassland Savannah preserve.

(a) Assess the adequacy of a homogeneous Poisson model.

(b) Kaluzny (1987) suggested fitting an inhomogeneous Poisson model by smoothing the counts as a function of location in the lattice. Compute such a fit, and assess its quality.

CHAPTER 5

Point processes

A point process is a model for events that occur separated in time or space. We will describe a clustering model for traffic density. The relation between nerve cells is analyzed using nonparametric inference. We look at some nonstationary models, including a periodic model of cyclonic storms and a counting process approach to a problem in medical sociology. Radar and rain gauge approaches to precipitation measurements are related using a marked point process. Methods for spatial point processes are used to assess some ecological theory of tropical forests.

5.1. A model of traffic patterns

Successful traffic engineering requires a thorough understanding of traffic patterns. The data in Figure 5.1 were recorded on a two-lane rural road in Sweden in the early 1960's. They consist of the times at which 129 successive vehicles passed a fixed position on the road. The total time from the first to the last vehicle is 2023.5 seconds, or just over half an hour. On average, 3.8 cars passed per minute. The figure shows a clear indication of bunching.

Figure 5.1. Vehicle passage times on a Swedish rural road.

A simple stochastic model for the traffic data was proposed by Bartlett (1963), who postulated a Poisson process of primary arrivals at rate ρ. For each primary arrival there is a cluster (possibly empty) of additional cars. The random size M of each cluster is assumed to be iid according to some discrete distribution $g_M(m)=\mathbf{P}(M=m)$, while the points of the cluster are strung out at points $\alpha_1, \alpha_1+\alpha_2, \ldots, \alpha_1+\cdots+\alpha_M$ where the α_i are iid according to some density $f(x)$. If $M=0$ the cluster is empty, and only the primary arrival is counted. Figure 5.2 is a schematic description of this construction.

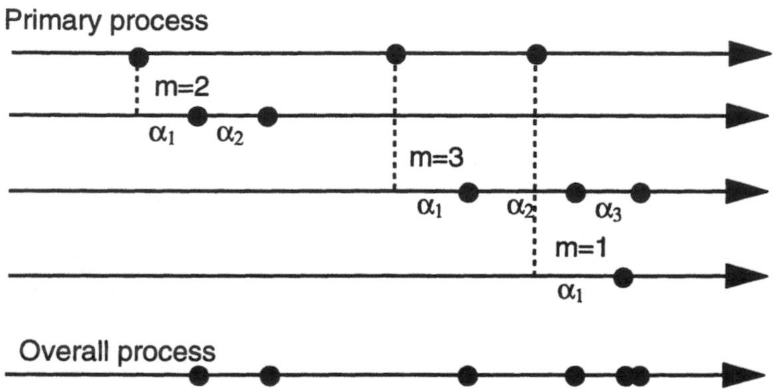

Figure 5.2. A schematic description of a Bartlett–Lewis clustering process.

Adopting this description of the traffic process, we first compute the rate of arrivals per unit time. Let $N(t,t+h]$ count the number of arrivals between time t and $t+h$. The rate is

$$p_1(t) = \lim_{h \downarrow 0} \frac{1}{h} \mathbf{E}N(t,t+h]. \tag{5.1}$$

In order to compute $\mathbf{E}N(t,t+h]$ consider the two different ways in which there can be a point in the interval: either it is a primary point, or an earlier primary point has a cluster with an event falling in $(t,t+h]$. For reasons that will be expanded upon in the next section, when h is small enough we can ignore the possibility of having more than one point in the interval. The ith cluster contribution from a cluster of size m with primary arrival at s falls in $(t,t+h]$ with probability $\int_t^{t+h} f^{*i}(u-s)du$, where f^{*i} is the i-fold convolution of f. Summing over the m cluster points, and unconditioning with respect to m and s yields

$$\int_{-\infty}^{t} \sum_{m=1}^{\infty} \int_{t}^{t+h} \sum_{i=1}^{m} f^{*i}(u-s)\, du\, g_M(m)\, \rho\, ds = \sum_{m=1}^{\infty} \int_{t}^{t+h} m\, \rho\, g_M(m)\, du$$

$$= \rho h \mathbf{E}(M), \tag{5.2}$$

so the rate corresponding to cluster points is $\rho\mathbf{E}(M)$, whence the overall rate is

$$p_1(t) \equiv p_1 = \rho(1+\mathbf{E}(M)). \tag{5.3}$$

Notice that this rate is not time-dependent. We say that this process is **first order stationary**. Generally, if the distribution of the process does not depend on the time origin we call the process **stationary**. A formal definition follows in the next section.

A simple way of assessing the dependence structure of the points is to estimate the conditional probability density of finding a point at time $t+u$, given that there was a point at t. This density is called the **autointensity function**. We can estimate it by the proportion of times that one vehicle was followed by another one nearly u time units later, say between $u-\beta/2$ and $u+\beta/2$, where β is a bandwidth parameter. Taking $\beta=10$, the function shown in Figure 5.3 results.

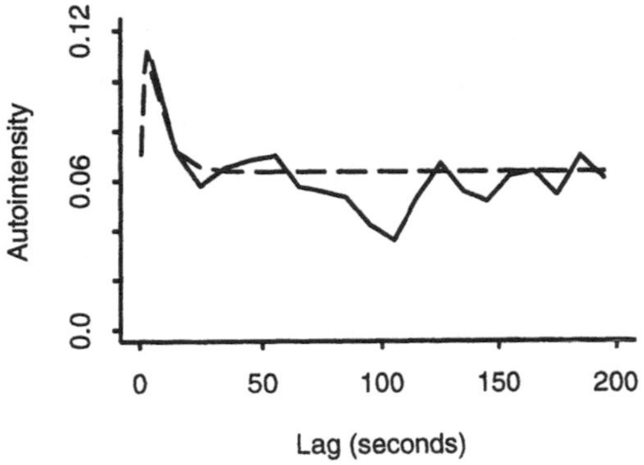

Figure 5.3. Estimated (solid line) and model-based (dashed line) autointensity functions for the Swedish vehicle data.

The clustering show up clearly in the peak at small lags, indicating a higher than average fraction of vehicles following another closely. The theoretical counterpart of the estimated autointensity will be derived in the next section. Assuming with Bartlett that the cluster point interarrival times have a $\Gamma(2,2\rho)$ distribution, and that the cluster size distribution is modified geometric with

$$g_M(i) = \begin{cases} 1-c & i=0 \\ cp\,(1-p)^{i-1} & i\geq 1 \end{cases}, \tag{5.4}$$

so $\mathbf{E}(M)=c/p$, the theoretical autointensity function is

$$h(u) = p_1 + \frac{1}{1+\mathbf{E}(M)} \sum_{i=1}^{\infty} f_i(u)(1-p)^{i-1} \tag{5.5}$$

where f_i is the density of a $\Gamma(2i,2\rho)$ distribution. Choosing $c=0.01$, $p=0.283$, and $\rho=0.966\hat{p}_1=0.0617$ cars per second, we obtain the dashed line in Figure 5.3. These parameter values were obtained by fitting the square roots of empirical and theoretical estimates using least squares (the reason for using square roots will be explained in section 5.3). We see that the general shape is correct, suggesting that the model adequately describes the traffic situation on this road as far as second order behavior (properties involving pairs of points) is concerned. The peak in the theoretical autointensity cannot be fully resolved in the empirical estimate, since the bandwidth is too large.

5.2. General concepts

Let τ_i, $i=1,2,...$ be locations (in time, space, etc.) of events and define the (set-valued) **counting process** $N(A) = \#\{i:\tau_i \in A\}$. When the events occur in time one often considers $N_t = N(0,t] = \#\{i:\tau_i \le t\}$. The process N_t is a jump process. Knowing N_t for all t, or more generally $N(A)$ for a large enough class of sets A, determines the locations of events, provided that events do not coincide. We will therefore use the point process (the location of the points) and the counting process interchangeably. One way of obtaining the separation of events in the temporal case is to require that the process is **orderly**,[1] meaning that

$$\mathbf{P}(N(t,t+\delta]>1) = o(\delta) \text{ as } \delta \to 0 \text{ for any } t. \tag{5.6}$$

We will concentrate on orderly temporal point processes until further notice. A consequence of the definition (5.6) is that $\mathbf{P}(\exists t: N(\{t\}>1))=0$ (Daley, 1974), i.e., that there are no multiple points.

Calculations in point process theory can often be done using an informal differential calculus. Write $dN_t = N(t,t+dt] = N_{t+dt} - N_t$ where dt is a small (infinitesimal) time increment. For an orderly process we have $\mathbf{E}\,dN_t = \mathbf{P}(dN_t = 1)$, since each infinitesimal interval will contain either 0 or 1 points, but not more. Furthermore,

$$\mathbf{Var}\,dN_t = \mathbf{E}\,dN_t^2 - (\mathbf{E}\,dN_t)^2 = \mathbf{E}\,dN_t + o(\mathbf{E}\,dN_t) = \mathbf{E}\,dN_t + o(dt) \tag{5.7}$$

indicating that the process is locally Poisson: for very small time intervals it behaves like a Poisson process at least as regards first- and second-order moments.

Formal computations can also be done using the differential notation. Recall that for a nondecreasing function $G(x)$ we can define the **Stieltjes integral** $\int f(x)\,dG(x)$ for suitable functions $f(x)$ by the limit of the Riemann–Stieltjes sums

$$\sum f(x_i)(G(x_{i+1}) - G(x_i)) \tag{5.8}$$

as the division points x_i get closer and closer. This allows us to define a **stochastic Stieltjes integral** with respect to a counting process N_t as

$$\int_A f(t)\,dN_t = \sum_{\{i:\tau_i \in A\}} f(\tau_i). \tag{5.9}$$

[1]The term orderly refers to the property that the points of the process come in a natural order. It is a translation of a Russian term that also can be interpreted as **ordinary**.

Relating temporal counting processes to the material in Chapter 3, we see that (N_t) is Markovian if and only if the **inter-event times** $Y_i=\tau_i-\tau_{i-1}$ are independent, exponentially distributed random variables, i.e., (N_t) is a pure birth process. Most interesting point processes are non-Markovian.

In order to describe a point process completely one must specify all probabilities

$$\mathbf{P}(N(A_1)=k_1, \ldots, N(A_n)=k_n) \tag{5.10}$$

for all n, $k_1, \ldots, k_n$, and all measurable sets $A_1, \ldots, A_n$. There are several equivalent ways of doing this. We will give three different quantities that determine the point process. The first is the **probability generating functional** (pgfl), defined by

$$G(\xi) = \mathbf{E}\prod_1^\infty \xi(\tau_i) = \mathbf{E}\left[\exp(\int_0^\infty \log\xi(u)dN_u)\right] \tag{5.11}$$

where $\xi(t)$ is a function such that $0\le\xi(t)\le1$ and $\xi\equiv1$ outside some bounded set. This is a generalization of the usual probability generating function.

Example (The Poisson process) Consider a Poisson process with (possibly nonstationary) rate $p_1(t)$. Suppose that ξ is identically one outside a bounded interval I. Let $P_1(I)=\int_I p_1(u)du$, so $N(I)\sim\text{Po}(P_1(I))$. Let $t_i=t_{i-1}+h$, $i=1,2,\ldots$ be a partition of the line. Then, writing $\theta_i=P_1(t_i,t_{i+1}]$, we calculate

$$\begin{aligned}
G(\xi) &= \lim_{h\downarrow0}\mathbf{E}\exp(\sum_{i=1}^\infty \log\xi(t_i)(N_{t_{i+1}}-N_{t_i})) \\
&= \lim_{h\downarrow0}\prod_{i=1}^\infty\sum_{k_i=0}^\infty \xi(t_i)^{k_i}\exp(-\theta_i)\theta_i^{k_i}/k_i! \\
&= \lim_{h\downarrow0}\exp(-\sum\theta_i(1-\xi(t_i))) = \exp(-\int_{\mathbb{R}}(1-\xi(t))p_1(t)dt.
\end{aligned} \tag{5.12}$$

In particular, choosing $\xi(t)=1-(1-s)1_A(t)$, we get

$$G(\xi) = \mathbf{E}s^{N(A)} = \exp(-\int_A(1-s)p_1(t)dt) = \exp(-(1-s)P_1(A)), \tag{5.13}$$

showing that (indeed) $N(A)\sim\text{Po}(P_1(A))$. □

Example (Superposition) Consider two independent point processes M^1 and M^2 with pgfl G^1 and G^2, respectively. The **superposition** $N(A)=M^1(A)+M^2(A)$ of the two processes, i.e., the collection of all points without regard for which process they came from, has pgfl

$$G(\xi) = \mathbf{E}\exp\int\log\xi(t)(dM_t^1 + dM_t^2)$$
$$= \mathbf{E}\exp\int\log\xi(t)dM_t^1 \times \mathbf{E}\exp\int\log\xi(t)dM_t^2 = G^1(\xi)G^2(\xi). \quad (5.14)$$

□

Example (Cluster processes) Consider a pair of point processes, M^c producing **primary** events (or **cluster centers**), and $M^s(\bullet \mid t)$ producing **secondary** events, or a cluster associated with the primary event at t. We assume that M^c generates independent clusters $M^s(\bullet \mid \tau_i)$. Denote the pgfl of M^c by $G^c(\xi)$, and that of $M^s(\bullet \mid t)$ by $G^s(\xi \mid t)$. The process of all secondary points is the superposition of the secondary processes corresponding to each of the cluster points. Thus, conditional upon the primary points the pgfl of all secondary points is $\prod_i G^s(\xi \mid \tau_i)$, so if the observed cluster process N consists of only secondary points it has pgfl

$$G(\xi) = G^c(G^s(\xi \mid \bullet)), \quad (5.15)$$

while if the observed process also counts the primary points it has pgfl

$$G(\xi) = G^c(\xi)G^c(G^s(\xi \mid \bullet)). \quad (5.16)$$

We sometimes assume that the clusters are identically distributed in the sense that $M^s(I \mid t) = M^s(I-t)$. Then

$$G^s(\xi \mid t) = \mathbf{E}\exp(\int\log\xi(u)M^s(du-t)) = \mathbf{E}\exp(\int\log\xi(u+t)dM^s(u))$$
$$= G^s(\xi(t+\bullet)). \quad (5.17)$$

The special case of **Poisson cluster processes**, where M^c is a Poisson process of rate $p^c(t)$ has pgfl

$$G(\xi) = \exp\left[\int(G^s(\xi \mid t)-1)p^c(t)\,dt\right], \quad (5.18)$$

provided that the primary points are not included.

The two main clustering models are due to Bartlett and Lewis (Bartlett, 1963, applied it to the traffic data in section 5.1, while Lewis, 1964, used it to analyze computer failure patterns) and to Neyman[1] and Scott (Neyman, 1939, described the migration of larvae on a field, while Neyman and Scott, 1959, modeled the large scale organization of the distribution of galaxies). Both models usually have stationary primary Poisson processes. Traditionally (because of the original applications) the Bartlett–Lewis model includes the primary points, while the Neyman–Scott model does not. As we saw in section

[1] Neyman, Jerzy (1894–1981). Polish statistician. Invented the theory of confidence intervals, and (with Egon Pearson) founded the theory of testing statistical hypothesis.

5.1, in the Bartlett–Lewis process the secondary events are generated as the successive cumulative sums of a random number M of iid random variables. For definiteness we assume that both M and the interval distribution f have finite means. The Neyman–Scott process differs from the Bartlett–Lewis process in that it generates M secondary points iid around a primary point.

In the Bartlett–Lewis case, we can write

$$G^s(\xi \mid t) = \sum_{m=0}^{\infty} g_M(m) \int f(v_1)\xi(t+v_1) \cdots$$

$$\int f(v_m)\xi(t+v_1+\cdots+v_m)dv_1 \cdots dv_m, \tag{5.19}$$

whereas in the Neyman–Scott case, we have

$$G^s(\xi \mid t) = \sum_{m=0}^{\infty} g_M(m) \int f(v_1)\xi(t+v_1) \cdots \int f(v_1)\xi(t+v_m)\, dv_1 \cdots dv_m$$

$$= P_M\Big(\int \xi(t+v) f(v) dv\Big), \tag{5.20}$$

where $P_M(s)$ is the pgf of M. □

Example (The output of an M/G/∞ queue) The arrivals to the queue form a Poisson process. Each customer is immediately served, with service times σ_i iid with density $f(x)$. Hence we can think of the output process as a Poisson process where each point τ is **randomly translated** to $\tau+\sigma$. But this is a Neyman–Scott Poisson cluster process in which each cluster has size 1 with probability 1. Hence the pgfl is

$$G(\xi) = G_c\Big(\int_{\mathbb{R}} \xi(\bullet+u) f(u) du = \exp\Big(-\int_{\mathbb{R}} (1-(\int_{\mathbb{R}} \xi(t+u) f(u) du)) p_1 dt\Big)$$

$$= \exp\Big(-\int_{\mathbb{R}^2} (1-\xi(t+u) p_1 dt) f(u) du\Big) \tag{5.21}$$

$$= \exp\Big(-p_1 \int_{\mathbb{R}^2} (1-\xi(v) dv) f(u) du$$

$$= \exp\Big(-p_1 \int_{\mathbb{R}} (1-\xi(v)) dv\Big) = G_c(\xi),$$

so the output process is again Poisson with the same rate as the input process. □

A different way to describe completely the distribution of a point process is (somewhat surprisingly, perhaps) the **zero probability function**

$$\zeta(A) = \mathbf{P}(N(A)=0). \tag{5.22}$$

This is obtained from G by setting $\xi = 1_{A^c}$ and, being a special case, would seem to contain much less information. The trick is that one must be able to specify ζ for all measurable sets.

Example (The Poisson process, continued) Setting $s=0$ in (5.13) we see that $\zeta(A) = \exp(-P_1(A))$. □

Example (The doubly stochastic Poisson process) Let $X(t)$ be a non-negative stochastic process. Given a realization x of X, generate a nonhomogeneous Poisson process with rate $x(t)$. This process was introduced by Cox (1955), and is called the **doubly stochastic Poisson process** (or the Cox process, although this name could properly be attached to a large variety of processes). The zero probability function is

$$\zeta(A) = \mathbf{P}(N(A)=0) = \mathbf{E}_X \exp(-\int_A X(t)\,dt). \tag{5.23}$$

A very simple case is when X is a two-state Markov process taking the values 0 and λ. This corresponds to a system were events are generated only under certain conditions (namely when $X>0$). Let $\pi(0)$ be the equilibrium probability that $X=0$. If X starts in equilibrium we have

$$\zeta(A) = \pi(0) + (1-\pi(0))\exp(-\lambda\,|A|) \tag{5.24}$$

where $|A|$ is the size (Lebesgue measure) of A. This model has been applied to rainfall (Smith and Karr, 1985). □

Example (Discretization) Rainfall processes are often recorded as hourly measurements in gauges. A simple approach to modeling is to assume that the hourly amounts, given that it rains, are conditionally independent from hour to hour, and think of the presence or absence of rainfall as a point process in discrete time (or **binary series**). We think of it as arising from a point process of storm events (rainfall occurrences) proceeding in continuous time. Let M_t again be an orderly, stationary point process on the line. Define the binary series

$$Z_k = 1(M(I_k)>0) \quad , \quad k=0,1,2,\cdots \tag{5.25}$$

where $I_k = \Delta(k, k+1]$, and Δ is the discretization interval. By using stationarity, it can be seen that

$$m = \mathbf{E}Z_k = \Pr\{M(I_k)>0\} = 1-\zeta(I_0). \tag{5.26}$$

A similar computation shows that

$$c_k = \mathrm{Cov}(Z_i, Z_{i+k}) = \zeta(I_0 \cup I_k) - \zeta^2(I_0). \tag{5.27}$$

Such moments are as useful to describe binary series derived from point processes as the corresponding product moment densities are for describing the continuous time point process itself. □

The third way of describing a point process is by using the **complete intensity function**

$$\lambda(t \mid \mathcal{N}_t) = \lim_{h \to 0} \frac{1}{h} \mathbf{P}(N(t, t+h] = 1 \mid \mathcal{N}_t) \tag{5.28}$$

where $\mathcal{N}_t$ is the history of the process up to time t. In general this will be a random function. We interpret it as the density of points at t, given the previous history.

Example (The Poisson process, continued) In section 3.3 we showed that the Poisson process has independent increments, so that, in particular,

$$\mathbf{E}(dN_t \mid \mathcal{N}_t) = \mathbf{E}\, dN_t = p_1(t)dt. \tag{5.29}$$

Hence the complete intensity function for a Poisson process is its rate function. Notice that it is deterministic, reflecting the independence of the past. □

Example (A self-exciting process) It is sometimes convenient to define a process by means of its complete intensity. The linear self-exciting process (Hawkes, 1971) is defined through the equation

$$\lambda(t \mid \mathcal{N}_t) = \gamma + \int_{-\infty}^{t} w(t-z)dN_z \tag{5.30}$$

where $\gamma > 0$ is a constant, and w is a non-negative function satisfying $w^* = \int_{\mathbb{R}} w(u)du < 1$. The rate satisfies

$$p_1 = \mathbf{E}\lambda(t \mid \mathcal{N}_t) = \gamma + \int_{-\infty}^{t} w(t-z) p_1\, dt = \gamma + w^* p_1, \tag{5.31}$$

so $p_1 = \gamma/(1-w^*)$. Hawkes introduced this process to describe earthquakes, and it was applied to microearthquakes in southern California by Udias and Rice (1975). □

Example (Likelihood) Let N_t be a Poisson process of rate $p_1(t,\theta)$ where θ is a parameter to estimate. The likelihood of a realization $\tau_1,\tau_2,\ldots,\tau_{N_T}$ can be obtained by discretizing $(0,T]$ into intervals of width h, and noting that events in disjoint intervals are independent. Hence, with $I_n=((n-1)h,nh]$, and writing $I(\tau_i)$ for the I_n containing τ_i, which is unique for small enough h, we have

$$\begin{aligned} L_T(\theta) &= \prod_{k=1}^{T/h} \frac{P_1(I_k)^{N(I_k)}}{N(I_k)!} \exp(-P_1(I_k)) \\ &= \left\{ \prod_{k=1}^{N_T} P_1(I(\tau_k)) \right\} \exp(-\sum_{k=1}^{T/h} P_1(I_k)) \\ &= h^{N_T} \left\{ \prod_{k=1}^{N_T} p_1(\tau_k;\theta) \right\} \exp(-\int_0^T p_1(u;\theta)du). \end{aligned} \tag{5.32}$$

Taking logarithms, and discarding the constant h^{N_T}, we can write

$$l_T(\theta) = \sum_{i=1}^{N_T} \log p_1(\tau_1;\theta) - \int_0^T p_1(u;\theta)du. \tag{5.33}$$

One of the important features of the complete intensity is that every point process that does not have points at predetermined times can be thought of as a Poisson process where time runs at a changed speed. Let M_t be a unit rate Poisson process, and let N_t have complete intensity $\lambda(t \mid \mathcal{N}_t)$. Write $\Lambda(s)=\int_0^s \lambda(t \mid \mathcal{N}_t)dt$. Then the point processes $N_{\Lambda}(t)$ and M_t have the same distribution. Consequently, the likelihood corresponding to points of N at $\tau_1,\ldots,\tau_{N_T}$ is the same as the likelihood of a unit rate Poisson process having points at $\Lambda(\tau_1),\ldots,\Lambda(\tau_{N_T})$. The same argument as above shows that if the complete intensity depends on a parameter θ, we have

$$l_T(\theta) = \sum \log \lambda(\tau_i;\theta \mid \mathcal{N}(\tau_i-)) - \int_0^T \lambda(u;\theta \mid \mathcal{N}_t)dt \tag{5.34}$$

where $\mathcal{N}(\tau_i-)$ is the history of events before τ_i. □

There are several parameters that provide a partial description of a point process, even though there may be many processes with the same parameters. The following are some of the more important ones:

- The **rate** $p_1(t)=\mathbf{P}(\text{point in } (t,t+dt])/dt=\lim_{h\to 0}\mathbf{P}(N(t+h)-N(t)=1)/h$. This describes the average number of points per time unit.
- The **autointensity function**

$$h(t,u) = \mathbf{P}(\text{point in } (u,u+du] \mid \text{point at } t)/du = \frac{p_2(t,u)}{p_1(t)} \tag{5.35}$$

where p_2 is given below. This is the conditional density of an event at u, given an event at t.

• The **kth order product moment**

$$p_k(t_1,\ldots,t_k) = \mathbf{P}(\text{points in } (t_i,t_i+dt_i\}, i=1,\ldots,k)/\prod_1^k dt_i. \tag{5.36}$$

Note that for an orderly process we can interpret $p_k(t_1,\ldots,t_k)\prod dt_i$ as $\mathbf{E}\prod_1^k dN_{t_i}$, whence the name product moment.

Example (The Poisson process, continued) For a Poisson process with rate $\lambda(t)$ we have (using l'Hôpital's rule)

$$p_1(t) = \mathbf{P}(dN_t=1)/dt = \lim_{h\to 0}(1-\exp(-\int_t^{t+h}\lambda(u)\,du))/h = \lambda(t). \tag{5.37}$$

By the independence of events in disjoint intervals,

$$p_k(t_1,\ldots,t_k) = \prod_1^k p_1(t_i), \tag{5.38}$$

so that, in particular,

$$h(t,u) = p_1(u). \tag{5.39}$$

□

As described in section 5.1, a point process is **stationary** if its distribution is independent of the origin, or, put differently, if for all $n, t_1,\ldots,t_n$,

$$\mathbf{P}(N(A_i+t_i)=k_i, i=1,\ldots,n) = \mathbf{P}(N(A_i)=k_i, i=1,\ldots,n) \tag{5.40}$$

where $A+t=\{x+t:x\in A\}$. The statistical analysis of stationary point processes is often simpler than the general case.

Example (The Poisson proces, continued) A stationary Poisson process has $p_1(t)\equiv p_1$, so that

$$p_k(t,t+t_1,t+t_2,\ldots,t+t_{k-1}) \equiv p_k(t_1,\ldots,t_{k-1}) = p_1^k \tag{5.41}$$

(notice that for stationary processes the product moment only depends on the $k-1$ differences between the arguments, so that the parameter function only needs $k-1$ arguments). □

Example (Poisson cluster processes) For any Poisson cluster model with stationary primary rate p^c the overall rate is (as we saw in section 5.1)

$$p_1 = p^c(\mathbf{E}M+1) \tag{5.42}$$

if primary points are included, and

$$p_1 = p^c \mathbf{E} M \tag{5.43}$$

if they are not.

To compute the second order product moment, note that two points at t and u either belong to separate clusters, in which case they occur independently, or belong to the same cluster with primary point at some v. For a Neyman–Scott model, we have

$$\mathbf{P}(dN_t = dN_u = 1) = (p^c \mathbf{E} M)^2 dt\, du + \{\mathbf{E} M(M-1) \int_{\mathbb{R}} f(u-v) f(t-v) p^c\, dv\}\, dt\, du, \tag{5.44}$$

so that the autointensity becomes

$$h(u) = p_1 + \frac{\mathbf{E} M(M-1)}{\mathbf{E} M} f_D(u) \tag{5.45}$$

where f_D is the density of the difference of two secondary displacements.

For the Bartlett–Lewis process, a similar computation yields

$$h(u) = p_1 + \frac{1}{\mathbf{E} M+1} \mathbf{E}\left(\sum_{k=1}^{M} (M-k) f^{*k}(u)\right). \tag{5.46}$$

Example (Discretization, continued) Many point process parameters have natural counterparts for the derived binary series. For example, the autointensity function $h(u)$ corresponds to the binary series parameter

$$h_k = \Pr\{Z_k=1 \mid Z_0=1\} = 1 - \frac{\zeta(I_0) - \zeta(I_0 \cup I_k)}{1-\zeta(I_0)} = m + \frac{c_k}{m}, \tag{5.47}$$

using equations (5.26) and (5.27). □

5.3. Estimating second-order parameters for stationary point processes

Let $(N_t, t \le T)$, or equivalently $(\tau_i, i \le N_T)$, be a stationary and orderly temporal point process. We are interested in estimating the second order product moment $p_2(u)$, or the autointensity function $h(u)$. For simplicity we will also assume that the process is **mixing**, which essentially means that events far apart are nearly independent. This implies that $p_2(u) \to p_1^2$ as $u \to \infty$. For the formal definition of mixing, see Brillinger (1975).

Let $I^T(u) = \#\{j,k : u-\beta/2 < \tau_j - \tau_k < u+\beta/2,\ j \ne k\}$. Here $\beta = \beta(T)$ is the binwidth. We divide the time axis into $[T/\beta]$ bins, and compute $I^T(u)$ for u at the midpoint of these bins. Computationally, we start from the first point, compute how far away the next point is, divide by β, take the integer part, and add one to that bin. Repeat for the distance between the first and the third point, and so on. In other words, we form a histogram of all differences (see Exercise C1). Estimate p_2 by

$$\hat{p}_2(k\beta) = I^T(k\beta)/\beta(T-k\beta). \tag{5.48}$$

We can write

$$I^T(u) = \iint\limits_{|s-t-u|<\beta/2, s\neq t, 0\leq s,t\leq T} dN_s dN_t \tag{5.49}$$

so

$$\mathbf{E}I^T(u) = \iint\limits_{|s-t-u|<\beta/2, s\neq t, 0\leq s,t\leq T} p_2(s-t)dsdt = \int_{r=u}^{T}\int_{v=u-\beta/2}^{u+\beta/2} p_2(v)dvdr$$

$$= (T-u)\int_{u-\beta/2}^{u+\beta/2} p_2(v)dv \doteq (T-u)\beta p_2(u) \tag{5.50}$$

if $p_2(\bullet)$ is smooth near u and β is small. Thus $\hat{p}_2(u)$ is approximately unbiased. A similar computation shows that $\mathbf{Var}I^T(u) \doteq \beta T p_2(u)$, suggesting that $I^T(u)$ may be approximately Poisson distributed. In fact, we have the following result from Brillinger (1975).

Theorem 5.1 If $p_2(u)$ is continuous, then

(i) if $\beta(T)=L/T$ for a fixed L, then

$$\hat{p}_2(u) \xrightarrow{d} \frac{1}{L}\mathrm{Po}(Lp_2(u)) \tag{5.51}$$

as $T\to\infty$, and estimates at points separated by at least $\beta(T)$ are asymptotically independent.

(ii) If $T\beta(T)\to\infty$, then $\hat{p}_2$ is asymptotically normally distributed, and estimates at points separated by at least $\beta(T)$ are asymptotically independent.

Remark (1) From result (i) we are led to apply a variance stabilizing transformation: $\hat{p}_2(u)^{1/2}$ has variance approximately $1/(4L)$. We are often interested in the autointensity function $h(u)$, which we estimate by

$$\hat{h}(u) = \frac{T\hat{p}_2(u)}{N_T}. \tag{5.52}$$

The asymptotic distribution of $\hat{h}(u)$ is $\mathrm{Po}(Lp_2(u))/Lp_1$, so $\hat{h}(u)^{1/2}$ has approximate variance $(4Lp_1)^{-1}$, which we can estimate by $(4LN_T/T)^{-1}=(\beta N_T)^{-1}$. For large L we can use the normal approximation to get pointwise confidence bands.

(2) In the Poisson case, Cox (1975) computes

$$\mathbf{E}I^T(u) = \beta(T-u)p_1^2; \tag{5.53}$$

$$\mathbf{Var}I^T(u) = \beta(T-u-2\beta)p_1^2 + O(\beta^3); \tag{5.54}$$

$$\mathbf{Cov}(I^T(u),I^T(v)) = -2p_1^2\beta^2 + O(\beta^3). \tag{5.55}$$

This can be used to test the hypothesis of a Poisson process. □

Application (Swedish traffic data) We will apply the asymptotic theory above to investigate the significance of the peaks in the autointensity function estimates of section 5.1. Figure 5.4 shows the square root of the autointensity function of the complete data, with a 95% pointwise confidence band, as developed above. It is clear that no straight line could fit in this band, so that the traffic data deviates from a Poisson process. On the other hand, the cluster model of Bartlett mostly fits inside the band, and is a good second order description. The only exception is at a lag of about 100. However, the confidence bands are pointwise confidence bands, and we should not be surprised to see occasional values outside the bands.

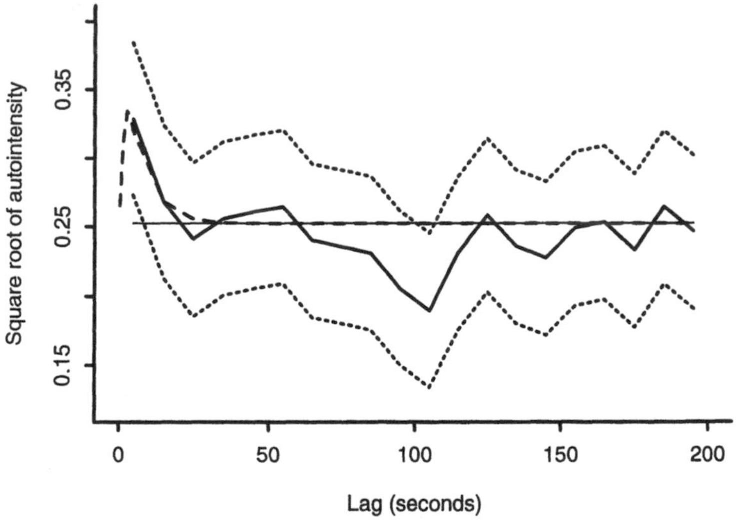

Figure 5.4. Square root of autointensity function estimate for the Swedish traffic data with 95% pointwise confidence bands (dotted lines). The dashed line is the fitted estimate based on the Bartlett–Lewis model, and the thin horizontal line corresponds to a Poisson process.

□

5.4. Relationships between processes

In section 3.8 we described the structure of nerve cells, and developed some models for the internal kinetics of them. The input to a nerve cell arrives through its dendrites, root-like strings that react to chemical stimuli (see Figure 3.10). The resulting current flow is converted to a membrane potential throughout the cell body. When the potential reaches a threshold, the neuron sends an action potential, or spike, along the axon to synapses, where the action potential is transmitted to the next neuron. Synaptic connections may be excitory or inhibitory, i.e., making the receiving cell more or less likely to fire. The voltage spikes corresponding to the action potential are usually very similar in size and shape. Figure 5.5 shows measured voltage fluctuations within cell R2 of *Aplysia californica* (the sea hare; a popular experimental animal since its neurons are easily identified).

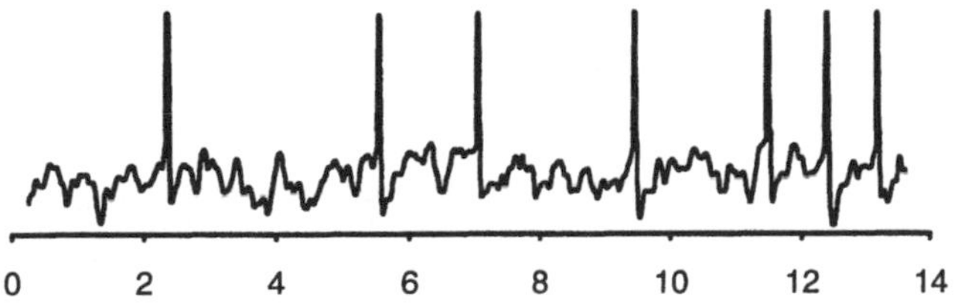

Figure 5.5. Transmembrane potential for R2 of *Aplysia.* Adapted from Brillinger (1992), with permission from the *Journal of the American Statistical Association.* Copyright 1992 by the American Statistical Association. All rights reserved.

It is quite reasonable to model the voltage spikes as events in a point process. Given some firings from a network of cells, we would like to infer the causal connections between the neurons. Figure 5.6 shows experimental data for two *Aplysia* neurons, L3 and L10. Let the firing times of L10 be denoted by σ_j, and those of L3 by τ_k. The plot, called a **rastor plot**, shows for each firing σ_j of L10 the time differences $\tau_k - \sigma_j$ plotted at vertical height j for all L3 firings within just over 4 seconds of the L10 firing.

There is a clear pattern in this plot, with a tendency for L3 to be inhibited by a firing of L10, and a rebound with higher rate of L3 firings about 0.5 seconds after an L10 firing. To present these observations more clearly, we introduce the **cross-intensity function** for two stationary point processes, $M=\{\sigma_j\}$ and $N=\{\tau_k\}$, by

$$h_{N|M}(u)du = \mathbf{P}(dN_{u+t}=1 \mid dM_t=1) \tag{5.56}$$

and estimate it, as for the autointensity functions, by a histogram type estimator

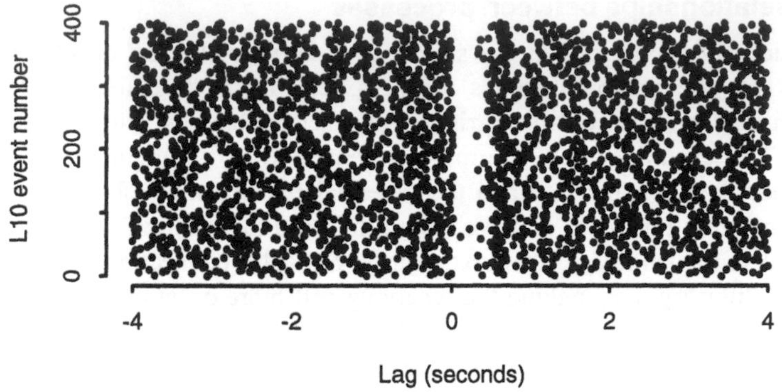

Figure 5.6. Rastor plot of L3 firings relative to L10 firings. Each line corresponds to a single L10 firing. Adapted from Brillinger (1992), with permission from the *Journal of the American Statistical Association.* Copyright 1992 by the American Statistical Association. All rights reserved.

$$\hat{h}_{N|M}(u) = \frac{\#\{u+\sigma_j-\beta/2<\tau_k\leq u+\sigma_j+\beta/2\}}{\beta M_T} = \frac{J_{MN}(u)}{\beta M_T}. \tag{5.57}$$

In effect we are counting the points in vertical strips of Figure 5.6. The properties of this estimator are given by Brillinger (1975), and are very similar to those of the autointensity estimator.

Theorem 5.2 Assume the mixing condition mentioned in section 5.3, and that $p_{MN}(u)=\mathbf{E}\, dM_t dN_{t+u}$ is a continuous function. Let $\beta=\beta_T$ depend on T. Then, for u_i separated by at least β_T we have as $T\to\infty$ with $\beta_T/T\to L$ that $J_{MN}(u_1),\ldots,J_{MN}(u_k)$ are independent Poisson random variables with means $\beta_T p_{MN}(u_k)$.

Remark It follows as in Proposition 5.1 that if $\beta_T\to 0$ with $T\beta_T\to\infty$, the $J_{MN}(u_i)$ are asymptotically normal. Likewise we see that $(\hat{h}_{N|M}(u_i))^{1/2}$ are asymptotically normal with variance that can be estimated consistently by $(4\beta M_T)^{-1}$. Thus we generally plot the cross-intensity on a square root scale. □

The estimate of $\hat{h}_{N|M}(u)$ shown in Figure 5.7 illustrates the strong refractory effect following a L10-firing. In addition, we notive an increased intensity of L3-firings just after the end of the refractory period.

In order to sharpen this qualitative description of the relationship between L3 and L10, we attempt a more structured model. To do so, we return to the idea that the neuron fires when voltage exceeds a threshold. This threshold depends on time, resets to a high level when the neuron fires (thereby prohibiting the cell from firing immediately after another firing), and then decays slowly. We first

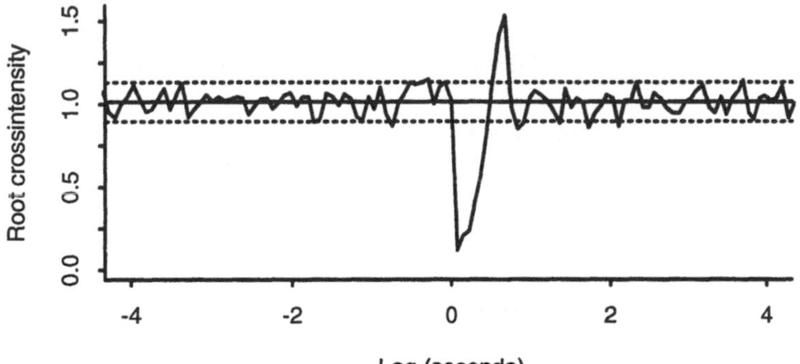

Figure 5.7. Square root of cross-intensity of L3 relative to L10. The dashed lines are asymptotic standard errors, displayed around the line of constant cross-intensity for readability. Adapted from Brillinger (1992), with permission from the *Journal of the American Statistical Association.* Copyright 1992 by the American Statistical Association. All rights reserved.

need to describe the voltage $U(t)$ in the cell body. We assume that each incoming pulse triggers a time-dependent potential, so that we can write

$$U(t) = \int_0^t a(t-u)dM(u). \tag{5.58}$$

The threshold function can be written

$$V^*(t) = b(\gamma_t) \tag{5.59}$$

where $\gamma_t = t - \tau_{N_t}$ is the **backward recurrence time**, i.e., the time since the last N-event. The idea is that the neuron fires when $U(t)$ equals $V^*(t)$. Generally this relation is not exact. Rather, we need to superimpose random noise over the threshold, yielding

$$V(t) = b(\gamma_t) + \varepsilon_t. \tag{5.60}$$

Let $\psi_t = U(t) - V(t)$, and write the conditional intensity

$$\lambda(t \mid \mathcal{H}_t) = \mathbf{P}(dN_t = 1 \mid \mathcal{H}_t)/dt = P(\psi_t) \tag{5.61}$$

where $\mathcal{H}_t$ is the joint history of the two processes up to time t, and P is the cdf of ε. From equation (5.34) we see that the the log likelihood based on observing the two processes in $(0,T)$ is

$$\int_0^T \log P(\psi_t)dN_t - \int_0^T P(\psi_t)dt, \tag{5.62}$$

which needs to be maximized. This is a tricky optimization problem, since we do not have parametric forms for the functions a and b, and we will attack it by discretization. Let $n_k = N_{k\Delta} - N_{(k-1)\Delta}$, where Δ is a small discretization interval, taken so that n_k is 0 or 1, and define m_k similarly from M. Write

$$u_k = \sum_0^K a_{k-l} m_l; \quad v_k = b_{k-\gamma_k} + \varepsilon_k: \quad \psi_k = u_k - v_k \tag{5.63}$$

where $\gamma_k = \max\{j \le k : n_j = 1\}$. The discrete approximation to the log likelihood becomes

$$\sum_{k=1}^K \log P(\psi_k) n_k - \Delta \sum_{k=1}^K P(\psi_k), \tag{5.64}$$

which can be maximized by iteratively reweighted least squares in the context of generalized linear models (McCullagh and Nelder, 1989).

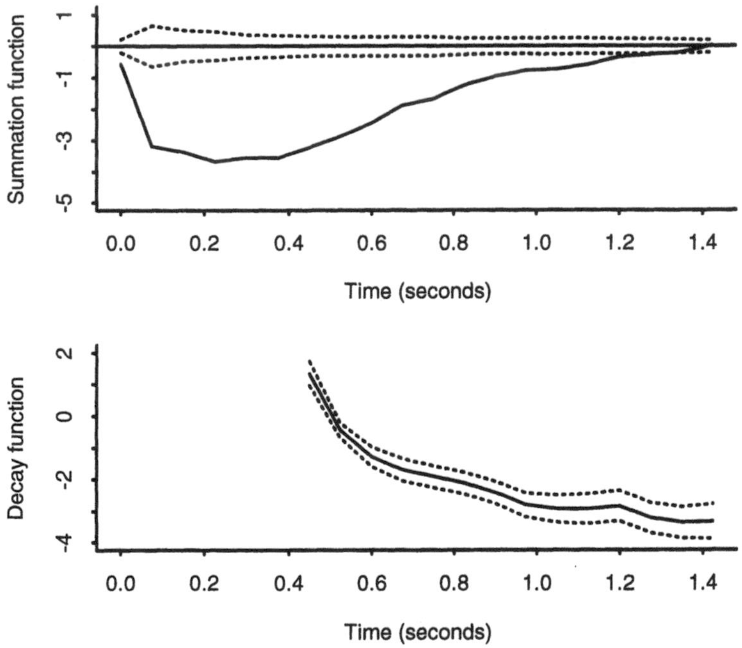

Figure 5.8. Estimates of the summation function a (upper panel) and the decay function b (lower panel) for the *Aplysia* L3 and L10 data. The standard error for a is exhibited around the horizontal axis (instead of around the estimate) for readability. Adapted from Brillinger (1992), with permission from the *Journal of the American Statistical Association*. Copyright 1992 by the American Statistical Association. All rights reserved.

Figure 5.8 shows the estimates of a_k and b_k with standard errors obtained from generalized linear model theory. Here P is taken to be Φ, the standard normal cdf, and Δ=0.75 s. We see that a_k immediately goes negative, producing the inhibition noted earlier. In other words, for up to 1.2 seconds after a firing of L10, the probability of L3 firing is relatively low. The decay function is infinite for the first five coefficients, since no pair of output spikes occurred closer than 0.49 s for this data set.

5.5. Modeling the complete intensity

Recall from section 5.2 that the complete intensity is defined as

$$\lambda(t \mid \mathcal{N}_t)\, dt = \mathbf{E}(dN_t \mid \mathcal{N}_t) \tag{5.65}$$

and completely determines the distribution of the process. Hence it is natural, particularly when a stationarity assumptions seems tenuous or incorrect, to try to write down explicit models for (5.65) based on the locations of previous points. A simple, yet very flexible, class of models is Aalen's (1978) **multiplicative intensity model** given by

$$\lambda(t \mid \mathcal{N}_t) = \alpha(t)\, Y(t) \tag{5.66}$$

where $Y(t)$ is a function of previous points, and hence a stochastic process in its own right, and $\alpha(t)$ is a time-dependent parameter.

Example (The linear BIDE process, continued) As in Chapter 3, let $(B(t),I(t),D(t),E(t))$ be the number of births, immigrants, deaths, and emigrants, respectively, during the time interval $(0,t]$. The observed BIDE process is

$$X(t) = X(0) + B(t) + I(t) - D(t) - E(t). \tag{5.67}$$

The conditional intensity for births is

$$\mathbf{E}(dB(t) \mid \mathcal{H}_t) = X(t)\lambda_t \tag{5.68}$$

where λ_t is the (possibly time-dependent) birth rate, and $\mathcal{H}_t$ is the joint history of the four-dimensional process. Similar equations can be written for the other components. □

In order to estimate $\alpha(t)$ from data, we first introduce

$$A(t) = \int_0^t \alpha(u)du. \tag{5.69}$$

Since $\mathbf{E}(dN_u \mid \mathcal{N}_t)=\alpha(u)Y(u)du$, a method of moments approach suggests estimating $A(u)$ by

$$\hat{A}(t) = \int_0^t \frac{dN(u)}{Y(u)} \tag{5.70}$$

where we interpret 0/0 as 0. The following result is proved by Aalen (1978).

Theorem 5.3 As $t \to \infty$, for any $0<\gamma<1$ $\hat{A}(\gamma t) - A(\gamma t)$ is asymptotically normal with mean 0 and variance $\sigma^2(\gamma t)$ which is consistently estimated by

$$\hat{\sigma}^2(\gamma t) = \int_0^{\gamma t} \frac{dN(s)}{Y(s)^2}. \tag{5.71}$$

Application (Psychiatric admissions) In 1972, Danish law changed to make induced abortion an elective right of all women during the first trimester of pregnancy. A research program was set up to study the effect of this liberalization of the law. Andersen and Rasmussen (1986) reported one such study, in which the Danish psychiatric registry was cross-referenced with the registries of birth and of induced abortion, so that all admissions to and discharges from psychiatric hospitals between October 1973 and December 1976 could be studied for all women with a birth or an induced abortion in 1975.

Let A denote the women with induced abortions during 1975, and B those who gave birth during 1975. There were n_A=26,657 individuals in group A, and n_B=71,378 in group B. Group A had 655 psychiatric admissions during the time of study, for a rate of 0.025, while group B had 624 admissions and a rate of 0.0087. Thus the admission rate was roughly three times larger in group A than in group B. The counting processes $N_A(t)$ and $N_B(t)$ are taken relative to the abortion/birth, rather than relative to calendar time. Thus $N_A(t)$ counts the number of admissions in group A up to t days after induced abortion. The analysis below focuses on the time intervals I_A = (–2 months, 12 months), where 2 months is the average gestation time at abortion, and I_B = (–9 months, 12 months).

Each individual in the study has her own counting process $N_g^i(t)$, $i=1,\ldots,n_g$, $g=A,B$. The overall counting process can be written

$$N_g(t) = \sum_{i=1}^{n_g} N_g^i(t). \tag{5.72}$$

A simple model of the intensity is of the form (5.66) where

$$Y_g(t) = \sum Y_g^i(t) = \sum 1(i\text{th individual not admitted at time } t) \tag{5.73}$$

counts the individuals **at risk** for psychiatric admission.

We estimate the parameters $A_g(t)$ using the Aalen estimators given in (5.70). The estimate $\hat{A}_A(t)$ is approximately linear, corresponding to a constant $\hat{\alpha}_A(t)$, while $\hat{A}_B(t)$ has a clear break immediately after t=0, indicating a higher risk of psychiatric admission immediately after giving birth. The slope of $\hat{A}_A$ is 3–4 times higher than that of $\hat{A}_B$, corresponding to the higher rate of admissions in the abortion group. □

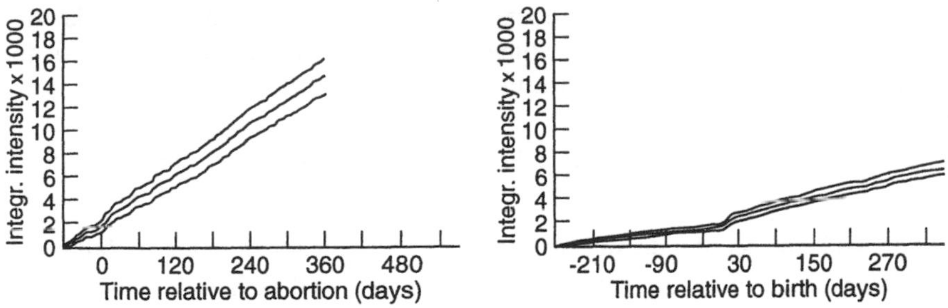

Figure 5.9. Aalen estimators of intensities of psychiatric admissions for women with induced abortions (left panel) and births (right panel). Reproduced with permission from Andersen and Rasmussen (1982).

It is frequently the case that the intensity of a counting process depends on covariates. Cox, in two parallel papers (1972a,b), introduced a simple, yet flexible way of modulating intensity functions, called the **proportional hazards** model (again, this model is frequently called the Cox model). The idea is to modify the intensity, while making sure that the result remains an intensity. For covariates $Z_i(t)$ this model is usually written

$$\lambda(t \mid \mathcal{N}_t) = \alpha(t)\exp(\sum \beta_i Z_i(t)). \tag{5.74}$$

Example (Some covariate models) The simplest time-dependent covariate is the linear trend, where $Z_1(t)=t$. An instantaneously self-exciting process has an intensity which is log-linear in $Z_1(t)=N(t)$. A periodic model is obtained when $Z_1(t)=\cos(\omega_0 t)$, $Z_2(t)=\sin(\omega_0 t)$.

An important class of time-independent covariates obtains when we have a sum process (as in the psychiatric admissions data above), and each individual has a qualitative covariate $Z_j^i=1$(individual i is of category j). If we have a binary categorization, the ratio of intensities for an individual with $Z^i=1$ to another with $Z^i=0$ is

$$\frac{\alpha(t)\exp(\beta)}{\alpha(t)\exp(0)} = \exp(\beta) \tag{5.75}$$

regardless of $\alpha(t)$, so the two intensities (or **hazards**) are proportional. □

The statistical analysis of the proportional hazards model in the more general counting process set-up is obtained by replacing $Y(t)$ in (5.70) by

$$\tilde{Y}(t;\boldsymbol{\beta}) = Y(t)\exp(\beta_1 Z_1(t) + \cdots + \beta_k Z_k(t)). \tag{5.76}$$

We first need to estimate the regression coefficients $\boldsymbol{\beta}$. When the intensity is additive, corresponding to independent individuals, survival analysis suggests the use of a **partial likelihood**, and conditions on the **baseline intensity** $\alpha(t)$.

For the case of a single, time-independent covariate the argument goes like this: consider n individuals, failing at time $\tau_1, \ldots, \tau_n$. Then the probability of the particular order $i_1, \ldots, i_n$ in which they failed can be written

$$p(i_1, \ldots, i_n) = \prod_{j=1}^{n} p_j(i_j \mid i_1, \ldots, i_{j-1}) = \prod_{j=1}^{n} \frac{\exp(\beta z_{i_j})}{\sum_{k \in R_j} \exp(\beta z_{i_k})} \tag{5.77}$$

where R_j is the set of individuals at risk (not failed) just before the jth failure, z_{i_j} is the covariate of the individual i_j that failed at τ_j, and each term in the product on the right-hand side of (5.77) is just the conditional probability of individual i_j failing, given that a member of R_j failed. We estimate $\boldsymbol{\beta}$ by maximizing $p(i_1, \ldots, i_n)$. The generalization to more than one covariate, and to time-dependent covariates, can be found in Cox and Oakes (1984, Chapters 7 and 8). Given an estimate of $\boldsymbol{\beta}$, we estimate the integrated baseline intensity by

$$\hat{A}(t) = \int_0^t \frac{dN_g(u)}{\sum_i \tilde{Y}^i(u;\hat{\boldsymbol{\beta}})}. \tag{5.78}$$

Application (Psychiatric admissions, continued) There are several important covariates for the psychiatric admissions. For example, the number of previous children appears to have a substantial effect on admission intensities. Dividing group B into subgroups based on the number of previous children (0, 1, 2 or ≥3), and performing the same analysis as before on each group separately yields Figure 5.10.

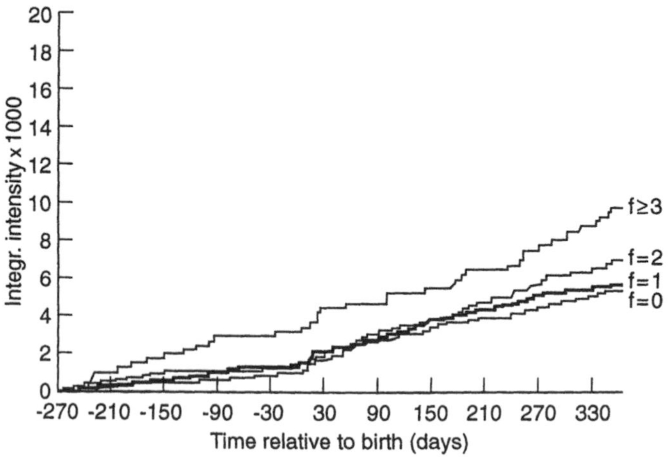

Figure 5.10. Estimated integrated admission intensities for subgroups of the women giving births, divided according to the number of previous children. Reproduced with permission from Andersen and Rasmussen (1982).

We see that the subgroup with at least 3 previous children has a substantially higher integrated intensity, particularly during the pregnancy, than the other groups. An assumption of proportional intensities appears reasonable.

Other factors that influence the intensity of admissions include marital status, age, and address (rural, urban, or cosmopolitan). We treat these demographic covariates as time-independent. In all cases there is no substantial deviation from proportional intensities, and a partial likelihood analysis yields Table 5.1.

Table 5.1 Regression coefficients

j	covariate z_j	$\hat{\beta}_{A_j}$	se	$\hat{\beta}_{B_j}$	se
1	not married	0.88	0.13	0.63	0.11
2	age < 19	–0.72	0.23	–0.052	0.27
3	age > 34	–0.36	0.16	0.70	0.18
4	no prev. children	–0.60	0.16	–0.23	0.12
5	2 prev. children	–0.10	0.16	0.22	0.15
6	≥ 3 prev. children	0.12	0.17	0.40	0.19
7	urban	–0.20	0.13	–0.41	0.11
8	rural	–0.48	0.15	–0.78	0.13

Marriage decreases the intensity in both groups, as indicated by a positive coefficient (increased intensity) for the complementary event of not married. The age effect is different between the groups. For the birth group the intensity increases with age, while for the abortion group those between 19 and 34 years of age (who would be most likely to carry a pregnancy to term) has the highest intensity. There are, of course, high correlations between the estimates of coefficients for age and parity, as well as for age and marital status, reflecting similar correlations between the corresponding covariates. The baseline hazards are very similar to those shown in Figure 5.9, but with substantially narrower standard errors. □

5.6. Marked point processes

It is frequently the case that there are random variables associated with the points τ_i of a point process. Such random variables are called **marks**, and the resulting point process is a **marked point process**. In the simplest marked point process the marks are iid, and independent of the points.

Example (M/G/∞ queue) One way of thinking of this queue is as a marked Poisson process, where the marks are the service times ε_i, assumed iid with density $f(x)$. Recall from section 3.3 that if we subsample a Poisson process with rate $\lambda(t)$ in such a fashion that a point at time t has probability $p(t)$ of being included in the sample, then the resulting process is a Poisson process of rate $\lambda(t)p(t)$. Furthermore the subsampled process is independent of the process of unsampled points. The following result is a generalization.

Theorem 5.4 Let (τ_i,δ_i) be a marked Poisson process of rate $\lambda(t)$ and iid marks with density $f(x)$. Then for any set $A\in\mathbb{R}^2$ the random variable $M(A)=\#\{i:(\tau_i,\delta_i)\in A\}$ has a Poisson distribution with mean $\iint_A \lambda(u)f(v)dvdu$.

Proof First let $A=I\times J$ be a rectangle. Partition $\mathbb{R}$ into intervals $J_i=(v_{i-1},v_i)$ where $v_i=v_{i-1}+h$. The the events $\delta_i\in J_j$ are independent, and $\{\tau_i:\delta_i\in J_j\}$ is a Poisson process with rate

$$\lambda(t)\mathbf{P}(\delta_i\in J_j) = \lambda(t)\int_{J_j} f(v)dv. \tag{5.79}$$

Writing $A=\sum I\times J_j$ we have $M(A)=\sum M(I\cap J_j)$ of independent Poisson variates, so it is itself Poisson with mean $\sum\int_I\int_{J_j} \lambda(t)f(v)dvdt$. The extension to general A is obtained by taking limits of sets of the above form, and using that counts in disjoint sets are independent. □

We use this result to deduce the number $X(t)$ of individuals in the system for an M/G/∞-queue. An individual is in the queue at time t if $\tau_i\le t$ and $\tau_i+\delta_i>t$. Hence $X(t)\sim\text{Po}(\mu_t)$ where

$$\mu_t = \iint_{A_t} \lambda(t)f(v)\,dv\,dt \tag{5.80}$$

where $A_t=\{(u,v):u\le t, u+v>t\}$. Hence we have, starting the process at time 0, that

$$\mu_t = \int_0^t \int_{t-u}^{\infty} f(v)\,dv\,\lambda(u)\,du = \int_0^t (1-F(w))\lambda(t-w)dw. \tag{5.81}$$

The equilibrium mean is obtained by letting $t\to\infty$. If we assume that the Poisson process is homogeneous, we get

$$\mu_\infty = \lambda\mathbf{E}(\varepsilon), \tag{5.82}$$

whence we have shown the following:

Corollary For an M/G/∞-queue, the number of customers in the system has a Poisson distribution with mean μ_t given by (5.81). In equilibrium, for a homogeneous input process, the number of customers is distributed $\text{Po}(\lambda\mathbf{E}(\varepsilon))$. □

Even the simple iid mark structure can yield interesting processes.

Application (Radar reflectivity of precipitation) Measurement of precipitation using weather radar is based on the reflectivity of precipitation, defined as the sixth power of raindrop diameter in a unit sample volume. This obtains from an expression of the backscattering cross section of a sphere as an infinite series, and ignoring terms of high order in the ratio between drop diameter and radar wavelength. This approximation is valid since typical weather radars operate on wavelengths between 3 and 10 cm, while raindrops are much smaller. Typically, the relation between radar reflectivity and precipitation amount as determined by rain gauges has been determined using empirical regression methods (Zawadzki, 1973). Theoretically, however, rain gauge volume is also a function of the drop size distribution; in this case proportional to the third power of drop diameter.

In order to measure drop size distributions a raindrop camera is utilized. It provides data on drop counts and sizes within a sample volume of about 1 m^3 at 1 minute time intervals. At the beginning of each minute the camera takes a sequence of 7 pictures, corresponding to equal parts of the sample volume. The observations are registered as drop counts in 65 equally spaced 0.1 mm size classes ranging from 0.5 mm to 7 mm. Figure 5.11 shows the drop arrival rate, mean drop diameter, and drop diameter coefficient of variation for sample data from a 75-minute storm in North Carolina on May 22, 1961. There is substantial temporal variability in these statistics.

The basic process describes the arrival times τ_i of raindrops at the top of the volume, and marks the arrivals with corresponding drop sizes δ_i. However, the distribution of drop diameter for the arrival process is not the same as the distribution of drop diameter in the sample volume, since drops of different size have different terminal velocities

$$v(\delta) = a\,\delta^b \tag{5.83}$$

where a=3.78 and b=0.67 (Atlas and Ulbrich, 1977). For example, a 0.5 mm drop takes 0.42 seconds to pass through a 1 m deep volume, while a 2.0 mm drop only takes 0.17 sec.

We assume that the arrival process $M(t)$ is a Poisson process of rate $\lambda(t)$, and a drop arriving at time t has size with density $f(x;\theta(t))$. Let $N(t)$ count the number of drops in the sample volume. We can write

$$N(t) = \sum_1^{M(t)} 1(\tau_i + v(\delta_i)^{-1} > t) \tag{5.84}$$

since $\tau_i + v(\delta_i)^{-1}$ is the exit time of the ith drop at the bottom of the sample volume. If we assume that the drop sizes are iid, we can think of the exit process as the output from an M/G/∞ queue, while $N(t)$ counts the number of customers in the system. Assuming that $\lambda(t)$ is approximately constant over time intervals

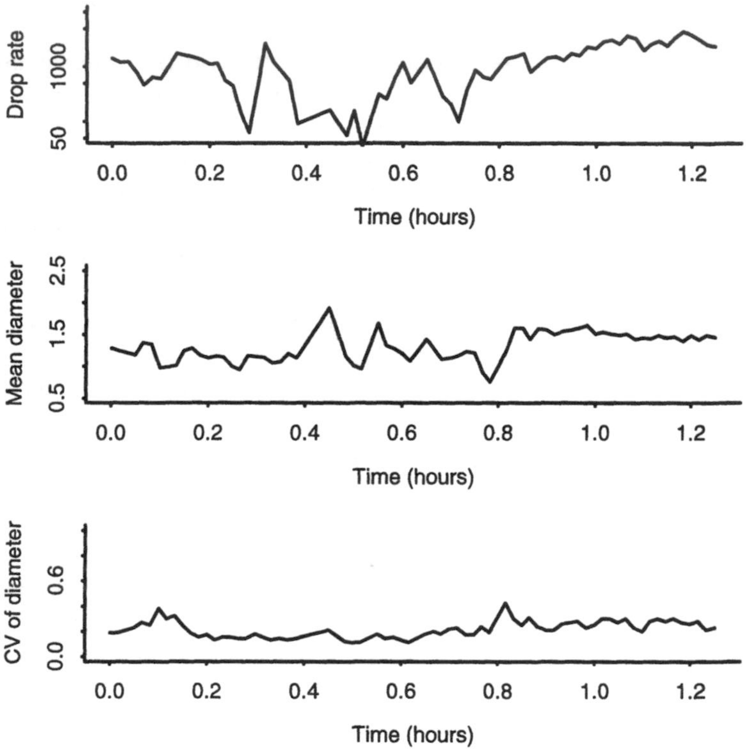

Figure 5.11. Drop arrival rate (top panel), mean sample drop diameter (middle panel), and coefficient of variation (lower panel) for the North Carolina storm. Adapted with permission from Smith (1993), published by the American Meteorological Society.

for which a drop is in the sample volume (less than half a second for the smallest drops observed), and using the corollary to Theorem 5.4 we find that this has a Poisson distribution with mean

$$\mu_t = \lambda(t)\mathbf{E}v(\delta(t))^{-1} = \frac{\lambda(t)}{a}\int_0^\infty y^{-b}f(y;\theta(t))dy. \tag{5.85}$$

We can write the radar reflectivity as

$$Z(t) = \sum_{i=1}^{M(t)} \delta_i^6 1(\tau_i + v(\delta_i)^{-1} > t). \tag{5.86}$$

The drop size distribution in the sample volume is seen to be

$$g(x;\theta(t),a,b) = \frac{x^{-b}f(x;\theta(t))}{\int_0^\infty y^{-b}f(y;\theta(t))dy}. \tag{5.87}$$

Taking the arrival drop size distribution f to be a gamma distribution with density

$$f(x;\theta(t)) = \frac{\theta_2(t)^{\theta_1(t)} x^{\theta_1(t)-1}}{\Gamma(\theta_1(t))} \exp(-y\theta_2(t)), \tag{5.88}$$

where $\theta_1(t)$ is the shape parameter at time t, and $\theta_2(t)$ is the scale parameter, we see that the observed drop size distribution is also gamma distributed, having the same scale parameter $\theta_2(t)$ as the arrival size distribution, but with a different shape parameter $\theta_1(t)-b$. We estimate the model parameter $\lambda(t)$ using the observed drop arrival rate for the North Carolina storm, and $\theta(t)$ using the method of moments applied to the observed drop sizes, yielding Figure 5.12.

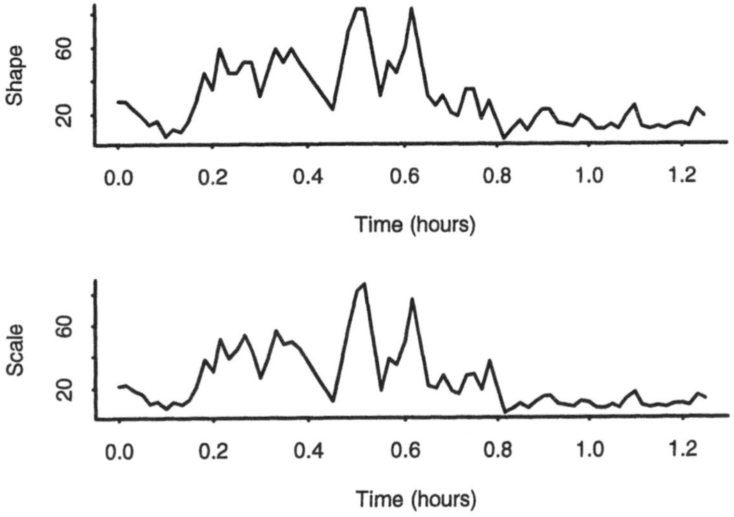

Figure 5.12. Parameter estimates for the North Carolina storm. The top panel is the estimated shape parameter while the bottom panel is the estimate of the scale parameter. Adapted with permission from Smith (1993), published by the American Meteorological Society.

How well does the model work? We can predict the expected reflectivity from the model, yielding

$$\mathbf{E}\, Z(t) = \frac{\lambda_t}{a} \frac{\Gamma(\theta_1(t)-b+6)}{\Gamma(\theta_1(t))} \theta_2(t)^{b-6}. \tag{5.89}$$

Figure 5.13 is a plot (on a dB scale) of observed reflectivity (computed as the sixth empirical moment of the observed drop size distribution) versus expected reflectivity (computed from the gamma model), showing substantial agreement.

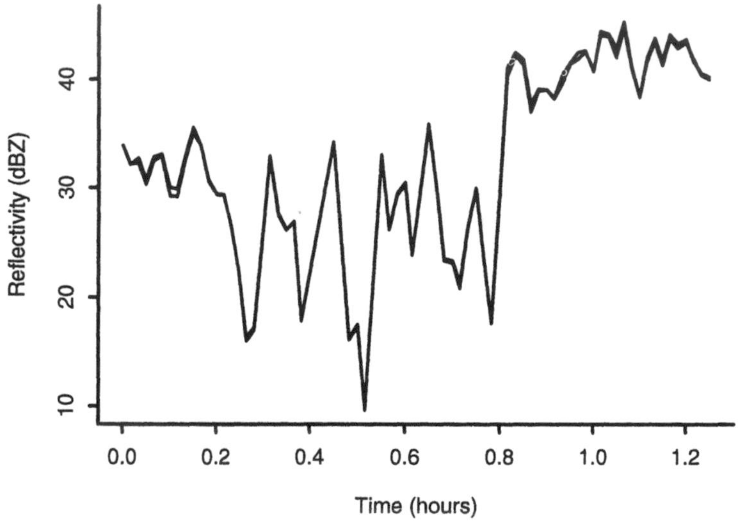

Figure 5.13. Observed (solid) and expected (dotted) reflectivity for the North Carolina storm. Adapted with permission from Smith (1993), published by the American Meteorological Society.

□

The iid marks model is often taken as a null model, which one attempts to reject in favor of more interesting structure. The following application illustrates an approach to this.

Application (A precipitation model) The MAP3S/PCN (Multistate Atmospheric Power Product Pollution Study / Precipitation Chemistry Network) network of nine monitoring stations in the northeastern United States was initiated in 1976 with the objective of creating a long-term, high quality database for the development of regional transport and deposition models. We shall only use data for station 3, located at the Pennsylvania State University campus in Middletown, Pennsylvania.

The data are event based, in the sense that there is at least 12 hours between successive events, although there may have been several precipitation instances during one event. Figure 5.14 shows the data. There were in all 465 events between October 1, 1977 and December 31, 1982, for an average occurrence of 0.24 events per day. The monthly variability is fairly small, from 0.21 events per day in September to 0.32 in June. The highest amounts occur in January (15 mm/event), and the lowest in December (6.6). The average event

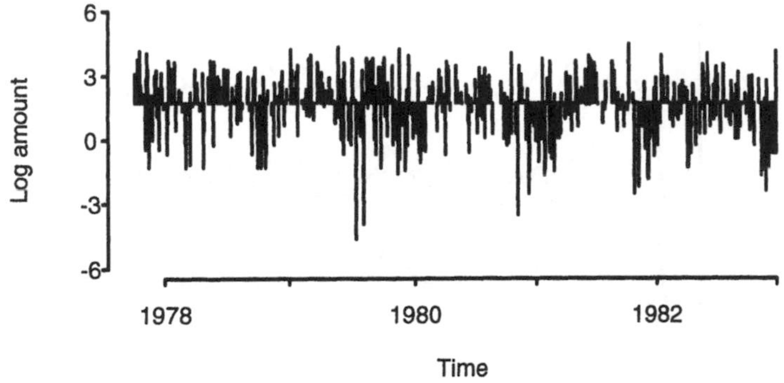

Figure 5.14. Precipitation data from MAP3S station 3. Amounts, on a log scale, originate at the median amount 1.87 (log mm), and are plotted at the time of event end. A line going up indicates above median amount. Adapted with permission from Guttorp (1988), published by the American Geophysical Union.

length is 6.7 hours. On a logarithmic scale, the median amount per event is 1.87.

The autointensity function (Figure 5.15) for the fall storm departure process shows (not surprisingly) some inhibition at 0.5 days, but apart from that not much structure. The construction of the sampling scheme enforces this inhibition. Thus, the arrivals are well described by a Poisson process with **dead time**, an inability to record events within 0.5 days of a previous event.

The distribution of amounts of precipitation (per storm) can be fitted by a variety of distributions. In the literature, gamma, Weibull, exponential, mixed exponential, and lognormal are among the suggested distributions. A Weibull distribution with shape parameter 0.85, estimated by maximum likelihood, provided an adequate fit to the amounts on the original scale (see Johnson and Kotz, 1970, Ch. 20, for a description of the Weibull distribution and estimation procedures for its parameters). In other words, if ξ denotes a rainfall amount for station 3, $\xi^{0.85}$ has approximately an exponential distribution.

Many researchers (Dovland and Mohn, 1975, Smith and Karr, 1983, Rodriguez-Iturbe et al., 1984, among others) have assumed that the amounts of precipitation are independent of the point process. In order to investigate this assumption for the MAP3S data, we look at the conditional probability density of amounts, given the history of the point process. Figure 5.16 shows the estimated densities of log amounts when the preceding 10 days had different numbers of events, as compared to the overall density, shown dashed. For example, the upper left-hand picture depicts the probability density for log amounts of only those events with no event in the preceding 10 days, as compared to the density of log amounts for all events. There is an indication of these

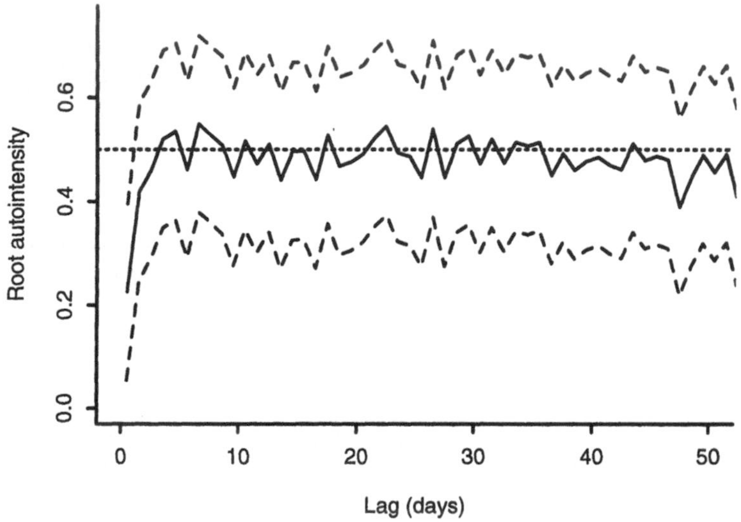

Figure 5.15. Autointensity function (on a square root scale) for the point process of storm departures (fall data only). The dashed lines are two standard errors above and below the estimate, while the dotted line is the rate. Adapted with permission from Guttorp (1988), published by the American Geophysical Union.

amounts being larger than average, while the amounts of events where there were four events in the preceding 10 days seem to be smaller than average. In order to assess the statistical significance of these indications, we need to develop some theory.

Silverman (1986) describes kernel density estimates of the type we use here. Let

$$f_n(x) = \frac{1}{\beta n}\sum_{i=1}^{n} w\left(\frac{\xi_i - x}{\beta}\right), \tag{5.90}$$

where w is a kernel function, integrating to one. We use a Gaussian kernel with bandwidth β=0.25. Assume we are estimating a density based on iid data $\xi_1, \ldots, \xi_n$, with density $f(x)$. If the amounts are independent of the underlying point process, we can choose indices for ξ_i according to the number of events in the previous time Δ, $Y_i = N_t - N_{t-\Delta}$. The conditional density estimate can then be written

$$f_n^*(x) = \frac{1}{\beta\sum_{j=1}^{n} Z_j}\sum_{i=1}^{n} Z_i w\left(\frac{\xi_i - x}{\beta}\right), \tag{5.91}$$

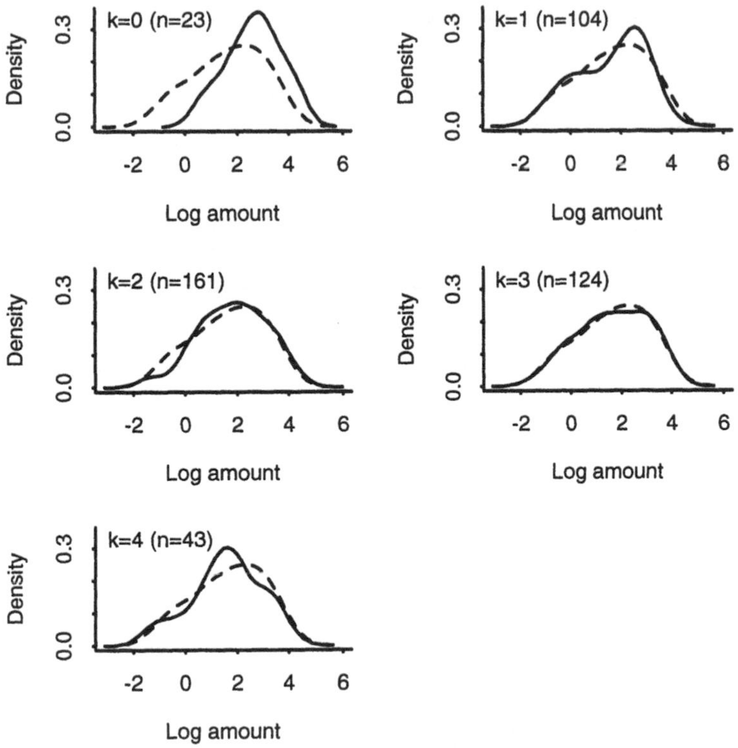

Figure 5.16. Estimated density of log amounts given the number k of events in the preceding 10 days (solid line) as compared to overall density (dashed line). The numbers in parentheses are sample sizes.

where $Z_i=1(Y_i\in A)$. Since the ξ's are independent of the Y's, it follows that Z_i ~Bin(n,$\mathbf{P}(Y\in A)$), and also independent of the ξ's.

Under suitable smoothness conditions on the underlying density f and on the kernel w, it is well known (e.g., Rosenblatt, 1971) that if $\int uw(u)du = 0$,

$$\mathbf{E}f_n(x) = f(x) + o(\beta^2), \tag{5.92}$$

and if also $\int w^2(u)<\infty$

$$\mathbf{Var}f_n(x) = \frac{f(x)}{n\beta}(\int w^2(u)du+o(\beta))+o(\frac{1}{n\beta}). \tag{5.93}$$

The Gaussian kernel satisfies this condition. Similarly one can derive (Guttorp, 1988)

$$\mathbf{E}f_n^*(x) \approx f(x) \tag{5.94}$$

and

$$\mathbf{Var}f_n^* \approx \frac{f(x)}{\beta}\int w^2(u)du\,\mathbf{E}\left[\frac{1}{\sum_{j=1}^n Z_j} \mid \sum_{j=1}^n Z_j>0\right] \tag{5.95}$$

where $\approx$ means that the difference between the right-hand side and the left-hand side goes to zero as $\beta \to 0$ and $n\beta \to \infty$.

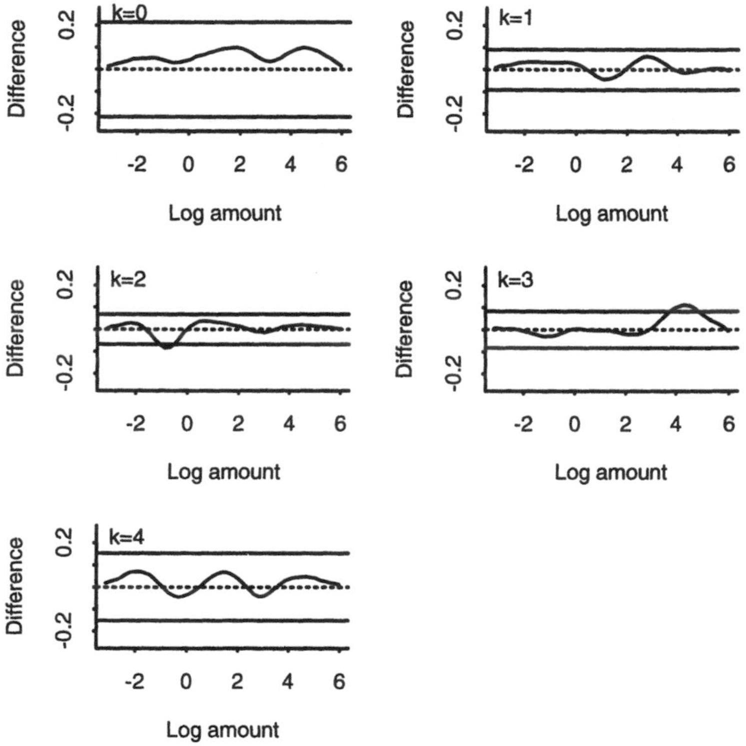

Figure 5.17. Difference of the square roots of the estimates in Figure 5.16. The vertical lines are approximately two standard errors from zero. If there were no difference between the conditional and the overall density, the plot would show random variation around zero (the dotted line).

In Figure 5.17 we standardize the comparison. It is a plot of the difference of the square roots of the densities shown in Figure 5.16. The square root transformation yields approximately constant variance of the density estimate, and we have supplied horizontal lines at two standard errors above and below zero. These standard errors were derived as follows. Since we are assuming that the Z_i are independent of the ξ_i, we can compute

$$\mathbf{Cov}(f_n(x), f_n^*(x))$$

$$= \frac{1}{n\beta^2}\sum_{i=1}^n \mathbf{Cov}\left[w\left(\frac{\xi_i - x}{\beta}\right), \frac{Z_i}{\sum_{j=1}^n Z_j} w\left(\frac{\xi_i - x}{\beta}\right) \mid \sum_{j=1}^n Z_j>0\right]$$

$$= \frac{1}{n\beta^2}\sum_{i=1}^{n}\mathbf{Var}\left[w(\frac{\xi_i-x}{\beta})\right]\mathbf{E}\left[\frac{Z_i}{\sum_{j=1}^{n}Z_j} \mid \sum_{j=1}^{n}Z_j>0\right] \quad (5.96)$$

$$\approx \frac{f(x)}{n\beta}\int w^2(u)du.$$

Therefore

$$\mathbf{Var}(f_n(x)-f_n^*(x)) \approx \frac{f(x)}{\beta}\int w^2(u)du \times \left[\mathbf{E}\{\frac{1}{\sum_{j=1}^{n}Z_j} \mid \sum_{j=1}^{n}Z_j>0\} - \frac{1}{n}\right]. \quad (5.97)$$

A simple computation shows that the expectation on the right-hand side is $(np)^{-1}(1-(1-p)^{n-1}) \approx (np)^{-1}$, where $p=\mathbf{P}(Y_i \in A)$. Using a Taylor expansion, it follows that

$$\mathbf{Var}(f_n^{1/2}(x)-f_n^{*1/2}(x)) \approx \frac{\int w^2(u)du}{4n\beta}\frac{1-p}{p}. \quad (5.98)$$

When p is unknown, we estimate it by $\sum_{j=1}^{n}Z_j/n$. The standard errors are computed pointwise, so about 5% of the points would be expected to fall outside the error lines. This happens for small amounts in the plot with two events in the preceding 10 days, and for large amounts in the plot with three events in the preceding 10 days. We conclude that the simple model that assumes that event amounts are independent of the point process of storm departures at least has some problems. □

5.7. Spatial point processes

A slightly different statistical theory has been developed around events that are located in space, rather than in time. By analogy to our classification of Markov processes, we should call these processes point fields, but that terminology does not seem to have caught on, and the processes are simply termed **spatial point processes** or **spatial patterns**.

As with other point processes, a convenient null hypothesis is that of **complete spatial randomness**, given by the Poisson process. Figure 5.18 shows a realization of a Poisson process.

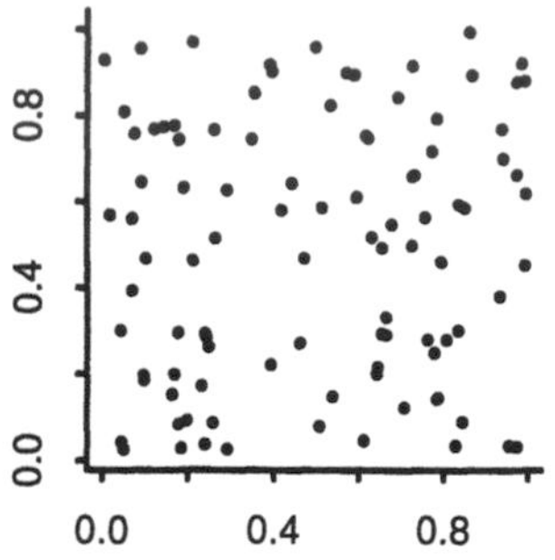

Figure 5.18. A realization of a spatial Poisson process.

In comparison to a Poisson process a spatial pattern can be viewed as **clustered** or **regular**. The Poisson cluster processes discussed in sections 5.1 and 5.2 generalize easily to the spatial context (in fact, the Neyman–Scott process was first introduced in the plane, and later applied to three-dimensional space). Whenever the clusters have more than one member, this yields a clustered pattern. On the opposite end, consider a Poisson process from which all pairs of points closer than a distance δ are removed. Such a process is called a **hard core rejection model**, and is regular compared to a Poisson process. This process was introduced by Matérn (1960). Figure 5.19 shows a clustered and a regular spatial pattern.

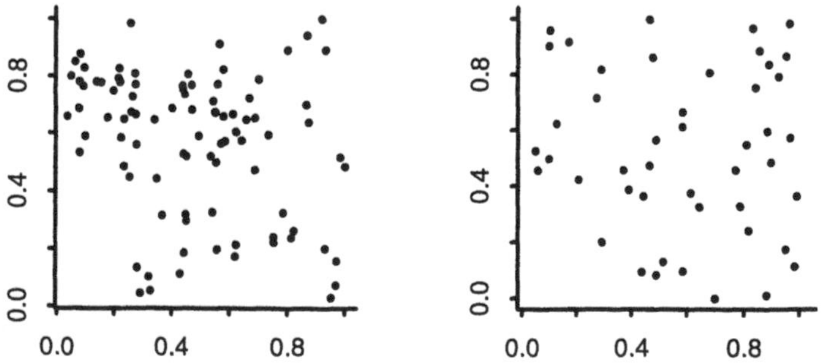

Figure 5.19. Realizations of a Neyman–Scott cluster process (left) and of a Matérn regular process (right).

Application (Tropical rain forest) The Atlantic Forest is a band of forest along the eastern coast of Brazil. This region has high precipitation and almost no dry season. The forest extends from sea level to elevations of almost 2,000 m. Batista (1994) describes a study made by Cia. Vale de Rio Dore, a state-owned mining company, in the Linhares Reserve north of the town of Linhares. This reserve remains in an unmanaged condition, and is dedicated to conservation and research.

Generally speaking, ecological theories of forest dynamic are focused on gaps, created by fall of limbs, disintegration of standing dead trees, fall of multiple trees due to wind storms, etc. The variation in the size of gaps can produce different environments, allowing species with different requirements to coexist in a forest stand. Tropical forests can be thought of as a mosaic of patches in different developmental stages, driven by how recently the corresponding gaps were formed.

One of the key features in the formation of a gap is that it induces a different light environment. These differences provide the most common mechanism used to explain forest dynamics. For example, a large clearing in a forest will enable shade-sensitive species to establish themselves. However, the light regime in a tropical forest varies vertically as well, and growing plants encounter different light environments during different stages of their growth.

The data consist of complete maps, breast height diameter, and species identification (if known) of all trees with breast height diameter at least 10 cm in each of five experimental plots. The plots are rectangles of 100×50 meters, located 75 m from roads. They have been mapped every three years from 1980 to 1989. The design of this experiment involved a variety of different management strategies. We focus on two extremes, namely the control group with no intervention, and the clear-cut treatment. While the dynamics of growth is very interesting, we restrict attention to data from year 6 after intervention. Figure 5.20 shows the spatial patterns, with the size of the circles corresponding to the breast height diameter of the tree. □

In order to test the hypothesis of complete spatial randomness, as well as to study particular parametric models for the mechanism generating the points, we need to develop an inferential framework. This has two components: a parameter function that can be estimated parametrically or nonparametrically, and a method of assessing variability. As in the temporal case we use a second order parameter (involving only the distribution of pairs of points). While the second order product moment

$$p_2(\mathbf{x},\mathbf{y}) = \lim_{|d\mathbf{x}|,|d\mathbf{y}|\to 0} \frac{\mathbf{E}dN(\mathbf{x})\,dN(\mathbf{y})}{d\mathbf{x}\,d\mathbf{y}} \tag{5.99}$$

is as valid in this context as in the temporal case, spatial point process analysis has focused on a slightly different parameter function. Consider **isotropic** (or

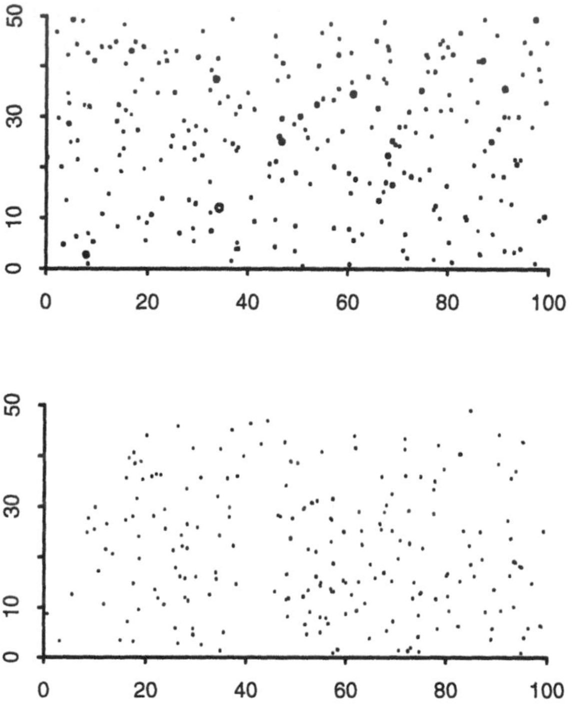

Figure 5.20. Control plot (top) and clear-cut plot (bottom) from the Linhares Reserve forest experiment, block A. The size of each tree is shown to scale.

rotation-invariant) stationary point patterns, having $p_2(\mathbf{x},\mathbf{y})=p_2(|\mathbf{x}-\mathbf{y}|)$. The K-function is an integrated version of the product moment function, namely

$$K(t) = \frac{2\pi}{p_1^2}\int_0^t u p_2(u)du. \tag{5.100}$$

To motivate this definition, first note that the conditional intensity of an event at $\mathbf{x}$, given one at $\mathbf{0}$, is $p_2(|\mathbf{x}|)/p_1$, so the expected number of further events within distance t of an arbitrary event is

$$\int_0^{2\pi}\int_0^t \frac{p_2(u)}{p_1} du = \frac{2\pi}{p_1}\int_0^t u p_2(u)du, \tag{5.101}$$

whence we interpret $K(t)$ as the expected number of further events within distance t of an arbitrary even, divided by the rate of events.

Example (The Poisson process, continued) As before we have $p_2(t)=p_1^2$, so $K(t)=\pi t^2$. □

Example (Neyman–Scott Poisson cluster process) Recall that $p_2(t)=p_1^2+\rho \mathbf{E}S(s-1)g(t)$, where ρ is the cluster center rate and g is the density of differences between dispersions. Since $p_1=\rho\mathbf{E}(S)$, we get

$$K(t) = \pi t^2 + \mathbf{E}S(S-1)G(t)/(\rho \mathbf{E}^2 S). \tag{5.102}$$

where G is the cdf corresponding to g. Figure 5.21 shows the K-function for the Neyman–Scott and Poisson processes.

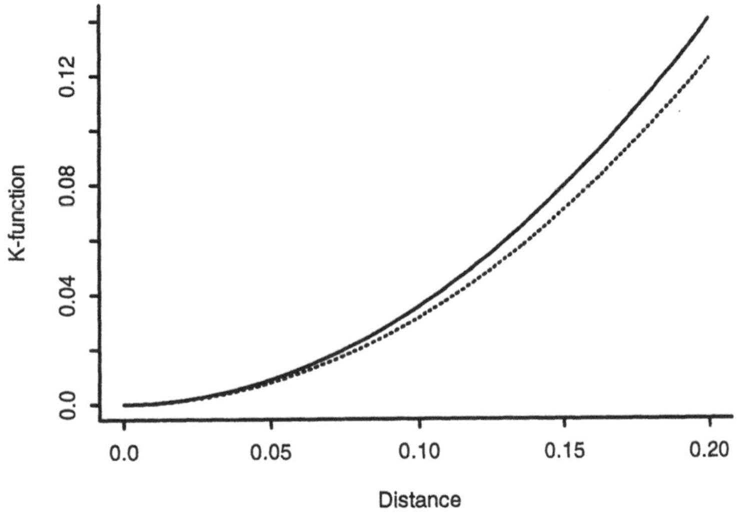

Figure 5.21. K-function for the Neyman–Scott process (solid line) and the Poisson process (dotted line).

□

Example (A Poisson process with hard core rejection) Assuming that the original Poisson process has rate ρ, the probability that a point survives the thinning is the probability of having no points within distance δ, or $\exp(-\rho\pi\delta^2)$, so the rate of the resulting process is $p_1=\rho\exp(-\rho\pi\delta^2)$. Similarly we have

$$p_2(t) = \begin{cases} 0 & 0<t<\delta \\ \rho^2\exp(-\delta A_\delta(t)) & t\geq\delta, \end{cases} \tag{5.103}$$

where $A_\delta(t)$ is the area of the union of two discs with equal radius δ and centers t apart. Elementary integration shows that

$$A_\delta(t) = \pi\delta^2 - t\delta(1-\frac{t^2}{4\delta^2})^{1/2} - 2\delta^2\arcsin(\frac{t}{2\delta}),\ 0\le t<2\delta. \qquad (5.104)$$

Hence $K(t)=2\pi\exp(2\rho\pi\delta^2)\int_0^t u\exp(-\rho A_\delta(u))du$, which can be computed numerically (Figure 5.22).

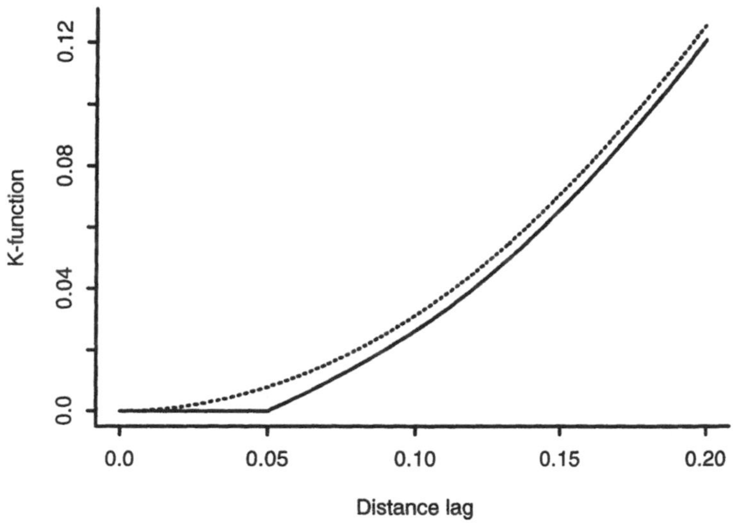

Figure 5.22. K-functions for the Matérn hard-core rejection model (solid line) and the Poisson process (dotted line).

□

In order to estimate $K(t)$ from data, given by events observed in a set A, count the number of event pairs of distance at most t apart. The expected value of this quantity is approximately $p_1^2 \mid A \mid K(t)$, where the approximation arises from the possibility that an event in A can be within distance t of the boundary of A, and hence have a neighbor falling outside of A. There are two ways of dealing with this: ignoring all points within a distance t of the boundary, or correcting for the bias due to boundary effects. We will do the latter here. Let $w(\boldsymbol{x},\boldsymbol{y})$ be the proportion of the circumference of a circle centered at $\boldsymbol{x}$ and going through $\boldsymbol{y}$ that falls inside A. Except for points in a strip around the edge this will be one. But for points near the edge we need to weight them more heavily (making them correct for unobserved points outside A). The appropriate weight given to a pair of points at $\boldsymbol{x}$ and $\boldsymbol{y}$ is $1/w(\boldsymbol{x},\boldsymbol{y})$. Notice that $w(\boldsymbol{x},\boldsymbol{y})\neq w(\boldsymbol{y},\boldsymbol{x})$. We estimate $K(t)$

by

$$\hat{K}(t) = \frac{|A|}{N(A)^2} \sum_{i \in A, j \in A} \frac{1}{w(\tau_i,\tau_j)} 1(d(\tau_i,\tau_j) \le t), \tag{5.105}$$

where d is distance in the plane. Ripley (1976) provides expressions for $w(\boldsymbol{x},\boldsymbol{y})$ for different shapes of A. As in the temporal time series case, a square root transformation stabilizes the variance. For comparison with a Poisson process it is visually useful to look at $\hat{k}(t)=\hat{K}(t)^{1/2}-\pi^{1/2}t$. For a Poisson process this should fluctuate around the x-axis, while regularity would show up as a systematic deviation below the axis, and clustering as a deviation above.

While the estimate $\hat{K}(t)$ does not depend on any particular parametric assumption, we need such an assumption in order to obtain confidence bands. The simplest assumption is that of complete spatial randomness. Using the Monte Carlo testing approach outlined in section 2.6, we simulate $m-1$ independent Poisson point processes, compute the estimate $\hat{K}^{(i)}(t)$ for each realization, and define **simulation envelopes**

$$U(t) = \max_{1 \le i \le m} \hat{K}^{(i)}(t) \quad \text{and} \quad L(t) = \min_{1 \le i \le m} \hat{K}^{(i)}(t) \tag{5.106}$$

where $\hat{K}^{(m)}(t) \equiv \hat{K}(t)$ is the estimate based on the observed data. If the data come from a Poisson process, the probability for each t that $\hat{K}(t)=U(t)$ is $1/m$. Likewise $\mathbf{P}(\hat{K}(t)=L(t))=1/m$. Note that if we view these bands as confidence bands for $\hat{K}$, they are not simultaneous. The procedure for producing confidence intervals for the square root transformation $\hat{k}(t)$ suggested above is to compare to $U(t)^{1/2}-\pi^{1/2}t$ and $L(t)^{1/2}-\pi^{1/2}t$.

Application (Tropical rain forest, continued) Figure 5.23 shows the estimate $\hat{k}(t)$ for the control plot (left) and the clear-cut plot (right), together with envelope functions obtained from a homogeneous Poisson process. In the control plot we see only slight deviation from complete spatial randomness in the direction of regularity. This probably results from competitive exclusion, reducing the frequency of trees close to a given one below what would be expected under the Poisson process hypothesis. The clear-cut plot, on the other hand, shows substantial clustering on both small and large scales. The small scale clustering probably results from new growth on stumps, while the clustering on larger scale could be due to micro-site variation in soil properties or the soil seed bank. □

Informally, we may also generate data from a fitted parametric model and compute envelope functions. If the parametric model were fully specified (i.e., if we knew the parameters), the Monte Carlo test approach to producing simulation envelopes would again be valid, but since we have estimated parameters from the data, we really need to simulate conditionally upon some sufficient statistics. This is rarely feasible. Nevertheless, the simulation envelopes provide

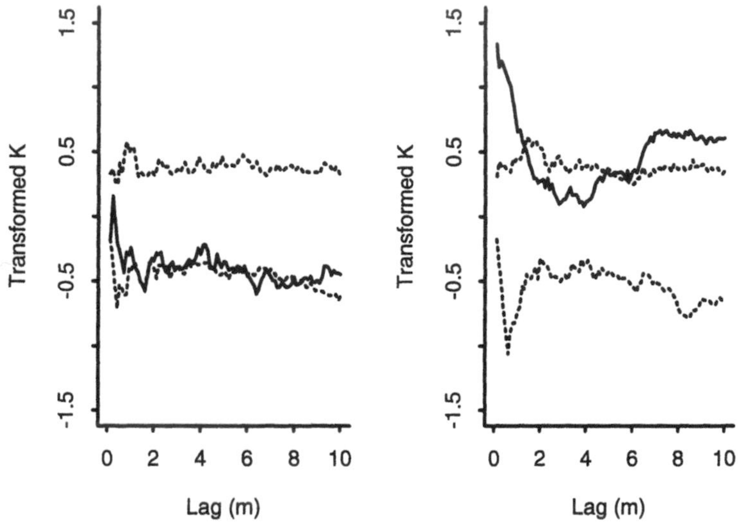

Figure 5.23. Estimates $\hat{k}(t)$ of the transformed K-function for the control plot (left) and the clear-cut plot (right). The dashed lines are Poisson process envelope functions based on 99 replications.

a rough guide to the goodness of the fit.

The extension to marked patterns is straightforward. Assume that we have a categorical mark i, $i=1,\ldots,I$, and compute

$$\hat{K}_{ij} = \frac{|A|}{N_i(A)N_j(A)} \sum_{k,l} 1(d(\tau_k^{(i)},\tau_l^{(j)}) \le t). \tag{5.107}$$

Application (Tropical rain forest, continued) According to ecological theory, inter-tree competition reaches its peak among trees with similar environmental requirements. We would expect that trees in the understory (i.e., trees whose tops are below the general forest canopy) to exploit openings in the canopy for their growth. In addition, we would expect trees in the canopy of the forest to exhibit a fairly regular pattern. There should also be a repulsion between overstory trees (whose tops protrude through the canopy) and understory trees. This repulsion should be the strongest for trees immediately beneath the canopy, since similar size trees tend to respond to the same type of light environment.

In order to test these hypotheses, we first divide the trees up according to their diameter at breast height. There is no indication in the data as to which trees are canopy, overstory, or understory, but diameter may be a proxy for this, so that, e.g., understory trees would be those with relatively small diameter.

Figure 5.24 shows the transformed K-function estimate for the control plot for two different diameter categories.

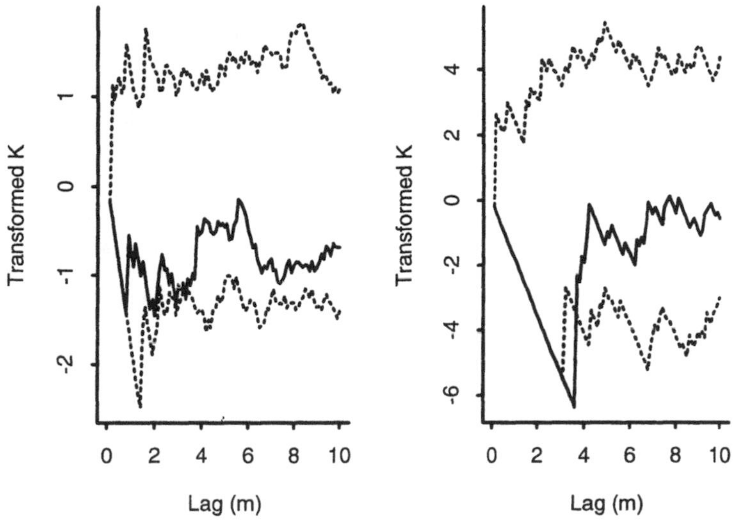

Figure 5.24. Transformed K-function estimate $\hat{k}(t)$ for the control plot trees with breast height diameter smaller than 30 cm (left) and larger than 30 cm (right).

The smaller category, corresponding to understory trees, conforms reasonably to complete spatial randomness. On the other hand, the category with diameter between 30 and 50 cm shows indications of regularity at small and intermediate scales. We expect these trees to be mainly canopy trees, and the regularity to result from similar size crowns that prohibit other trees of the same height in the vicinity, as predicted by the inter-tree competition hypothesis.

Considering now the clear-cut area, there is clustering at all size classes (Figure 5.25). We also look at the interaction between size classes (Figure 5.26). The control plot shows no tendency for any particular relationship (exclusion or clustering) between the two size classes, while for the clear-cut plots there is a tendency for small trees to cluster near large ones. This may hold some information about the early dynamics of this type of forests. □

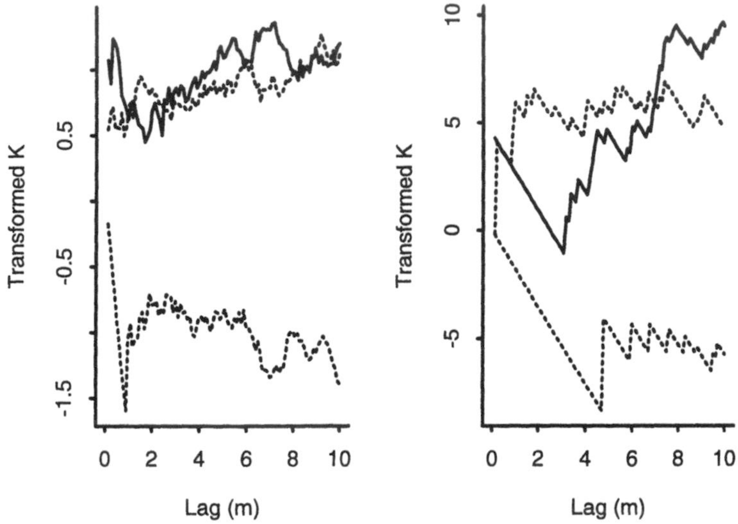

Figure 5.25. Transformed K-function estimate $\hat{k}(t)$ for the clear-cut plot in two different diameter categories: less than 15 cm (left) and over 20 cm (right).

5.8. Bibliographic remarks

The Swedish car data are discussed at length in Cox and Lewis (1966), as well as in the paper by Bartlett (1963). Generalities about point processes can be found in Cox and Isham (1980), and with considerably more mathematical rigor in Daley and Vere-Jones (1988).

Statistical analysis of point processes was the subject of the monograph by Cox and Lewis (1966). A more recent treatment is Brillinger (1978), and Brillinger's Wald lecture (Brillinger, 1988) has several success stories of scientific applications of point process analysis. Particular attention to doubly stochastic Poisson processes is given by Grandell (1976) and Karr (1991). The latter monograph also has much material on inference based on the complete intensity function.

The neurophysiology analysis follows Brillinger (1992). A general reference for neurophysiology is Abeles (1982). The Aalen model is described and developed in detail in Andersen et al. (1992). A point process approach to proportional hazards models is the subject of Fleming and Harrington (1991). The marked precipitation models are due to Smith (1993) for the radar reflectivity analysis and Guttorp (1988) for the relation between storm amounts and intensity of arrivals. Hanisch and Stoyan (1979) develop the theory of second order analysis of marked point processes.

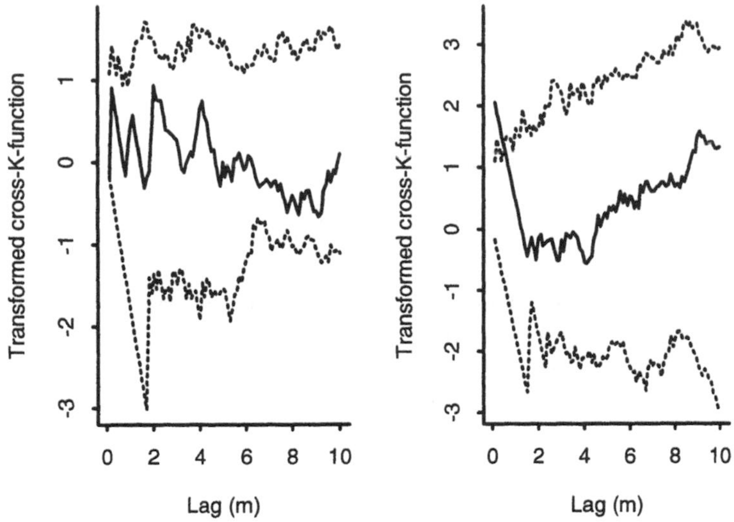

Figure 5.26. Cross-K-function estimates for the interaction with large and small trees in the control plot (left) and the clear-cut plot (right). Diameter categories as in the two previous figures.

Much of the statistical analysis of spatial point processes originated with the seminal dissertation of Matérn (1960; reprinted 1986). Bartlett (1975), Diggle (1983) and Ripley (1988) contain the modern statistical theory. Baddeley and Møller (1989) relate the theory of Markov point processes to stochastic geometry. The presentation here is essentially based on Diggle's book, which also contains several data sets. The application originates in the PhD dissertation by Batista (1994), although the results given here ignore the interesting temporal development of the experiment. A more parametrically oriented approach to dynamic forest modeling (in a different climate) is that of Rathbun and Cressie (1994).

5.9. Exercises

Theoretical exercises

1. Show that for a doubly stochastic Poisson process we have

$$\mathbf{Var}N(a,b) > \int_a^b \mathbf{E}\Lambda(u)du,$$

so that the counts are overdispersed compared to a regular Poisson process.

2. Let N_t be a doubly stochastic Poisson process with stationary rate function $\Lambda(t)$ having $\mathbf{E}\Lambda(t)=\mu$, $\mathbf{Cov}(\Lambda(t),\Lambda(s))=\gamma(t-s)$. Show that the autointensity function of N_t is $h(t)=\mu+\gamma(t)/\mu$.

3. Let

$$Y_t=\int_{-\infty}^{\infty} h(t-u)dN(u)$$

where N is a Poisson process of rate λ, and $h(t)=\mathrm{sgn}(t)\,|t|^{-1}$. Show that Y_t has a Poisson distribution.

4. Let X_i be a stationary Markov chain on $\{0,1\}$. Let N_t be a stationary, orderly point process of rate ρ, and create a thinned process M_t by retaining the ith point if $X_i=1$ and deleting it otherwise. To maintain the rate we rescale M_t so that points occur at times $\pi\tau_i$ for i such that $X_i=1$, where π is the stationary probability of a 1 for the Markov chain.

(a) Show that M_t has rate ρ.

(b) Show that

$$\mathbf{Var}M_t=\pi^2\mathbf{Var}N_{t/\pi}+\frac{(1-\pi)(1+\beta)}{1-\beta}\rho t-\frac{2\pi(1-\pi)}{(1-\beta)^2}\beta(1-\mathbf{E}\beta^{N_{t/\pi}})$$

where $\beta=(p_{11}-\pi)/(1-\pi)$.

(c) Discuss the cases $|\beta|=1$ and $\beta=0$. For example, what type of sampling is this?

(d) Show that the autointensity function is given by

$$h_M(t)=\rho+\frac{1}{2\rho}\frac{d^2}{dt^2}\mathbf{Var}\,M_t.$$

5. For a stationary, orderly point process N_t, subjected to iid displacements with density f, show that the resulting autointensity function is $\int h(u-v)f_D(v)dv$, where f_D is the density of the difference of two iid random variables with density f. What happens when the variance of the displacement gets large?
Hint: Write $f_D(x)=\sigma f(x/\sigma)$, and take σ large.

6. Consider k independent stationary orderly point processes $N_1,\ldots,N_k$ with rates ρ_i and autointensity functions h_i. Let N be the **superposition** of the processes, i.e., the collection of events irrespective of which process they belong to.

(a) Show that N has rate $\rho=\sum_1^k\rho_i$.

(b) Show that N has autointensity $h(u)=\rho+\rho^{-1}\sum_1^k\rho_i(h_i(u)-\rho_i)$.

7. Let N_t be a doubly stochastic Poisson process with rate function $\Lambda(t)=\int_{-\infty}^{t} w(t-u)dN_c(u)$ where N_c is a Poisson process and w a weight function. Show that this process is a Neyman–Scott process and identify its cluster size distribution and displacement density.

8.Let Y_k, $k=1,\ldots,T$ be a time series obtained from sampling a stationary, orderly point counting process $(M_t)_{t\geq 0}$ at discrete, equispaced times, i.e.,

$$Y_k = M_{k\Delta}-M_{(k-1)\Delta} = \int_{(k-1)\Delta}^{k\Delta} dM_t,$$

where Δ is the sampling window.

(a) Show that $\mathbf{E}(Y_k)=\Delta p_1$ where p_1 is the rate of M.

(b) Show that

$$\mathbf{E}(Y_k Y_{k+u})=\Delta^2 \int_{-\infty}^{\infty} g_\Delta(r-u\Delta)p_2(r)dr$$

where $g_\Delta(s)=\Delta^{-1}(1-|s|/\Delta)1_{\{-\Delta\leq s\leq\Delta\}}$, is the triangular window function.

(c) How would you use the time series data to estimate the second order product moment of the point process?

9. Consider a binary series (cf. section 5.2) obtained from a Neyman–Scott cluster process by observing only the presence or absence of points in equispaced intervals of unit length.

(a) Suppose that the cluster process has Poisson cluster starts, exponential dispersion with mean $1/\beta$, and geometric cluster sizes with parameter p. Let (a_i) and (b_i) be sequences of numbers such that $b_i \leq a_{i+1}$. Show that the zero probability function of unions of intervals is

$$\log \zeta(\bigcup_1^k(a_i,b_i)) = \lambda\{b_n + q\sum_{j=1}^{n}(b_j-a_j) - \frac{1}{\beta}\sum_{j=0}^{n-1}\log(\frac{qC_{j+1}+pd_j}{qC_{j+1}+pf_{j+1}})$$
$$-\frac{1}{\beta}\sum_{j=1}^{n}\log(\frac{q(C_{j+1}-d_j)+f_j}{q(C_{j+1}-d_j)+d_j})$$
$$-\frac{q}{\beta}\sum_{j=1}^{n}\log(\frac{1+q(C_{j+1}-d_j)/d_j}{1+q(C_{j+1}-d_j)/f_j})\},$$

where $q=1-p$, $d_j=\exp(-\beta b_j)$, $f_j=\exp(-\beta a_j)$, $C_j=\sum_{i=j}^{n}(f_i-d_i)$, and $C_{n+1}=0$. *Hint:* Use the pgfl.

(b) Show that the discrete autointensity function (5.47) can be written

$$h_k = 1 - \frac{\zeta(I_0)-\zeta(I_0\cup I_k)}{1-\zeta(I_0)},$$

and write it in terms of the parameters of the cluster process in (a).

10. Let N_t be a Poisson process with periodic mean function $p_1\lambda(t)$ (so $\lambda(t+k)=\lambda(t)$ for integer k). Let $\Lambda(t)=\int_0^t\lambda(u)du$, and assume that $\Lambda(1)=1$. Let $f(t)=[t]+\Lambda^{-1}(t-[t])$, where $[y]$ is the integer part of y. Show that $N_{f(t)}$ is a homogeneous Poisson process.

11. Consider a stationary point process $M_t, t\leq T$, subjected to time-dependent thinning so that a point occurring at time τ is retained with probability $\pi(\tau)$, where π is a smooth function, with $\pi(T)=1$. Call the process of retained points N_t.

(a) Show that N_t has rate $\pi(t)p_M$ where p_M is the rate of M_t.

(b) Show that N_t has autointensity $h(t,t+u)=\pi(t+u)h_M(u)$.

(c) Suppose the function π is known. How would you estimate p_M and h_M from observations on $N_t, t\leq T$?

12. In marketing research buyers are sometimes classified into "innovators" and "imitators." The innovators are not influenced in their decision to buy a new product by the number of people who have bought it already, while the imitators are. Suppose that there are n potential buyers, and that the conditional intensity of purchasing a new product for the first time at time t is (Böker, 1987)

$$\lambda(t \mid \mathcal{N}_t) = a\,(n-N(t-)) + b\,N(t-)\,(n-N(t-)).$$

(a) Estimate a when $b=0$, and interpret the estimator.

(b) Estimate b when $a=0$, and interpret the estimator.

(c) Estimate a and b jointly.

(d) How would you assess the standard errors of these estimators?

13. Consider an insect with crop-destroying larvae, such as the potato beetle. Suppose that a beetle lays a clump of eggs at a randomly chosen spot on a field. These egg masses can be considered primary events in a cluster process. If we assume that different beetles act independently, the primary process will be a Poisson process with rate λ, say. In due course the larvae hatch and begin to crawl around to look for food. If the larvae have no social instinct, they will move independently, and at least to a first approximation the distance they move within a given period of time may be identically distributed for all larvae. At some fixed time we take a census of the experimental field E. Assume that the distance moved is uniform over a bounded set of accessibility A, and that the number of surviving eggs in each egg mass has a Poisson distribution with mean m.

(a) Verify that the total number $N(E)$ of larvae in E has pgf

$$P(s) = \exp(\lambda \int_{\mathbb{R}^2} [\exp(-m(1-s)\frac{|\cap A|}{|A|})-1]dt)$$

$$\approx \exp(-\lambda\,|E|\,[1-\exp(-m(1-s))]).$$

(b) Deduce that

$$p_{n+1} = \mathbf{P}(N(E)=n+1) = \frac{\lambda m}{n+1}\sum_{k=0}^{n} p_{n-k}\frac{m^k}{k!}e^{-m}$$

and that $p_0=\exp(-\lambda(1-\exp(-m)))$.

(c) Under what circumstances would $N(E)$ have a Poisson distribution?

14. Show that for an isotropic, stationary point pattern we have

$$p_2(t) = \frac{p_1^2}{2\pi t}K'(t).$$

Computing exercises

C1. How can you use a histogram routine, allowing you to specify binwidth and/or number of classes, to estimate autointensity and crossintensity functions?

C2. Generate a Neyman–Scott Poisson cluster process, with geometric cluster size and parameters of your choice. Using the simulation, estimate the parameters, and assess the variability of your estimates.

C3. For simulated paths from a Neyman–Scott Poisson cluster process, consider observing the corresponding binary series as in Exercise 9, and estimating the parameters of the process (as outlined in that Exercise) for different discretization intervals Δ. Assess how the quality of estimation is affected by the coarseness of discretization.

C4. (a) Using Exercise 3.13 (b), write a routine to generate a Poisson process with rate $\lambda(t)$.

(b) When the rate $\lambda(t)$ is complicated, one can use a thinning method due to Lewis and Shedler (1979). Let $\lambda^*(t)\geq\lambda(t)$, and generate (using the method in (a)) points $\tau_1, \cdots \tau_n$. Each point is deleted with probability $1-\lambda(t)/\lambda^*(t)$. Show that this yields a process with the intended rate. The dominating rate λ^* is chosen so that generation using (a) is as easy as possible, while keeping the rejection rate small (i.e., keeping λ^* close to λ).

(c) Generate a Poisson process with rate $\lambda(t)=t/(1-t)$, $0<t<1$.
Hint: Start away from 0 and end before 1. Why?

Data exercises

D1. Lake Constance, bordered by Austria, Germany and Switzerland is Europe's third largest lake. Records of freezes of major parts of the lake has been kept since the year 875, and a total of 38 major freezes have been reported since then (Steinijans, 1976). The years are given in Table 5.2. Analyze these data.

Table 5.2 Major freezes of Lake Constance

875	895	928	1074	1076	1108	1217	1227
1277	1323	1325	1378	1379	1383	1409	1431
1435	1460	1465	1470	1479	1497	1512	1553
1560	1564	1565	1571	1573	1684	1695	1763
1776	1788	1796	1830	1880	1963		

D2. The data in Table 5.3 are historical record of large earthquakes in the S. Kanto area of Japan (Vere-Jones and Ozaki, 1982).

Table 5.3 Japanese earthquakes

818	841	878	1096	1213	1227	1240	1241
1257	1293	1433	1498	1525	1590	1605	1633
1647	1648	1649	1670	1697	1703	1782	1812
1853	1854	1955	1905	1922	1923	1924	1933

Investigate the possibility of a periodicity in these data.

D3. Data set 9 has the times of arrival of severe cyclonic storms striking the Bay of Bengal coast during the period 1877–1977 (Mooley, 1981). For the North Indian tropical cyclone basin, severe cyclonic storms are define as storm systems with wind velocities of 88–116 km/h. Systems with velocities above 116 km/h are called hurricanes. They are quite rare in this area. The data consist of all recorded severe cyclonic storms and hurricanes.

(a) Consider the within-year behavior of the cyclonic storm arrivals. Are they uniform over the year?

(b) Do the storm arrivals show evidence of clustering?

(c) Fit the model outlined in Exercise 10, and test the transformed points for Poisson behavior.

D4. Table 5.4 contains years of major flooding on the Rio Negro River at Maraus, Brazil.

Table 5.4 Rio Negro floods

1892	1895	1898	1904	1908	1909	1913
1918	1920	1921	1922	1944	1953	1955
1971	1972	1973	1975	1976	1982	1989

A major flood is defined a having a river level exceeding 28.5 m sometime during the year.

(a) Estimate the autointensity function for this process.

(b) Due to deforestation, there is expected to be an increase in flooding over the last two decades. Assess this hypothesis.

D5. Data set 10 contains the record of all known volcanic eruptions in Iceland between 1550 and 1978 (Gudmundsson and Saemundsson, 1980). Especially the early data are not particularly precise, sometimes only giving the year. Use these data to assess any regularity among the eruptions.

D6. Table 5.5 contains the centers of 42 biological cells (see Diggle, 1983, for the source). Estimate the K-function, and compare it to a Matérn inhibition process.

Table 5.5 Locations of cell centers

0.350	0.025	0.637	0.050	0.825	0.125
0.237	0.150	0.575	0.212	0.062	0.362
0.325	0.287	0.650	0.362	0.337	0.462
0.600	0.475	0.938	0.400	0.350	0.600
0.725	0.512	0.987	0.512	0.175	0.650
0.462	0.750	0.737	0.687	0.237	0.787
0.775	0.850	0.175	0.912	0.462	0.900
0.487	0.087	0.775	0.025	0.087	0.187
0.400	0.162	0.737	0.237	0.212	0.337
0.450	0.287	0.900	0.262	0.462	0.425
0.800	0.387	0.150	0.500	0.562	0.575
0.862	0.525	0.062	0.750	0.337	0.750
0.525	0.650	0.862	0.637	0.637	0.812
0.900	0.775	0.350	0.962	0.625	0.950

CHAPTER 6

Brownian motion and diffusion

The random motion of pollen particles in water (or dust in sunlight) leads to models for continuous state processes in continuous time. We shall think of the outcomes of such processes as randomly chosen functions. The class of second order processes is perhaps the largest for which any structural statements about these random functions (such as continuity) can be verified simply. The Brownian motion process is an important tool in building more complicated processes. We will briefly touch upon likelihood theory for stochastic differential equations, and discuss some applications.

6.1. Brownian motion

In 1827, the botanist Robert Brown noticed that when pollen is dispersed in water, the individual particles did not settle down, but kept dancing around in a lively, unpredictable manner. As the phenomenon repeated itself with all sorts of different organic substances, Brown believed that he had found the basic unit of living matter (called a **molecule** by Buffon). When he later found that inorganic substances presented the same phenomenon, he deduced that he had isolated the primitive molecule of all matter. In 1865, Cantoni and Oehl found that the motion continued unchanged for a year when the liquid was sealed between two glass plates. Various explanations of the phenomenon were put forward: it was thought to be caused by irregular heating by incident light, by electrical forces, or by temperature differences in the liquid. In 1877 Delsaux first expressed the theory that the motion was caused by impacts of the molecules of the liquid on the immersed particles. Basically, since a heavy particle is immersed in a fluid consisting of light molecules, which are in constant motion due to the heat of the liquid, its velocity will vary due to a large number of very small bumps each time the particle runs into a molecule.

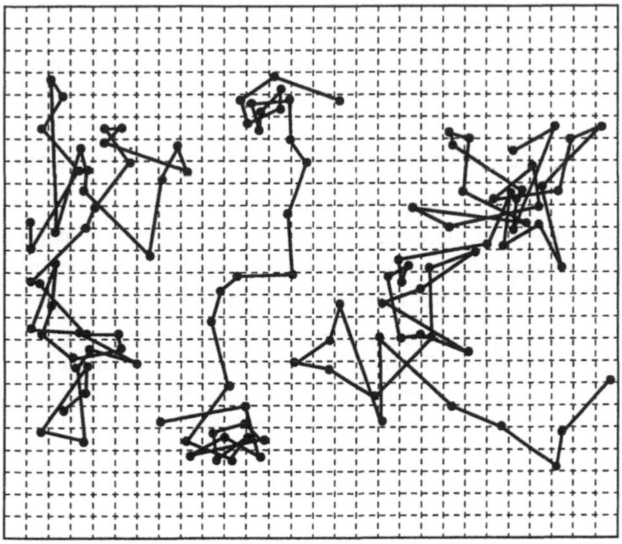

Figure 6.1. A graph of three paths of Brownian motion of the same mastic grain (radius 0.53 microns) as seen through the microscope of Perrin (1913). The scale is 16 divisions to 50 microns, and positions are marked every 30 sec. Reproduced from *Atoms* by J. Perrin, reprinted by Ox Bow Press, 1990.

If, for simplicity, we think of the motion as one-dimensional, we first note that at a given velocity v, the particle will on the average have more collisions in front (slowing the particle down) than from behind (speeding it up). Hence the change dv in velocity over a small time increment dt depends on v, but not on earlier velocity values. Thus the process should be Markovian. When the system is in equilibrium, we are looking at a stationary process. Translating this into physics, if we assume that the collisions are elastic, the molecule bouncing off the pollen particle will change its momentum (velocity $\times$ mass) by twice its original value, since its velocity will be multiplied by -1. In order to conserve momentum (the physical analog of stationarity) the particle must receive this change in momentum, so that the velocity change of the particle is twice the speed of the molecule times the ratio of the molecular to pollen particle mass. The predictions produced by this reasoning, however, yielded most varying and inconsistent results, and the phenomenon seemed as mysterious as ever. The breakthrough occurred in 1905, when Einstein[1] and Smoluchowsky independently realized that the measurement of results of *individual* bounces was not the motion that had been observed experimentally. Instead, the process is

[1] Einstein, Albert (1879–1955). German physicist. Fundamental contributions to the theory of special and general relativity. Nobel prize 1921 for his work on the photoelectric effect.

observed at discrete time points, and between the times of observation it has undergone many variations in velocity. What we observe is the net displacement. If, as seems reasonable, we can assume that the individual collisions are independent, the net displacement should be normally distributed. Furthermore, the displacement is independent of previous locations, so the position of the particle is also a Markov process (at least at the coarse time scale imposed by observations).

The importance of the theory of Brownian motion was as a verification of the molecular theory, which at the turn of the century was not universally accepted. Einstein's first paper contains a caveat:

> In this paper it will be shown that according to the molecular-kinetic theory of heat, bodies of microscopically visible size suspended in a liquid will perform movements of such magnitude that they can be easily observed in a microscope, on account of the molecular motions of heat. It is possible that the movements to be discussed here are identical with the so-called "Brownian molecular motion"; however, the information available to me regarding the latter is so lacking in precision, that I can form no judgement in the matter.

The motion was indeed identical with the Brownian motion, and empirical verification of Einstein's and Smoluchowsky's theory was immediately forthcoming. From a physical point of view, the most important finding in Einstein's paper was the identification of the diffusion coefficient. In this section, we will outline the derivation of this result. From the point of view of stochastic modeling, however, the derivation of the probability density of the location of a particle undergoing Brownian motion was far more important, and we will return to this in section 6.3.

Suppose that suspended particles are irregularly dispersed in a liquid. There are two opposing forces acting on the particles. Since the concentration is not constant, there is osmotic pressure towards a constant concentration, and, according to the kinetic theory of heat, there is also diffusion because the particles are hit by the heat-induced movement of the molecules making up the liquid. If the system is in equilibrium, these two forces have to be equal and opposing.

If we look at a small volume of solution between x and $x+dx$, and denote the osmotic pressure at x by $p(x)$, then the pressure on this small volume from the rest of the solution is $p(x)-p(x+dx)$. Therefore the osmotic force acting on unit volume is $-p'(x)$. Let $q(x)$ be the density of particles per unit volume at x. The osmotic pressure of dissolved substances follows van't Hoff's law, which states that the osmotic pressure of a dissolved substance equals the pressure it would exert if it were an ideal gas occupying the same volume as that of the solution, or in equation form

$$p(x) = \frac{RTq(x)}{N} \tag{6.1}$$

where $R=8.31$ is the gas equation constant, T is the (absolute) temperature in

Kelvin, and N is Avogadro's number 6.02×10^{23}. Thus the osmotic force on a unit volume of solution is

$$-p'(x) = -\frac{RT}{N}q'(x). \tag{6.2}$$

By the osmotic force acting on them, the particles will be moving in the solution. If we assume that the particles are spherical of radius r, then by Stoke's law a force K imparts on a single particle the velocity $K/6\pi r\eta$, where η is the viscosity of the liquid. Therefore, in unit time $K/6\pi r\eta$ particles will pass a unit volume. Now consider the movement of particles by diffusion, produced by the thermal molecular movements. This is described by Fick's law, which says that the flux of particles is $-Ddq/dx$, where D is the diffusion coefficient. If the system is in equilibrium, these two fluxes must be equal, yielding

$$\frac{K}{6\pi r\eta} = -D\frac{dq(x)}{dx}. \tag{6.3}$$

Using the expression (6.2) for K, we see that

$$-\frac{RT}{6\pi rN\eta}q'(x) = -Dq'(x) \tag{6.4}$$

so

$$D = \frac{RT}{6\pi rN\eta}. \tag{6.5}$$

This formula can be used to compute, for example, the size of a molecule for which the diffusion coefficient is known. If the size of the molecule is known, it can further be used to estimate the size of Avogadro's number N, in particular since there is a simple relation between the dispersion of Brownian motion and the diffusion coefficient, as we shall presently see.

In order to estimate the diffusion coefficient, Einstein considered the motion of particles in solution over a short time interval τ. During this time the displacement of a particle by irregular thermal motion is a random variable Y. The magnitude of Y will be different for different particles, but since we presuppose a dilute solution, this displacement is only controlled by the surrounding solvent. Hence, in portions of the solution of different concentrations these displacements Y will be on the average of equal magnitude, and just as frequently positive as negative. Thus μ should be 0. Now simplify the situation by assuming that Y takes the values $\pm y$ with equal probabilities. If we look at the motion across the point x, only particles starting in $(x-y,x)$ can pass from left to right across x in time τ. Only half of those molecules will in fact move to the right, or about $\frac{1}{2}q(x-y/2)y$ molecules. Similarly, the number of molecules moving to the left is about $\frac{1}{2}q(x+y/2)y$, so the net number is $\frac{1}{2}y(q(x-y/2)-q(x+y/2))$. But this difference is very nearly $-y^2q'(x)$, so the quantity of the substance which diffuses across x in unit time is $-\frac{1}{2}y^2q'(x)/\tau$. Thus we have a second expression for the diffusion coefficient, namely $D=y^2/2\tau$. The displacement over unit time will be the sum of $1/\tau$ iid random variables, so would have variance $\sigma^2=\mathbf{Var}(Y)/\tau$. In other words, $\mathbf{Var}(Y)=\sigma^2\tau=y^2$, so $D=\sigma^2/2$. Thus, in order to estimate D, we need to estimate σ^2 from data.

The estimation of Avogadro's number from experiments on Brownian motion was performed by Perrin, an achievement for which he was awarded the Nobel prize in 1926. Guttorp (1991, section 5.2) contains a derivation of a confidence interval for N based on observations by Westgren (1916) of gold particles in a colloidal solution. The experimental confirmation of Einstein's theory of Brownian motion yielded the definitive verification of the molecular theory.

6.2. Second-order processes

Consider $X(t), t\geq 0$, a stochastic process such that $\mathbf{Var}X(t)<\infty$ for all t. Such processes are called **second order** processes, and we shall now look at some sample path properties for such processes. But before we study these properties, some general facts from probability theory are needed. As we saw in the previous chapter, it is not always possible to specify uniquely a probability measure by its distribution at a finite number of points. However, such **finite dimensional distributions** (fdd's) are the cornerstones of stochastic process theory. As we pointed out in Chapter 1, Kolmogorov's consistency condition guarantees the existence of at least one process with the given set of fdd's. However, there may be many different processes with the same fdd's.

The main problem with Kolmogorov's construction is that it does not assign probabilities to important sets such as the probability that X has continuous paths. Call two processes X and Y **equivalent** if $\mathbf{P}(X(t)=Y(t))=1$ for all t. If time is discrete, we have that

$$\begin{aligned}\mathbf{P}(X_k=Y_k,\text{ all }k) &= \mathbf{P}(\bigcap_k X_k=Y_k) = 1-\mathbf{P}(\bigcup_k X_k\neq Y_k)\\ &\geq 1-\sum \mathbf{P}(X_k\neq Y_k) = 1,\end{aligned} \tag{6.6}$$

so two equivalent discrete time processes are equal with probability one. In the continuous time case the sum in the penultimate term is uncountable, so it need not be zero. Therefore equivalent processes need not be equal. On the other hand, it is frequently the case that we can pick a **version** with some desired property, such as continuous paths. All our statements to follow in this section will deal with the question whether a version exists with a given regularity condition satisfied for its paths.

Assume that X is a second order process. Our first problem is to determine when X has continuous paths. Here is a general criterion due to Cramér and Leadbetter (1967).

Theorem 6.1 Let $X(t)$ be a stochastic process on $0 \le t \le 1$. Suppose that for all $t, t+h \in [0,1]$

$$\mathbf{P}(|X(t+h)-X(t)| \ge g(h)) \le q(h) \tag{6.7}$$

where $g(h)$ and $q(h)$ are even functions, nonincreasing as $h \to 0+$, and such that

$$\sum_{n\ge1} g(2^{-n}) < \infty \quad \text{and} \quad \sum_{n\ge1} 2^n q(2^{-n}) < \infty. \tag{6.8}$$

Then there is a version of X with continuous paths.

This theorem is difficult to apply in its present form. A simplification occurs if we use Chebyshev's inequality.

Corollary If

$$\mathbf{E}|X(t+h)-X(t)|^2 \le \frac{K|h|}{|\log|h||^q} \tag{6.9}$$

for some $K>0$, $q>3$, then X has a version with continuous paths.

Proof Let b be a positive number to be chosen below, and use Chebyshev's inequality to estimate

$$\begin{aligned}\mathbf{P}(|X(t+h)-X(t)| \ge |\log|h||^{-b}) &\le \frac{\mathbf{E}|X(t+h)-X(t)|^2}{|\log|h||^{-2b}} \\ &\le \frac{K|h|}{|\log|h||^{q-2b}}.\end{aligned} \tag{6.10}$$

We use $g(2^{-n}) = (n\log 2)^{-b}$, which is summable whenever $b>1$, and $2^n g(2^{-n}) = K|n\log 2|^{-(q-2b)}$, which is summable whenever $q-2b>1$. Combining the two requirements, we need $1<b<(q-1)/2$, which is possible since $q>3$. □

Example (The Poisson process) For the Poisson process, $X(t+h)-X(t)$ is Poisson distributed with mean λh if $h>0$, so

$$\mathbf{E}|X(t+h)-X(t)|^2 = \lambda h + \lambda^2 h^2, \tag{6.11}$$

which does not go to zero fast enough. This does not prove that the Poisson process has no version with continuous paths, but is suggestive of this fact. □

Let $R(s,t) = \mathbf{Cov}(X(s), X(t))$. Then

$$\begin{aligned}\mathbf{E}|X(t+h)-X(t)|^2 &= R(t+h,t+h) - R(t+h,t) \\ &\quad - R(t,t+h) + R(t,t).\end{aligned} \tag{6.12}$$

Using the difference notation

$$\Delta_h \Delta_k f(s,t) = f(s+h,t+k) - f(s+h,t) - f(s,t+h) + f(s,t) \tag{6.13}$$

we can write the condition of the corollary

$$\Delta_h \Delta_h R(t,t) < \frac{K|h|}{|\log |h||^q}. \tag{6.14}$$

In particular, if $|\Delta_h \Delta_h R(t,t)/h^2|$ is bounded for small h, we have a version with continuous paths. This condition is implied by $R(s,t)$ being differentiable on the diagonal $s=t$.

Arguments very similar to those needed to prove Theorem 6.1 can be used to produce conditions for other classes of paths. Here are two examples.

Proposition 6.1 If

$$\Delta_h^2 \Delta_h^2 R(t,t) < \frac{K|h|^3}{|\log |h||^q} \tag{6.15}$$

for some $K>0$, $q>3$, then there is a version with differentiable paths. In particular, this is the case if R is twice differentiable on the diagonal.

Proposition 6.2 If for $t_1<t_2<t_3$

$$\mathbf{E}|(X(t_3)-X(t_2))(X(t_2)-X(t_1)|^p < K|h|^{1+r} \tag{6.16}$$

for $K,p,r>0$ where $h=t_3-t_1$, then X has a version with paths that have at most jump discontinuities.

Example (The Poisson process, continued) We know that the Poisson process has only jump discontinuities, so it should satisfy the criterion in Proposition 6.2. To see this, take $p=2$ and $r=1$. We will just consider the stationary case. Since the Poisson process has independent increments, we have

$$\begin{aligned}\mathbf{E}|(X(t+h)-X(t))|^2 &= \mathbf{E}^2|X(t+h)-X(t)|^2 \\ &= (\lambda h+\lambda^2 h^2)^2 \le Kh^2 \end{aligned} \tag{6.17}$$

for small h by choosing K large enough. □

The most important class of second order processes are the **Gaussian** (or normal) processes. A process X is Gaussian if all its fdd's are multivariate normal. It is **stationary** if its distribution does not depend on the time origin. If X is stationary, then $\mathbf{E}X(t)\equiv m$, which we will assume to be 0. We will also scale the random variables so that $R(t,t)=r(0)=\mathbf{Var}X(t)=1$. A stationary Gaussian process has continuous paths if

$$r(h) = 1-O(|\log |h||^{-a}) \tag{6.18}$$

as $h\to 0$ for some $a>3$, where $r(h)=R(t,t+h)$. The proof is an application of Theorem 6.1 (Exercise 1). The class of Gaussian processes have interesting paths. The following result is due to Dobrushin (1960).

Theorem 6.2 For all stationary Gaussian processes, either the sample paths are continuous with probability one, or the sample paths have discontinuities of the second kind at every point (i.e., no limits from the left or the right) and are in fact unbounded on every interval.

6.3. The Brownian motion process

We now turn to a probabilistic description of Brownian motion. Let $X(t)$ denote the position at time t of a particle undergoing Brownian motion. Assume that $X(t)$ has:

(i) **Independent increments:** $X(t+\Delta t)-X(t)$ is independent of $(X(s), s \le t)$;

(ii) **Stationary increments:** the distribution of $X(t+\Delta t)-X(t)$ is independent of t;

(iii) **Continuity:** $\lim_{\Delta t \to 0+} \mathbf{P}(|X(t+\Delta t)-X(t)| \ge \delta)/\Delta t \to 0$ for all $\delta > 0$.

Remark

Assumption (i) is actually not quite adequate: we said in section 6.1 that the change in momentum, rather than the change in position (integrated velocity), imparted on the particle in $(t, t+\Delta t)$ can reasonably be assumed independent of the history. Assuming independent position change only makes sense if the displacement of the particle due to its initial velocity at the beginning of the interval $(t, t+\Delta t)$ is small compared to the displacements it suffers as a result of molecular momentum exchange (by collision) in the same interval. Thus there needs to be a lot of collisions, so Δt needs to be relatively large. The consequences of this will become clear later.

Assumption (ii) just means that the medium is infinite in extent and does not have a preferred origin.

The idea behind (iii) is that we would like the process to have continuous paths. This assumption does not quite require that directly. However, if B has continuous paths, it is uniformly continuous on bounded intervals, so the **modulus of continuity**

$$h(1/n) = \max_{1 \le k \le n} |B(k/n) - B((k-1)/n)| \tag{6.19}$$

must go to zero as $n \to \infty$. Let Y_k be the kth increment. By (i) the Y_k are independent, and by (ii) they have identical distributions. Thus

$$\mathbf{P}(h(1/n) \ge \delta) = 1 - \mathbf{P}(\max_{1 \le k \le n} Y_k < \delta) = 1 - (\mathbf{P}(Y_1 < \delta))^n \tag{6.20}$$

$$= 1 - (1 - \mathbf{P}(Y_1 \ge \delta))^n \approx 1 - \exp(-n\mathbf{P}(Y_1 \ge \delta)),$$

so $\mathbf{P}(h(1/n) \ge \delta) \to 0$ iff $n\mathbf{P}(Y_1 \ge \delta) \to 0$, i.e., condition (iii). □

Theorem 6.3 Any random process $X(t)$ satisfying (i), (ii), and (iii), with $X(0)=0$, has $X(t) \sim N(\mu t, \sigma^2 t)$.

Sketch of proof We write

$$Y_k = X(\frac{kt}{n}) - X(\frac{(k-1)t}{n}), \tag{6.21}$$

so that $X(t)=\sum_1^n Y_k$. The Y_k are again iid, and hence their sum has an infinitely divisible distribution (these are the limiting distributions for scaled and centered sums of iid random variables). This distribution is normal iff $M_n=\max_{1\le k\le n} Y_k$ converges in probability to zero (Breiman, 1968, Proposition 9.6). But that is just the calculation in (6.20). To compute the mean, let $\phi(t)=\mathbf{E}X(t)$. Then

$$\phi(t+\tau)=\mathbf{E}(X(t+\tau)-X(t))+\mathbf{E}X(t)=\phi(\tau)+\phi(t). \tag{6.22}$$

The only (measurable) solution to this equation is linear. Likewise the variance is seen to be linear in t. □

Assume that $s<t$. Then the covariance function is

$$\begin{aligned} R(s,t) &= \mathbf{E}(X(t)-\mu t)(X(s)-\mu s) \\ &= \mathbf{E}([X(t)-X(s)-\mu(t-s)] + X(s)-\mu s)(X(s)-\mu s) \\ &= \mathbf{Var}\, X(s) = \sigma^2 s, \end{aligned} \tag{6.23}$$

so that for general s,t we have $R(s,t)=\sigma^2\min(s,t)$. In order to verify that the process has continuous paths, we use Theorem 6.1. We assume for simplicity that $\mu=0$ and that $\sigma^2=1$. Compute

$$\mathbf{P}(|X(t+h)-X(t)|>|h|^a) = 2(1-\Phi(|h|^{a-\frac{1}{2}}) \le 2|h|^{\frac{1}{2}-a}\phi(|h|^{a-\frac{1}{2}}) \tag{6.24}$$

where we used that

$$1-\Phi(x)\le\frac{1}{x}\phi(x) \quad \text{for } x>0, \tag{6.25}$$

and where $\phi(x)=(2\pi)^{-\frac{1}{2}}\exp(-x^2/2)$ is the standard normal density and $\Phi(x)$, as before, is the corresponding cdf. We need $a>0$ to make $g(2^{-n})$ summable. Write

$$2^n q(2^{-n}) = 2^{n(1+2a)/2}\phi(2^{n(1-2a)/2})\phi(x)(2\pi)^{-\frac{1}{2}}\exp(-x^2/2) \tag{6.26}$$

which is summable whenever $a<\frac{1}{2}$. Thus the Brownian motion process has continuous paths. Note that we cannot use the corollary to Theorem 6.1, since $\Delta_h\Delta_h R(t,t)=h$.

Wiener[1] (1923) first showed the continuity of paths of the Brownian motion process, and it is often called the Wiener (or Wiener-Bachelier) process.

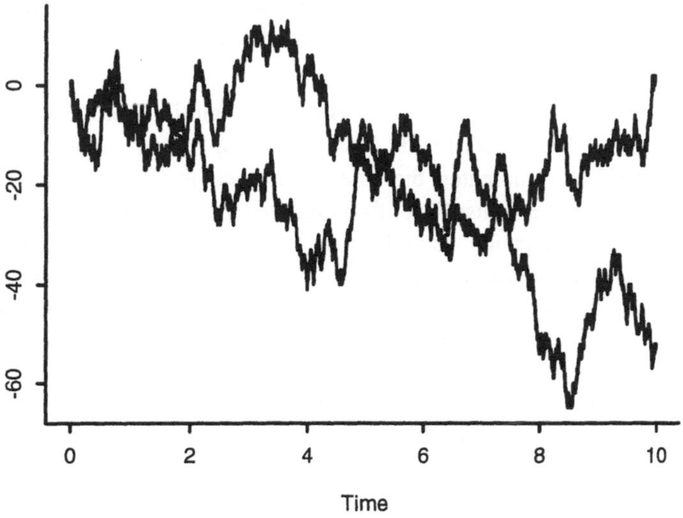

Figure 6.2. Two simulated Brownian motion process paths.

Figure 6.2 is an illustration of two simulated Brownian motion process paths. Although they are continuous, they do vary rather irregularly. In fact, the paths are quite weird (when viewed as real functions). For example, the paths are not of bounded variation. Recall that a function $f(t)$ is of **bounded variation** if there is a finite C such that

$$\sum_{i=1}^{n} |f(t_i)-f(t_{i-1})| < C \tag{6.27}$$

for all ordered finite sequences $t_0<t_1<\cdots<t_n$ in the domain of definition of f. This is a necessary condition in order to be able to define Riemann–Stieltjes integrals of Brownian motion. To compute the variation of a Brownian motion path, let $t_i=i2^{-N}$ for $i=0,\ldots,2^N$. Then

$$2^N S_{2^N} = 2^N \sum_{i=0}^{2^N-1} (X(t_{i+1})-X(t_i))^2 \sim \chi^2(2^N). \tag{6.28}$$

By the law of large numbers, $S_{2^N}\to 1$ a.s. We want to show that $\sum_0^{2^N-1} |X(t_{i+1})-X(t_i)| \to\infty$. Write

[1]Wiener, Norbert (1894–1964). American mathematician. Introduced probability measure on function spaces, developed generalized harmonic analysis and prediction theory for time series, and introduced cybernetics.

$$\sum |X(t_{i+1})-X(t_i)|^2 \le \max_i |X(t_{i+1})-X(t_i)| \sum |X(t_{i+1})-X(t_i)|. \quad (6.29)$$

The left-hand side goes to one, whereas the first term on the right-hand side goes to zero, since X has continuous paths, and therefore uniformly continuous paths on a bounded interval. Therefore the second term on the right-hand side must grow unboundedly.

In fact, the paths of the Brownian motion process are nowhere differentiable. The proof of this is complicated, but the fact that $\Delta_h^2\Delta_h^2 R(t,t)=2|h|$ indicates that the covariance is too weak, so that nearby points on the path are insufficiently correlated for the path to be differentiable. The rate of convergence according to Proposition 6.1 should be faster than $|h|^3$. This is a serious problem for a model for actual Brownian motion. The derivative of the location of a particle is its velocity, so we have defined a process in which velocity is undefined. The model has $\Delta X(t)$ of order $\Delta t^{½}$, so that the velocity $\Delta X(t)/\Delta t \to \infty$ with probability 1. This indicates that the model is not adequate for very small time intervals, on the order of magnitude of the time intervals between collisions. This will be remedied in the next section.

Returning now to Einstein's derivation of the distribution of Brownian motion, we now make that argument a little more precise. Consider a particle starting at the origin. In time τ it jumps an amount Y, with $\mathbf{P}(Y=y)=1-\mathbf{P}(Y=-y)=\pi$. The moment generating function of Y is

$$\mathbf{E}e^{\theta Y} = \pi e^{\theta y}+(1-\pi)e^{-\theta y}. \quad (6.30)$$

In time t the particle takes t/τ independent steps, so that the total displacement $X(t) = \sum_1^{t/\tau} Y_i$ has moment generating function

$$\mathbf{E}e^{\theta X(t)} = \left[\pi e^{\theta y}+(1-\pi)e^{-\theta y}\right]^{t/\tau} \quad (6.31)$$

with mean $\frac{t}{\tau}(2\pi-1)y$ and variance $\frac{t}{\tau}4\pi(1-\pi)y^2$. We are going to let $y,\tau\to 0$ so as to produce a sensible limit; namely mean converging to μt and variance converging to $\sigma^2 t$. To get that, we need

$$y=\sigma\tau^{½} \quad \text{and} \quad \pi=½(1+\frac{\mu\tau^{½}}{\sigma}). \quad (6.32)$$

We notice that $y=O(\tau^{½})$, and $\pi=½+O(\tau^{½})$, in other words that the displacement y must be of larger order of magnitude than τ, and that π must be near ½. Substituting in (6.31) we obtain

$$\mathbf{E}e^{\theta X(t)} = \left[\frac{1}{2}(1+\frac{\mu\tau^{½}}{\sigma})\exp(\theta\sigma\tau^{½})+\frac{1}{2}(1-\frac{\mu\tau^{½}}{\sigma})\exp(-\theta\sigma\tau^{½})\right]^{t/\tau} \quad (6.33)$$

We now take logarithms and expand $\exp(\theta\sigma\tau^{½})$ as a power series to get

$$\frac{t}{\tau}\log\left[\frac{1}{2}(1+\frac{\mu\tau^{½}}{\sigma})\exp(\theta\sigma\tau^{½})+\frac{1}{2}(1-\frac{\mu\tau^{½}}{\sigma})\exp(-\theta\sigma\tau^{½})\right]$$

$$= \frac{t}{\tau}\log\left[1+\frac{\theta^2\sigma^2\tau}{2}+\mu\theta\tau+O(\tau^2)\right] = \frac{\theta^2\sigma^2 t}{2}+\mu\theta t+O(\tau). \qquad (6.34)$$

It follows that $\mathbf{E}\exp(\theta X(t))\to\exp(\mu\theta t+\theta^2\sigma^2 t/2)$, the moment generating function for $N(\mu t,\sigma^2 t)$. μ is called the **drift** of the process.

We now use this limiting process to determine the Kolmogorov equations. The forward equation for a simple random walk with steps ± 1 with probability π of going up is

$$p_{jk}^{(n)} = \pi p_{j,k-1}^{(n-1)} + (1-\pi)p_{j,k+1}^{(n-1)}. \qquad (6.35)$$

We rescale this process to take steps of magnitude Δx in time intervals of length Δt. Let $p(x_0,x;t)$ denote the probability that the process is at x at time t, given that it started at x_0 at time 0. With $x_0=j\Delta x$, $x=k\Delta x$, and $t=n\Delta t$, we get

$$p(x_0,x;t) = \pi p(x_0,x-\Delta x;t-\Delta t) + (1-\pi)p(x_0,x+\Delta x;t-\Delta t). \qquad (6.36)$$

Now assume that $p(x_0,x;t)$ is smooth; specifically that we can write a second order Taylor expansion

$$p(x_0,x+\Delta x;t-\Delta t) = p(x_0,x;t)-\Delta t\frac{\partial p}{\partial t} + \Delta x\frac{\partial p}{\partial x}$$
$$+ \frac{1}{2}(\Delta x)^2\frac{\partial^2 p}{\partial x^2} + \Delta x\Delta t\frac{\partial^2 p}{\partial x\partial t} + (\Delta t)^2\frac{\partial^2 p}{\partial t^2} + \cdots \qquad (6.37)$$

Then

$$0 = (1-2\pi)\Delta x\frac{\partial p}{\partial x}-\Delta t\partial\frac{p}{\partial}t + \tfrac{1}{2}(\Delta x)^2\frac{\partial^2 p}{\partial x^2} \qquad (6.38)$$
$$+ (2\pi-1)\Delta x\Delta t\frac{\partial^2 p}{\partial x\partial t} + \tfrac{1}{2}(\Delta t)^2\frac{\partial^2 p}{\partial t^2} + \cdots$$

Letting $\tau=\Delta t$, $\pi=\tfrac{1}{2}(1+\frac{\mu\tau^{1/2}}{\sigma})$ and $\Delta x=\sigma\tau^{1/2}$ as before, we obtain

$$\frac{1}{2}\sigma^2\frac{\partial^2}{\partial x^2}p(x_0,x;t) - \mu\frac{\partial}{\partial x}p(x_0,x;t) = \frac{\partial}{\partial t}p(x_0,x;t). \qquad (6.39)$$

We have derived the Kolmogorov forward equation for a Brownian motion with drift. For $\mu=0$ this is the **heat equation**. Generally, it is a Fokker–Planck equation for one-dimensional diffusion with an external force field. Note that we already know the solution to this equation. Things get more interesting (or more complicated, if you will) when μ and σ depend on x. We will return to that situation in section 6.5.

Example (Exercising a stock option) Suppose that we have the option of buying, at some time in the future, one unit of a stock at a fixed log price C, independent of its current market log price, which we (by subtracting it from all quantities considered) will take to be 0. Suppose that the log price changes according to a Brownian motion process with negative drift coefficient $-\mu$. When, if ever, should we exercise our option? Consider a policy that exercises the option when the market log price is y. Its expected gain is $P(y)(y-C)$, where $P(y)$ is the probability that the process will ever reach y. To determine $P(y)$, we first study $R(x)=\mathbf{P}(X(t)$ hits A before $-B \mid X(0)=x)$. Let $Y=X(h)-X(0)$. Then

$$R(x) = \mathbf{E}(R(x+Y))+o(h) \tag{6.40}$$

where the error term corresponds to the probability of hitting any of the boundaries before time h. Expanding R in a Taylor series yields

$$R(x) = \mathbf{E}(R(x)+YR'(x)+\tfrac{1}{2}Y^2R''(x)+\cdots)+o(h). \tag{6.41}$$

Since Y is $\mathrm{N}(-\mu h,h)$ we see that

$$R(x) = R(x) - R'(x)\mu h + R''(x)\frac{\mu^2h^2+h}{2} + o(h), \tag{6.42}$$

so that

$$-R'(x)\mu + \tfrac{1}{2}R''(x) = 0 \tag{6.43}$$

which has solution

$$R(x) = \frac{e^{-2\mu B}-e^{2\mu x}}{e^{-2\mu B}-e^{2\mu A}}. \tag{6.44}$$

The quantity of interest to us is obtained by letting $B\to\infty$ and setting $x=0, A=y$, to get $P(y)=\exp(-2\mu y)$. Thus the optimal y maximizes $(y-C)\exp(-2\mu y)$, and is seen to be

$$y = C + \frac{1}{2\mu}. \tag{6.45}$$

□

Now consider a Brownian motion process $B(t)$ and a fixed time s. Then the process $X(t)=B(t+s)-B(s)$ is again a Brownian motion which is independent of $B(t), t\leq s$. The **strong Markov property** asserts that the same is true if we pick s as a random time, provided that it depends only on the path up to that time, and not on anything happening later. Two examples of such **stopping times** are

$$T_1 = \inf\{t: B(t) \geq 1\};$$

$$T_2 = \inf\{t\geq a: B(t) = 0\}. \tag{6.46}$$

6.4. A more realistic model of Brownian motion

Since the Brownian motion process is inadequate as a description of physical Brownian motion for very short time intervals, efforts were soon underway to improve the model. In 1930, Uhlenbeck and Ornstein introduced a new model, assuming that the change in momentum, rather than in position, is independent of the past. Let $U(t)$ be the velocity of a particle (so we compute the position by integrating the velocity). The basic equation is due to Langevin (1908):

$$m\frac{dU(t)}{dt} = -\beta\, U(t) + K(t) \tag{6.47}$$

and states that the rate of change in momentum is decreased by friction (β^* is the coefficient of friction; the friction includes the Doppler effect associated with a particle being hit more often from the front than from the back), and also affected by a fluctuating force, due to the impact of the molecules. The force K has mean zero, and a covariance function which decreases extremely rapidly. Basically, except possibly at molecular distances, the impacts of the molecular shocks are independent. We will divide through by m. This changes β^* to $\beta=\beta^*/m$. It is physically reasonable to assume that the fluctuating force K is independent of the state u of the particle. A formal solution to (6.47) is obtained by multiplying the equation by $\exp(\beta t)$, so that

$$\frac{d}{dt}(e^{\beta t}U(t)) = e^{\beta t}K(t). \tag{6.48}$$

Integrating from 0 to s we get that

$$U(s) = U(0)e^{-\beta s} + \int_0^s e^{-\beta(s-t)}K(t)dt. \tag{6.49}$$

If K is continuous, the last integral makes sense as a pathwise Riemann integral. However, it is quite doubtful that K would be continuous, since it essentially is built from independent random variables. In order to define the integral, write (again formally) $K(t)=dB(t)/dt$, so that (6.49) becomes

$$U(s) = U(0)e^{-\beta s} + \int_0^s e^{-\beta(s-t)}dB(t). \tag{6.50}$$

We shall assume that B is a second order process that has stationary and independent increments. In fact, we will assume that B is a Brownian motion process with variance σ^2 and drift 0. Since a Brownian motion process does not have differentiable paths, we have somehow to make sense of the statement that $K(t)=dB(t)/dt$. The way we will do that is by defining $\int f(t)K(t)dt$ to mean the same thing as $\int f(t)dB(t)$, leaving us to try to make sense of the latter integral. The process B has mean zero and variance $\sigma^2 t$. In order to define $\int_a^b f(t)dB(t)$, we consider the Riemann–Stieltjes sum

$$S_m = \sum_1^m f(t_j)[B(t_{j+1}) - B(t_j)] \tag{6.51}$$

where $a=t_1<t_2<\cdots<t_{m+1}=b$. Clearly $\mathbf{E}S=0$. In defining the integral we will let $m\to\infty$ while $\max(t_{j+1}-t_j)\to 0$. In order to make the Riemann–Stieltjes sum converge for all paths of B, the process must have paths with bounded variation. We saw earlier that the Brownian motion process does not have such paths. Therefore we just require that the sum S converges in **mean square**, rather than with probability one. A sequence X_n converges in mean square to X, written $\text{l.i.m.}X_n=X$, if $\mathbf{E}(X_n-X)^2\to 0$. This is equivalent to saying that $\mathbf{E}X_nX_m\to c$ as $n,m\to\infty$ (Exercise 2). By Chebyshev's inequality mean square convergence implies convergence in probability. The following is a useful condition:

Proposition 6.3 Let $R(t,u)=\mathbf{Cov}(X(t),X(u))$. If $R(t,u)$ is of bounded variation, and f is such that

$$Q = \int_a^b\int_a^b f(t)f(u)dR(t,u) < \infty \tag{6.52}$$

then the stochastic Riemann–Stieltjes integral

$$J = \text{l.i.m.}\sum_1^m f(t_j)(X(t_{j+1}) - X(t_j)) \tag{6.53}$$

exists with $\mathbf{E}J=0$ and $\mathbf{Var}\,J=Q$. We write $J=\int_a^b f(t)dX(t)$.

Remark In a similar fashion we can define $\int f(s)X(s)ds$ in a mean square sense. A sufficient condition for existence is that $R(t,u)$ is continuous, and $\int_a^b\int_a^b f(t)f(u)R(t,u)dtdu<\infty$. An integration by parts formula holds for these integrals:

$$\int_a^b f(t)dX(t) = f(b)X(b)-f(a)X(a)-\int_a^b f'(t)X(t)dt. \tag{6.54}$$

The integrals can be defined for all second order processes with adequately smooth covariance functions. If a pathwise integral exists it is the same as the mean square integral. □

Example (Poisson process integrals) Let X be a Poisson process. Then

$$Y = \int_a^b f(s)dX(s) = \text{l.i.m.}\sum f(t_i)(X(t_{i+1})-X(t_i)) = \sum_{a\le\tau_i\le b} f(\tau_i). \tag{6.55}$$

In fact, as we saw in the previous chapter, this holds pathwise, and not only in mean square. We have that

$$\mathbf{E}Y = \mathbf{E}(f(b)X(b)-f(a)X(a)-\int_a^b f'(t)X(t)dt)$$

$$= f(b)\lambda b - f(a)\lambda a - \lambda\int_a^b tf'(t)dt \tag{6.56}$$

$$= f(b)\lambda b - f(a)\lambda a - \lambda bf(b) + \lambda af(a) + \lambda\int_a^b f(t)dt = \lambda\int_a^b f(t)dt.$$

In particular, $\mathbf{E}(X(b)-X(a))=\lambda(b-a)$. Since $R(t,u)=\lambda\min(t,u)$, Proposition 6.3 yields that

$$\mathbf{Var}\, Y = \lambda\int_a^b f^2(t)dt \tag{6.57}$$

(after a lot of algebra). This can be seen more easily by looking at

$$\mathbf{Var}\sum f(t_i)(X(t_{i+1})-X(t_i)) = \sum f^2(t_i)\lambda(t_{i+1}-t_i) \to \lambda\int f^2(t)dt. \tag{6.58}$$

Since the covariance function for Brownian motion is of the same form as that of a Poisson process, the same formula holds for stochastic integrals in the mean square sense with respect to Brownian motion. □

We now return to (6.50). Note that as $t\to\infty$, the effect of $U(0)$ disappears. Since the stochastic integral is defined as the limit of linear combinations of normal random variables, this limit must, too, be normally distributed. Its variance is

$$\sigma^2\int_0^s e^{-2\beta(s-t)}dt = \sigma^2\frac{1-e^{-2\beta s}}{2\beta}. \tag{6.59}$$

Maxwell's law for the frequency distribution of the velocity of particles in a gaseous medium in equilibrium requires that they be independent normal variables with mean zero and variance kT/m, where k is Boltzmann's constant, T is the absolute temperature, and m is the mass of the particle. If we want the process to be in equilibrium, we choose $U(0)$ to follow the limiting distribution. If we also take $U(0)$ independent of the Brownian motion, we obtain the equation for variance

$$\frac{kT}{m} = e^{-2\beta s}\frac{kT}{m} + \sigma^2\frac{1-e^{-2\beta s}}{2\beta}, \tag{6.60}$$

from which we get

$$\sigma^2 = \frac{2\beta kT}{m}. \tag{6.61}$$

Now we integrate U to compute the displacement X, namely

$$X(t) = X(0) + \frac{1-e^{-\beta t}}{\beta}u(0) + \frac{1}{\beta}\int_0^t (1-e^{-\beta(t-\tau)})dB(\tau). \tag{6.62}$$

Thus X is normally distributed with mean $X(0)$ and variance

$$\frac{2kT}{m\beta^2}\left[e^{-\beta t}-1+\beta t\right]. \tag{6.63}$$

From Stoke's law we have $\beta=6\pi r\eta/m$. For large values of t (compared with β) we can ignore the exponential and constant terms in the variance, yielding a variance of $kTt/3\pi r\eta=2Dt$ as in Einstein's derivation. For small values of t, however, the variance behaves like kTt^2/m. Hence the velocity exists (as it should, since $X(t)$ is integrated velocity).

To specify fully the distribution of the Ornstein–Uhlenbeck process we must compute its covariance function. First note by changing the order of integration that

$$\mathbf{E}\int f(t)dB(t)\int g(u)dB(u) = \iint f(t)g(u)\sigma^2(\min(t,u)dtdu. \tag{6.64}$$

Thus

$$\begin{aligned}\mathbf{Cov}(U(t),U(s)) &= e^{-\beta(t+s)}\mathbf{Var}\,U(0) + \sigma^2\int_0^t\int_0^s e^{-\beta(t+s-v-u)}\min(v,u)dvdu \\ &= e^{-\beta(t+s)}\frac{kT}{m} + \frac{2\beta kT}{m}\frac{e^{-\beta|s-t|}-e^{-\beta(t+s)}}{2\beta} \\ &= \frac{kT}{m}e^{-\beta|s-t|}.\end{aligned} \tag{6.65}$$

We have verified that $U(t)$ is a second order stationary process and, being a normal process, also strictly stationary. The Ornstein–Uhlenbeck process is the only stationary normal Markov process in continuous time.

The corresponding computation for the integrated Ornstein–Uhlenbeck process yields

$$\begin{aligned}\mathbf{Cov}(X(t),X(u)) &= \mathbf{Cov}\left(\int_0^t U(s)ds\,,\int_0^t U(v)dv\right) \\ &= \int_0^t\int_0^u \mathbf{Cov}(U(s),U(v))dsdv \\ &= \frac{\sigma^2}{2\beta^3}\left[2\beta\min(t,u)+e^{-\beta t}+e^{-\beta u}-1-e^{-\beta|t-u|}\right].\end{aligned} \tag{6.66}$$

We see that the Brownian motion process obtains when $\beta\to\infty$ in such a fashion that σ^2/β^2 converges to a constant.

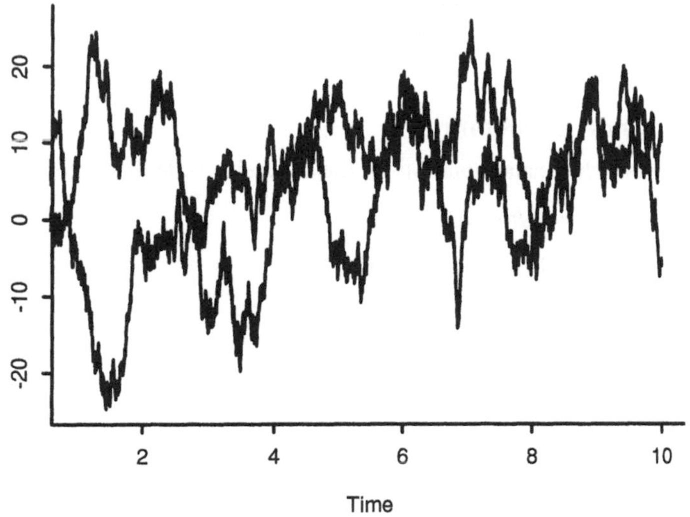

Figure 6.3. Two simulated paths of an Ornstein–Uhlenbeck process.

Application (Modeling the progression of AIDS) In biostatistical applications dependent data often arise in the context of **longitudinal studies**, i.e., repeated observations on the same individual. Frequently the measurements are taken at unequal times, and the number of observations can vary between subjects. There are two main approaches to modeling this type of data. One is a regression model with **random effects**, where the coefficients are realizations of a random variable. The common distribution of these random parameters induces correlation across subjects. The other approach is to view each subject as the realization of a stochastic process.

In AIDS study an important aspect of disease progression is the change in CD4 T-cell numbers, a critical indicator of the quality of the immune system, with small values corresponding to severe immunodeficiency. Immunological theory suggests that the rate of change in CD4 numbers "tracks" in the sense that individuals maintain their rate of change for long periods of time. This would correspond to a random effects model. For example, De Gruttola et al. (1991) suggest that the population mean decreases at a linear rate on a square root scale. An alternative hypothesis is that CD4 values evolve in a Markovian fashion. A feasible parametric model, having both a random effects and a stochastic process part, allows us to discriminate between these hypotheses. Let the response Y_{it} (taken to be the fourth root of CD4 counts to achieve homogeneity of variances) of individual i at time t satisfy

$$Y_{it} = a_i + b_i t + X_i(t) + \varepsilon_{it} \tag{6.67}$$

where (a_i, b_i) are independent bivariate normal vectors, $X_i(t)$ are integrated Ornstein–Uhlenbeck processes, and ε_{it} are iid measurement errors. □

6.5. Diffusion equations

A convenient shorthand notation for the Ornstein–Uhlenbeck process is

$$dU = Udt + dB \tag{6.68}$$

which is interpreted in integral form

$$U(t) - U(0) = \int_0^t U(s)ds + B(t) - B(0). \tag{6.69}$$

A more general form of this kind of **stochastic differential equation** is

$$dY = \mu(Y)dt + \sigma(Y)dB. \tag{6.70}$$

The way to think about this intuitively is in terms of differences:

$$\Delta_h Y(t) = \mu(Y(t))\Delta_h t + \sigma(Y(t))\Delta_h B(t) \tag{6.71}$$

where $\Delta_h f(t) = f(t+h) - f(t)$ as before. This process must be Markovian, since B has independent increments. Furthermore the paths should be continuous, since ΔY is a linear piece plus an increment of a continuous process. The process behaves locally as Brownian motion with drift. In particular

$$\mathbf{E}(\Delta_h Y(t) \mid Y(t)=y) = \mu(y)h \tag{6.72}$$

and

$$\begin{aligned} \mathbf{E}((\Delta_h Y(t))^2 \mid Y(t)=y) &= \sigma^2(y)\mathbf{E}(\Delta_h B(t))^2 + o(h) \\ &= \sigma^2(y)h + o(h). \end{aligned} \tag{6.73}$$

The coefficients μ and σ^2 are called the **infinitesimal mean** (or drift coefficient) and the **infinitesimal variance** (or diffusion coefficient), respectively. They may depend on t as well as $Y(t)$. A stochastic process is called a **diffusion** if it is a continuous time Markov process with continuous paths.

Example (Models for Brownian motion) If $\mu=0$ and $\sigma^2=1$ we have the standard Brownian motion process. The Brownian motion process with drift has a non-zero constant μ. The Ornstein–Uhlenbeck process is obtained by using $\mu(y)=-\beta y$ and $\sigma^2(y)\equiv\sigma^2$. This process has a tendency to go back towards zero, and the tendency is stronger the farther away from zero the process has drifted. □

Let $p_{s,t}(y,x)$ be the transition probability density for a transition from $Y(s)=y$ to $Y(t)=x$. Then, using the Chapman–Kolmogorov equation, we have that

$$\begin{aligned}
p_{s-ds,t}(y,x)-p_{s,t}(y,x) &= \int p_{s-ds,s}(y,z)(p_{s,t}(z,x)-p_{s,t}(y,x))dz \\
&= \int p_{s-ds,s}(y,z)\{(z-y)\frac{\partial}{\partial y}p_{s,t}(y,x) \\
&\quad + \tfrac{1}{2}(z-y)^2\frac{\partial^2}{\partial y^2}p_{s,t}(y,x) + \cdots\}dz \qquad (6.74)\\
&\approx \frac{\partial p_{s,t}(y,x)}{\partial y}\int(z-y)p_{s-ds,s}(y,z)dz \\
&\quad + \tfrac{1}{2}\frac{\partial^2 p_{s,t}(y,x)}{\partial y^2}\int(z-y)^2 p_{s-ds,s}(y,z)dz.
\end{aligned}$$

But

$$\int(z-y)p_{s-ds,s}(y,z)dz = \mathbf{E}(\Delta_{ds}Y(s-ds)\mid Y(s-ds)=y) = \mu(y)ds \qquad (6.75)$$

and

$$\int(z-y)^2 p_{s-ds,s}(y,z)dz = \sigma^2(y)ds, \qquad (6.76)$$

whence

$$\frac{\partial p_{s,t}(y,x)}{\partial s} = \mu(y)\frac{\partial p_{s,t}(y,x)}{\partial y} + \tfrac{1}{2}\sigma^2(y)\frac{\partial^2 p_{s,t}(y,x)}{\partial y^2}. \qquad (6.77)$$

This is the backward equation for a diffusion with infinitesimal parameters μ and σ^2. The infinitesimal parameters determine the distribution of the process in the interior of its state space, but boundary conditions may be needed on the state space boundary. Two diffusions may have the same infinitesimal parameters, but different boundary conditions, and therefore different behavior. We can compute the forward equation in a similar fashion; it is

$$\frac{\partial p_{s,t}(y,x)}{\partial t} = -\frac{\partial}{\partial x}(\mu(x)p_{s,t}(y,x)) + \tfrac{1}{2}\frac{\partial^2}{\partial x^2}(\sigma^2(x)p_{s,t}(y,x)). \qquad (6.78)$$

If a diffusion has a stationary distribution, we will need $\partial p_{0,t}/\partial t \to 0$ as $t\to\infty$. Then the forward equation yields that a stationary solution $\pi(x) = \lim_{t\to\infty}p_{0,t}(0,x)$ must satisfy

$$0 = \frac{d}{dx}\left[-\mu(x)\pi(x) + \tfrac{1}{2}\frac{d}{dx}(\sigma^2(x)\pi(x))\right]. \qquad (6.79)$$

We integrate this to obtain the differential equation

$$\frac{d}{dx}\log\pi(x) = \frac{2\mu(x)-\frac{d}{dx}\sigma^2(x)}{\sigma^2(x)}. \qquad (6.80)$$

For example, if we study the Ornstein–Uhlenbeck process, we see that

$$\pi(x) \propto e^{-\frac{\beta x^2}{\sigma^2}}, \tag{6.81}$$

so that the stationary distribution is Maxwell's normal distribution for velocities in a gas in equilibrium. On the other hand, Brownian motion does not have a stationary distribution since, in the standard case for example, the stationary distribution would have to be uniform over the reals.

Example (Branching process) An important use for diffusion processes is to approximate processes that more naturally evolve in discrete time and/or space, but have complicated behavior. To illustrate this we shall look at a Bienaymé-Galton-Watson process with offspring mean μ and variance σ^2. Then

$$\mathbf{E}(Z_{n+1}-Z_n \mid Z_n=k) = k\mu-k = k(\mu-1) \tag{6.82}$$

and

$$\mathbf{Var}(Z_{n+1}-Z_n \mid Z_n=k) = k\sigma^2. \tag{6.83}$$

In order to find an approximating diffusion we use infinitesimal parameters that are proportional to the current population size, namely

$$\mu(y) = \beta y \quad \sigma^2(y) = \alpha y. \tag{6.84}$$

The forward equation, writing p_t for $p_{0,t}$, is

$$\tfrac{1}{2}\alpha\frac{\partial^2}{\partial x^2}(xp_t(x_0,x)) - \beta\frac{\partial}{\partial x}(xp_t(x_0,x)) = \frac{\partial}{\partial t}p_t(x_0,x). \tag{6.85}$$

If we try to establish a stationary distribution for this diffusion, the method above yields

$$\frac{d}{dx}\log\pi(x) = 2\frac{\beta}{\alpha}-\frac{1}{x} \tag{6.86}$$

so that

$$\pi(x) \propto xe^{2\beta x/\alpha}. \tag{6.87}$$

Since this is not integrable over the positive axis, there is no stationary distribution. This is in accordance with the behavior of the branching process.

Let $\phi(\theta,t) = \mathbf{E}\exp(X(t)\theta)$, where $X(t)$ is the diffusion process that will approximate Z_n. Multiplying the forward equation by $\exp(\theta x)$ and integrating over x we get

$$\frac{\partial\phi(\theta,t)}{\partial t} = (\beta\theta - \tfrac{1}{2}\alpha\theta^2)\frac{\partial\phi(\theta,t)}{\partial\theta} \tag{6.88}$$

which has solution

$$\phi(\theta,t) = \exp\left[\frac{-x_0e^{\beta t}\theta}{1+\frac{\alpha}{2\beta}(e^{\beta t}-1)}\theta\right]. \tag{6.89}$$

By differentiating and setting $\theta=0$ we see that

$$\mathbf{E}X(t) = x_0 e^{\beta t} \tag{6.90}$$

and

$$\mathbf{Var}X(t) = \frac{\alpha x_0}{\beta} e^{\beta t}(e^{\beta t}-1) \tag{6.91}$$

which can be compared to the branching process moments

$$\mathbf{E}Z_n = z_0\mu^n \quad \text{and} \quad \mathbf{Var}Z_n = \frac{z_0\sigma^2}{\mu(\mu-1)}\mu^n(\mu^n-1). \tag{6.92}$$

In order to obtain convergence we must have $n=t/\tau$, $\mu-1=\beta\tau$, and $\sigma^2=\alpha\tau$. The limit is obtained as $\tau\to 0$. If we let $\tau=1/N$, we are considering a sequence Z^N of branching processes, with the Nth process having offspring mean $1+\beta/N$, and offspring variance α/N. We then define a sequence of continuous time processes $X^N(t)=Z^N_{[t/N]}$. This corresponds to looking at a very large population, measured in units of N individuals, and having Nt generations pass in time t. Since a single individual has weight $1/N$, the average population increase due to one individual, or $m-1$, should be of the same order. Likewise, the offspring variance should correspond to the weight of an individual. Then all fdd's of X^N converge to those of X. □

The stochastic differential equation (6.70) is interpreted as

$$Y(t)-Y(0) = \int_0^t \mu(Y(t))dt + \int_0^t \sigma(Y(t))dB(t). \tag{6.93}$$

Integrals of the form $\int f(t)dB(t)$ have so far only been defined for non-random functions f. In order to make sense of the integral with a random f we proceed in a similar fashion, and define the integral as the quadratic mean limit of the approximating Stieltjes sums.

Example (Integrating Brownian motion) Let $X(t) = \int_0^t B(s)dB(s)$. Standard integration by parts would yield $X(t)=B(t)^2/2$, provided that $B(0)=0$. We define $X(t)$ as the q.m. limit of

$$S_n = \sum_1^n B(t_{i-1})(B(t_i)-B(t_{i-1})) = \frac{B(t)^2}{2} - \tfrac{1}{2}\sum_1^n (B(t_i)-B(t_{i-1}))^2. \tag{6.94}$$

Now,

$$\sum_1^n (B(t_i)-B(t_{i-1}))^2 \overset{\text{q.m.}}{\rightarrow} t \tag{6.95}$$

by our earlier computations. Thus

$$\text{l.i.m.}\, S_n = B(t)^2/2 - t/2. \tag{6.96}$$

The usual integration by parts formula is no longer valid. We can write this result in differential notation:

$$d(B(t)^2) = 2B(t)dB(t)+dt. \tag{6.97}$$

By Taylor expansion of a twice differentiable function f we have

$$df(x) = f'(x)dx + \tfrac{1}{2}f''(x)(dx)^2. \tag{6.98}$$

Using $f(x)=x^2$ and $x=B(t)$ we see that

$$d(B(t)^2) = 2B(t)dB(t) + (dB(t))^2, \tag{6.99}$$

from which we see that we must have $(dB(t))^2=dt$. For an ordinary integral the square of the differential element is negligible, but this is not the case for stochastic integrals. □

The reasoning in the example above works in some generality. Here is a general change of variables formula for stochastic differential equations, called **Ito's formula.**

Theorem 6.4 Suppose that $Y(t)$ satisfies the stochastic differential equation

$$dY = \mu(Y)dt + \sigma(Y)dB. \tag{6.100}$$

If f is a twice differentiable function, and $X(t)=f(Y(t))$, then X satisfies the stochastic differential equation

$$dX = (f'(Y)\mu(Y) + \tfrac{1}{2}f''(Y)\sigma^2(Y))dt + f'(Y)\sigma(Y)dB \tag{6.101}$$

where the $B(t)$-processes are the same.

Sketch of proof In essence this is just an expansion of the differential to two terms:

$$\begin{aligned} df(Y) &= f'(Y)dY + \tfrac{1}{2}f''(Y)(dY)^2 \\ &= f'(Y)\{\mu(Y)dt+\sigma(Y)dB\}+\tfrac{1}{2}f''(Y)\{\mu^2(Y)dt^2 \\ &\quad + 2\mu(Y)\sigma(Y)dtdB + \sigma^2(dB)^2\}. \end{aligned} \tag{6.102}$$

We now ignore all terms having orders higher than dt, which are the dt^2 and the $dtdB$ terms. But the $(dB)^2$ term is of order dt, and it is not negligible. □

Example (Geometric Brownian motion) In the modeling of stock price options in the previous section, we were describing the log prices as Brownian motion with drift. In other words, if $Y(t)$ describes the log price of a stock at time t we have that $dY(t)=\mu dt+\sigma dB(t)$, and the price $X=\exp(Y)$. From

Ito's formula we obtain

$$dX = e^{Y}((\mu+\sigma^2)dt + \sigma dB) = X((\mu+\sigma^2)dt + \sigma dB). \tag{6.103}$$

A solution to this stochastic differential equation is given by

$$X(t) = \exp(\mu t+\sigma B(t)). \tag{6.104}$$

If (6.103) were an ordinary differential equation, its solution would be

$$X(t) = \exp((\mu+\sigma^2)t+\sigma B(t)). \tag{6.105}$$

□

Application (Monitoring hormone concentrations in pregnancy) In evaluating the result of a laboratory test, a reference interval based on data from an appropriate group of healthy subjects is often used. For monitoring purposes, much is gained by using the subject as his or her own reference, to eliminate the inter-individual variation from the comparison. This requires the presence of one or more observations obtained when the subject was in "good health", as well as a model for the temporal development, within an individual, of the quantity monitored.

The concentration of plasma progesterone during pregnancy is a factor that is easily monitored. During later stages (after week 12) of a pregnancy the plasma progesterone value reflects the size of the placenta. In a normal pregnancy, the growth rate of the placenta is proportional to its current size. There is some indication that this growth rate may be fluctuating randomly in time for a given individual (see Figure 6.4). Thus a reasonable model may be

$$dX(t) = X(t)\,a(t)\,dt \tag{6.106}$$

where $a(t)$ is a white noise process with fixed mean α. We can write this $a(t)dt=\alpha dt+\sigma dB(t)$. From the previous example we see that a solution is $X(t)=\exp(\sigma B(t)+(\alpha-\sigma^2)t/2)$, in other words that $\log X(t)$ is a Brownian motion process with drift. □

Example (Optimal harvesting of untended forest stands) In order to determine the value of a biological asset, such as a stand of trees, we need to take into account at least two aspects of random variability: in the growth of the asset, and in its price. Let $P(t)$ be the unit price of the lumber, and $X(t)$ the size of the asset, both at time t. If we harvest at time t, the revenue is $R(t)=P(t)X(t)$, which is called the **aggregate intrinsic value** of the forest stand. More generally, let the **market value** of the asset be $V(x,p,t)$ when $X(t)=x$ and $P(t)=p$. This value is the expected present value of revenues earned, given that the harvest takes place at the (random) time τ that maximizes the expected present value. Assuming a constant instantaneous discount rate δ,

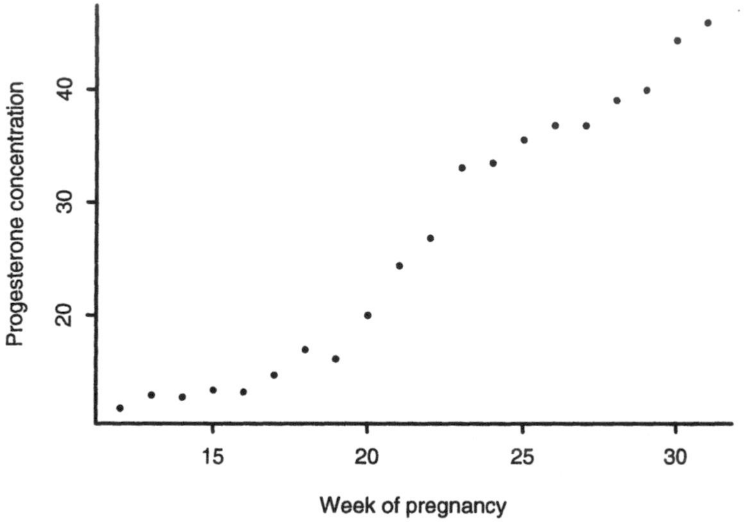

Figure 6.4. Concentration of plasma progesterone for a pregnant woman. Adapted from P. Guttorp, *Statistical Inference for Branching Processes*, with permission of John Wiley & Sons, Inc. (1991).

we have (letting the present time be 0)

$$V(x,p,0) \equiv V(x,p) = \sup_{\tau \geq 0} \mathbf{E}(\exp(-\delta\tau)P(\tau)X(\tau) \mid P(0)=p, X(0)=x). \quad (6.107)$$

The optimal harvest problem is to determine the stopping rule τ which maximizes the right-hand size of (6.107).

We assume that the price change is proportional to current price, and thus follow a geometric Brownian motion process satisfying

$$dP = \mu P dt + \sigma_1 P dW_1. \quad (6.108)$$

Letting $Q(t)=\log P(t)$ we get by Ito's formula

$$dQ = (\mu - \tfrac{1}{2}\sigma_1^2)dt + \sigma_1 dW_1. \quad (6.109)$$

In specifying a stochastic model for biological growth, we again assume that the growth is proportional to current size, writing

$$dX = Xf(X)\,dt + \sigma_2 X\,dW_2, \quad (6.110)$$

where f is a decreasing growth rate (so we have **compensatory density-dependent growth**). The special case where f is a linear function yields logistic growth (see Exercise 12 for further discussion of this). Taking logs in (6.110) we get, with $Y(t)=\log X(t)$, that

$$dY = (f(e^Y) - \tfrac{1}{2}\sigma_2^2)dt + \sigma_2 dW_2. \tag{6.111}$$

It turns out (Reed and Clarke, 1990) that the optimal harvest rule is of the form: stop whenever Y exceeds a barrier b (or, equivalently, when the asset size hits a target value $x = \exp(b)$). Let v be the value corresponding to this stopping rule, namely

$$v = \mathbf{E}\exp(-\delta T_{y,b} + Q(T_{y,b}) + y \mid Q(0)=q, Y(0)=y) \tag{6.112}$$

where $T_{y,b} = \inf\{t > 0 : Y(t) = b \mid Y(0) = y\}$. Using the solution (6.104) to (6.108) and conditioning on $T_{y,b}$, we see that

$$v = \exp(q+b)\,\mathbf{E}^y \exp(-BT_{y,b}) \tag{6.113}$$

where $B = \delta - \mu - \tfrac{1}{2}\sigma_1^2$. Write

$$M(y,b;B) = \mathbf{E}^y \exp(-BT_{b,y}). \tag{6.114}$$

As a function of B, this is the Laplace transform of the distribution of $T_{y,b}$. Consider an infinitesimal increase in the starting value y, and write

$$M(y,b;B) = \mathbf{E}^y\{\exp(-Bdt)M(y+dy,b;B)\} + o(dt) \tag{6.115}$$

$$= (1 - Bdt)(M + \frac{\partial M}{\partial y}\mathbf{E}dy + \tfrac{1}{2}\frac{\partial^2 M}{\partial y^2}\mathbf{E}dy^2 + o(dt))$$

using a Taylor expansion of the exponential function as well as of the Laplace transform. Simplifying, dividing by dt and letting $dt \to 0$ we see that

$$\tfrac{1}{2}\sigma_2^2 \frac{\partial^2 M}{\partial y^2} + f(exp(y))\frac{\partial M}{\partial y} - BM = 0. \tag{6.116}$$

Write $h(y) = f(\exp(y))$ and $\phi(y) = (\partial/\partial y)\log M$. Then (6.116) can be written

$$\phi'(y) = -\phi^2(y) - \frac{2}{\sigma_2^2}(h(y)\phi(y) - B), \tag{6.117}$$

which is a Riccati equation. The boundary condition is $\phi(b)=1$, since on the stopping boundary the value must equal the intrinsic value. This equation can be solved numerically. □

6.6. Likelihood inference for stochastic differential equations

In order to utilize the model for placenta monitoring in the previous section, we must develop some likelihood theory so as to be able to estimate parameters of the process. The first step is to write down the likelihood. This is quite difficult, and we will first give a heuristic derivation of the likelihood for a Brownian motion process with drift.

Example (Brownian motion with drift) Consider the equation $dX(t)=\mu dt+\sigma dB(t)$ for a Brownian motion process with drift. Thinking of it as a difference, we have

$$\Delta_h X(t) = \mu h + \sigma \Delta_h B(t). \tag{6.118}$$

Since $\Delta_h B$ has a normal distribution with mean 0 and variance h, independently of the development of the process up to time t, the additional likelihood from the observation in $(t,t+h)$ is

$$(2\pi\sigma h)^{-1/2}\exp(-(\Delta_h x(t)-\mu h)^2/2\sigma^2 h). \tag{6.119}$$

If we observe the process from 0 to T, and multiply together all these independent pieces, we get a log likelihood proportional to

$$-\tfrac{1}{2}\sigma\sum\log h - \frac{1}{\sigma^2}\mu^2\sum h + \sum\Delta_h x(t)\frac{\mu}{\sigma^2}. \tag{6.120}$$

Ignoring the first term and letting $h\to 0$, this converges to

$$-\frac{\mu^2 T}{\sigma^2} + \frac{\mu}{\sigma^2}\int_0^T dx(t). \tag{6.121}$$

The reason for ignoring the first term is that it is common to all processes driven by the same Brownian motion process with the same variance function, and therefore does not contribute to our ability to distinguish different values of μ. Technically, this amounts to computing the likelihood with respect to a process solving

$$dZ(t)=\sigma dB(t). \tag{6.122}$$

Taking partial derivatives in (6.121) we find the maximum likelihood estimate

$$\hat{\mu} = \frac{1}{T}\int_0^T dx(t) = \frac{x(T)-x(0)}{T}. \tag{6.123}$$

To estimate the variance we use the fact that the quadratic variation of a Brownian motion with drift is linear with slope σ^2, which formally can be written

$$(dX)^2 = \sigma^2 dt. \tag{6.124}$$

This suggests an estimate obtained by averaging the observed quadratic variation, such as

$$\hat{\sigma}^2 = \frac{1}{T}\sum_{1}^{T2^n}(x(i2^{-n})-x((i-1)2^{-n}))^2. \tag{6.125}$$

In fact, as $n\to\infty$, this estimate converges to σ^2. In other words, we can *compute* σ^2 from a completely observed path. This is the reason that we can drop the first term in (6.121). In reality, one would usually not observe the path continuously, so the approximation using observed quadratic variation would be needed. □

The argument in the example above generalizes in a fairly straightforward fashion to more general diffusions. Suppose that Y solves

$$dY = \mu(Y;\boldsymbol{\theta})dt + \sigma(Y)dB \tag{6.126}$$

where $\boldsymbol{\theta}$ is a parameter of interest. Then the likelihood of an observed path (with respect to the diffusion given by $dZ=\sigma(Z)dB$) is

$$L(\boldsymbol{\theta}) = \exp\Bigl(\int_0^T \frac{\mu(y(s);\boldsymbol{\theta})}{\sigma(y(s))}dy(s) - \tfrac{1}{2}\int_0^T \frac{\mu^2(y(s);\boldsymbol{\theta})}{\sigma(y(s))}ds\Bigr). \tag{6.127}$$

Here we are assuming that σ is a known function. If the expression for σ contains unknown parameters, we can use Ito's formula to find a transformed process $X(t) = f(Y(t))$, where f must satisfy $f'(x)\sigma(x)\equiv 1$. This will be a diffusion with drift coefficient $\mu/\sigma+\tfrac{1}{2}f''\sigma^2$.

Under regularity conditions (to ensure appropriate behavior of the second derivative of the log likelihood), an estimate of $\boldsymbol{\theta}$ based on maximizing the logarithm of (6.127) is consistent and asymptotically normal with covariance matrix computed from the second derivative of the log likelihood in the usual fashion.

Application (Monitoring hormone concentrations in pregnancy, continued) In the previous section we derived the stochastic differential equation

$$dX(t) = X(t)(\alpha dt+\sigma dB(t)). \tag{6.128}$$

We want to estimate α. From (6.127) the log likelihood is

$$\frac{\alpha}{\sigma}\int_0^T dx(s) - \tfrac{1}{2}\frac{\alpha^2}{\sigma}\int_0^T x(s)ds = \frac{\alpha}{\sigma}(\alpha(x(T)-x(0))-\tfrac{1}{2}\int_0^T x(s)ds, \tag{6.129}$$

whence the maximum likelihood estimate of α is

$$\hat{\alpha} = \frac{x(T)-x(0)}{\int_0^T x(s)ds} \doteq \frac{x_T-x_0}{\sum_1^T x_i} = 0.064. \tag{6.130}$$

This approximation is a Riemann sum using the observed values $x_1, \ldots, x_T$. The approximate variance of $\hat{\alpha}$ is given by $\sigma^2/\int_0^T x(s)ds$. In order to estimate σ, consider $\log X$, which is a diffusion with differential equation

$$dY = (\alpha-\tfrac{1}{2}\sigma^2)dt + \sigma dB, \tag{6.131}$$

so that we can compute σ^2 from a path using the quadratic variation as in the previous example:

$$\hat{\sigma}^2 = \frac{1}{T}\sum_1^{T2^n}\log^2\left[\frac{x(i2^{-n})}{x((i-1)2^{-n})}\right]. \tag{6.132}$$

The data are only observed to precision $n=0$, yielding the approximate estimate

$$\hat{\sigma}^2 \doteq \frac{1}{T}\sum_1^{T}\log^2(x_i/x_{i+1}) = 0.0102. \tag{6.133}$$

□

Application (Modeling the progression of AIDS, continued) We consider data from a cohort study of 1637 homosexual and bisexual men in Los Angeles starting in 1984–5. Here we shall only look at the **seroconverters**, the 87 individuals observed to change from HIV negative to HIV positive during the study. The midpoint between the last HIV negative and the first HIV positive measurements was taken to be the time of infection, and data for the 6 months following infection were ignored, since CD4 counts decrease dramatically immediately following infection. Due to the influence of the AZT treatment on this cohort no data after January 1 1990 are used.

Since we do not have continuous or equispaced measurements, we use the mean and covariance structure for the observed data to compute the likelihood. Recalling the model (6.67) we have $(a_i,b_i)\sim N(\boldsymbol{\mu},\Sigma)$. The integrated Ornstein–Uhlenbeck process $X_i(t)$ is assumed to start at $X_i(0)=0$, and has parameters β and σ^2, while the measurement error has variance σ_m^2. Hence the mean of individual i at time t is

$$\mathbf{E}Y_{it} = \mu_a + \mu_b t, \tag{6.134}$$

while the variance is

$$\mathbf{Var}Y_{it} = \sigma_a^2 + \sigma_b^2 t^2 + 2\sigma_{ab}t + \frac{\sigma^2}{\beta^3}(e^{-\beta t} - 1 + \beta t) + \sigma_m^2, \tag{6.135}$$

and the covariance is

$$\mathbf{Cov}(Y_{it}, Y_{is}) = \sigma_a^2 + \sigma_b^2 ts + \sigma_{ab}(t+s) \tag{6.136}$$

$$= \frac{\sigma^2}{2\beta^3}(2\beta\min(t,s) + e^{-\beta t} + e^{-\beta s} - 1 - e^{-\beta|t-s|}).$$

Different individuals are independent. The contribution of the ith individual, with data at $t_1, \ldots, t_n$, to the log likelihood, is then given by a normal likelihood having the mean and covariance structure given in the equations (6.134)–(6.136).

The mle of β is 2.5, with a 95% **profile likelihood** interval of $(0.8, \infty)$. This is obtained by computing the likelihood as a function of β only, keeping all the other parameters at their maximum likelihood estimates. The random effects means μ_a and μ_b were estimated to be 5.17 (standard error 0.054) and −0.221 (0.026), respectively. The likelihood is very flat for $\beta > 2.5$. The large value of β indicates that the derivatives only "track" for a short period of time. The estimated correlation between observations 3 months apart on the same individual is 0.54, while for observations 1 year apart it is 0.08.

The population median of CD4 10 years after infection is estimated to be about 77. Clinical AIDS develops for CD4 around 100, and the median time to AIDS is about 10 years, indicating that the estimated rate of decline is reasonable. The median time to AIDS predicted by this model is about 9 years. □

6.7. The Wright–Fisher model of diploid populations

Consider a population with two types of genes, A and a. We assume that there is a fixed population of N individuals, each having 2 genes, for a total gene population of $2N$. Let $X(t)$ be the number of A-genes in the population at generation t. We saw in Chapter 2, when discussing the Hardy–Weinberg laws, that (in the absence of mutation and selection, at least) gene frequencies tend to remain steady. Then $X(t+1)$ should be a random variable with mean $X(t)$. Fisher (1930) and Wright (1931) assumed that $X(t+1) \sim \text{Bin}(2N, X(t)/2N)$. This is a Markov chain with transition probabilities

$$p_{ij} = \binom{2N}{j} \left[\frac{i}{2N}\right]^j \left[1 - \frac{i}{2N}\right]^{2N-j}. \tag{6.137}$$

This Markov chain has two absorbing states, 0 and $2N$. If it reaches either of these states, the population has lost its genetic variation at this locus, and we say that **fixation** has occurred. An important problem is to determine the probability of eventual fixation of the A-gene, say, as well as the expected time to fixation. One-step conditional arguments yield answers to some aspects of these questions.

To make things a little more complicated, consider selection and mutation. If $X(t)=i$, the expected frequencies of AA, Aa, and aa gene pairs are $(i/2N)^2$, $2(i/2N)(1-i/2N)$, and $(1-i/2N)^2$, respectively. If these zygotes reach maturity and in turn produce genes for the next generation in the ratios $1+s_1 : 1+s_2 : 1$, where s_1 and s_2 are small selective (dis)advantages, then the expected frequency of A at the next generation is

$$\pi_i^* = \frac{(1+s_1)i^2 + (1+s_2)i(2N-i)}{(1+s_1)i^2 + 2(1+s_2)i(2N-i) + (2N-i)^2}. \tag{6.138}$$

Suppose in addition that A-genes mutate to a with probability u, and a to A with probability v. Then the expected frequency of A is

$$\pi_i = (1-u)\pi_i^* + v(1-\pi_i^*). \tag{6.139}$$

Applying the Wright–Fisher model to this situation, we get the transition probabilities

$$p_{ij} = \binom{2N}{j} \pi_i^j (1-\pi_i)^{2N-j}. \tag{6.140}$$

It is extremely difficult to compute exact answers in this model, except that when u and v are both zero, fixation is certain, whereas when u and v are both positive, the chain has a stationary distribution.

We can determine the conditional moments of the model (6.140). Clearly

$$\mathbf{E}(X(t+1)-X(t) \mid X(t)=i) = \tag{6.141}$$
$$v+(1-u-v)\frac{(1+s_1)i^2+(1+s_2)i(2N-i)}{(2N)^2+s_1 i^2+2s_2 i(2N-i)} - i.$$

Suppose that $s_1=\alpha_1/2N$, $s_2=\alpha_2/2N$, $u=\beta_1/2N$, and $v=\beta_2/2N$. Consider the fraction $Y(t)=X(t)/2N$ of A-genes in the population at time t. We have that

$$\mathbf{E}(Y(t+1)-Y(t) \mid Y(t)=y) = v+(1-u-v)\frac{y+s_1 y^2+s_2 y(1-y)}{1+s_1 y^2+2s_2 y(1-y)} - y$$
$$= v(1-y) - uy + y + s_1 y^2 + s_2 y(1-y) - y(1+s_1 y^2 \tag{6.142}$$
$$+ 2s_2 y(1-y)) + O(N^{-2})$$
$$= (-\beta_1 y+\beta_2(1-y))/2N + x(1-x)(\alpha_1 x+\alpha_2(1-2x))/2N + O(N^{-2}).$$

A similar computation shows that

$$\mathbf{Var}(Y(t+1)-Y(t) \mid Y(t)=y) = y(1-y)/2N + O(N^{-2}). \tag{6.143}$$

In order to approximate the Markov chain described by (6.140) by a diffusion process, we need to measure time in units of size $2N$ generations. Let $Z^N(t) = X([2Nt])/2N$. Then Z^N converges to a diffusion process Z with infinitesimal mean

$$\mu(x) = x(1-x)(\alpha_1 x+\alpha_2(1-2x))-\beta_1 x+\beta_2(1-x) \tag{6.144}$$

and infinitesimal variance

$$\sigma^2(x) = x(1-x). \tag{6.145}$$

We will first use this diffusion to find expressions for the expected time to fixation, and for the probability of fixation of the A-gene. In this case we must assume that $\beta_i=0$, i.e., that there is no mutation. The problem of computing the fixation probability has already been solved in a previous section: let $u(x)=\mathbf{P}(Z$ hits 1 before $0 \mid Z(0)=x)$. Then the technique corresponding to a one-step conditional argument works:

$$u(x) = E(u(Z(h))) + o(h), \tag{6.146}$$

yielding (by Taylor expansion) a differential equation for u with the solution

$$u(x)=\frac{\int_0^x \psi(y)dy}{\int_0^1 \psi(y)dy} \tag{6.147}$$

where

$$\psi(y)=\exp(-2\int_0^x \frac{\mu(y)}{\sigma^2(y)}dy). \tag{6.148}$$

We may simplify matters by assuming that the selectivity is due to the A-gene only, so that $s_1=2s_2=2\alpha/2N$. Then $\mu(x)=\alpha x(1-x)$, and we have

$$u(x)=\frac{1-e^{-2\alpha x}}{1-e^{-2\alpha}}. \tag{6.149}$$

Suppose that we have a population with $N=10^6$, $x=\frac{1}{2}$, and $s_2=10^{-5}$. Then $u(\frac{1}{2})\doteq 1-\exp(-20)$. In other words, even for a relatively small population of one million, a gene with a selective advantage of 10^{-5} will eventually become fixed. It is perhaps even more surprising when one notes that the mean increase per generation in the number of A genes is five, and the chance of a decrease between successive generations is about $\Phi(-0.01)=0.496$. In other words, in the long run the selective advantage dominates the random fluctuations.

Suppose we want to infer the selective advantage α from observations of the frequency of A-genes in a population. The likelihood for a continuously observed path is

$$\alpha\int_0^T dz(t)-\frac{\alpha^2}{2}\int_0^T z(t)(1-z(t))dt, \tag{6.150}$$

so the maximum likelihood estimate of α is

$$\hat{\alpha}=\frac{z_T-z_0}{\int_0^T z(t)(1-z(t))dt}. \tag{6.151}$$

Its variance is approximately $(\int_0^T z(1-z)dt)^{-1}$. Returning to the original time scale, we would actually compute $\hat{\alpha}$ from observations on $2NT$ generations as

$$\tilde{\alpha}=\frac{X(NT)-X(0)}{\frac{1}{N}\sum_0^{NT} X(i)(1-X(i))}. \tag{6.152}$$

A different question is the chance of fixation of a newly mutated gene. If $x=1/2N$ we see that

$$u(1/2N) \doteq \frac{2s_2}{1-e^{-4Ns_2}} \tag{6.153}$$

which for $s_2 \gg 1/N$ becomes $u(1/2N) \doteq 2s_2$, independent of N.

Let T be the time to fixation (at 0 or 1). In order to compute its expected value, let

$$v(x) = \mathbf{E}(T \mid Z(0)=x) \tag{6.154}$$

and construct a differential equation by writing

$$v(x) = \mathbf{E}(h+\mathbf{E}(T \mid Z(h)) \mid Z(0)=x). \tag{6.155}$$

Expand to get

$$-1 = \mu(x)v'(x) + \tfrac{1}{2}\sigma^2(x)v''(x). \tag{6.156}$$

In the simple case of no selection and no mutation (the original Wright–Fisher model) we have

$$v(x) = -2(x \log x + (1-x)\log(1-x)) \tag{6.157}$$

with a maximum at $x=\frac{1}{2}$ equal to $2\log 2$. In other words, if the population starts with equal proportions A and a, fixation of A will occur on the average after about $2.8N$ generations, which is an extremely long time. If we look at the time to fixation following a new mutation we see that

$$v(1/2N) \doteq \frac{1+\log 2N}{N}. \tag{6.158}$$

In terms of generations it takes on the average $2+2\log 2N$ generations until either the mutation or the original gene is lost.

Going back to the model (6.140), when we have mutation there will be a stationary distribution, which as before we obtain from the forward equation. It satisfies

$$-\mu(x)\pi(x) + \tfrac{1}{2}\frac{d}{dx}(\sigma^2(x)\pi(x)) = 0. \tag{6.159}$$

The solution is

$$\pi(x) \propto x^{2\beta_2-1}(1-x)^{2\beta_1-1}\exp(2\alpha_2 x + (\alpha_1 - 2\alpha_2)x^2). \tag{6.160}$$

If, for example, β_1 is very small, so that mutation from A to a is unlikely, the stationary distribution approaches infinity at 1, indicating that it is very likely that at any given time the gene A is fixed, or nearly fixed. If both β_1 and β_2 are moderate in size, extreme values of the frequency of A are unlikely. When there is a selective advantage for A, the density concentrates near 1. A selective advantage for the heterozygote, i.e., for the Aa combination, will have α_1 small compared to α_2, and will have a local maximum at ½, with local minima on either side of ½ (Figure 6.5).

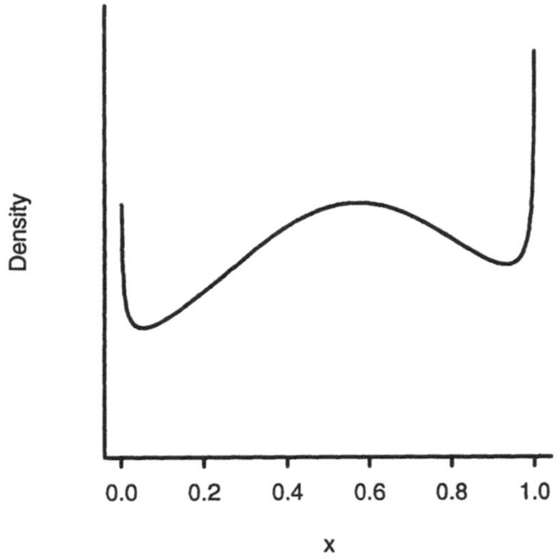

Figure 6.5. Density of A-gene with selective advantage for the heterozygote.

In some situations, we may want to consider models where the mutation or selection rates are not necessarily of the same order of magnitude. Assume that there is no selection, but that $u=\gamma_1/(2N)^d$ and $v=\gamma_2/(2N)$, $0<d<1$. The rate of mutation from A to a is of much larger order than the reverse rate. Consider

$$Z_N(t) = \frac{X([N^d t])}{N^d}. \tag{6.161}$$

The limiting diffusion Z has infinitesimal moments $\mu(x)=\gamma_2-\gamma_1 x$ and $\sigma^2(x)=x$. This diffusion is akin to that we used to approximate a Bienaymé-Galton-Watson process, and much of the analysis done above can indeed be carried out successfully by assuming that the Markov chain is approximated by a branching process instead of a diffusion process.

In order to attack more complicated genetics problem than the single locus diallelic ones discussed above, a variety of models have been proposed. Consider a population of N individuals, each having d observable characteristics, measured as integer multiples of a standard unit m. This is just a convenient way of keeping a discrete state space, and is not an essential assumption. The **type** of an individual is the set of its characteristics. Let$N_{\mathbf{k}}(t)$ count the number of individuals of type **k** at time t. The distribution of types over the population is a quantity of considerable interest. Let

$$p(t;\mathbf{k}) = N_{\mathbf{k}}(t)/N \tag{6.162}$$

be the **empirical distribution of types** at time t. We want to describe the development over time of the type distribution. To do that, assume that when the current type distribution is p, an individual of type $\mathbf{i}$ is replaced by an individual of type $\mathbf{j}$ at rate $g_{\mathbf{ij}}(p)$. The Ohta–Kimura model (Ohta and Kimura, 1973) assumes that

$$g_{\mathbf{ij}}(p) = \gamma p(\mathbf{i})p(\mathbf{j}) + Dp(\mathbf{i})\theta_{\mathbf{ij}} \tag{6.163}$$

where $\gamma>0$ is the rate at which individuals mate and are replaced by an offspring, while $D>0$ is a mutation rate and

$$\theta_{\mathbf{ij}} = \begin{cases} 1 & \text{if } |\mathbf{i}-\mathbf{j}|=1 \\ -2d & \text{if } \mathbf{i}=\mathbf{j}. \end{cases} \tag{6.164}$$

Diffusion approximations to this type of processes are much more complicated. Fleming and Viot (1978) allowed for the effect of a mutation to decrease with the population size, adjusting the time scale appropriately, in order to study

$$Y_N(t,A) = \sum_{bj/\sqrt{N}\in A} p(N^2t;bfj) \tag{6.165}$$

Notice that for fixed t, $Y_N(t,\bullet)$ is a **random measure**: it is an additive set function whose value depends on the realizations $(N_{\mathbf{k}}(t);\mathbf{k}\in\mathbf{Z}^d)$. It can be shown that this process converges to a limit, the **Fleming–Viot process**, which is a **measure-valued diffusion process**. The structure of this limiting process is quite interesting, but requires a bit of an excursion. Consider n particles moving in $\mathbb{R}^d$ according to independent Brownian motion processes. At a constant rate, one particle disappears and another simultaneously splits in two. Two particles resulting from such a binary fission are said to be **siblings**, and the particle from which they split is the **parent**. The **descendants** of a particle are all particles which can be traced backwards through a sequence of binary fissions to that particle, who is called their **common ancestor**.

The process counting the number of discontinuities (fissions) is a pure birth process with rate $\lambda_k=\gamma k(k-1)$. Going backwards in time we will eventually find a common ancestor to all individuals present at time t. The time of the first splitting of the common ancestor is the age of the system of particles alive at time t. Now let $n\to\infty$, resulting in an **infinite particle system**. The empirical distribution of these particles is the random measure called the Fleming–Viot process.

The probability theory of this type of processes, called **superprocesses**, has seen explosive growth recently. Much current work in statistical genetics involves tracing back to common ancestors, and trying to determine, given data on the genetic structure of current populations, likely paths of evolution.

6.8. Bibliographic remarks

The description of physical Brownian motion is from the papers in Einstein (1926), and the historical background is mainly from the Notes in that volume, written by Fürth.

The path properties of second order processes are presented by Cramér and Leadbetter (1967). The presentation of the Brownian motion and Ornstein–Uhlenbeck processes largely follows that of Breiman (1968). The application to AIDS modeling is due to Taylor et al. (1994), while the progesterone concentration model is from Winkel et al. (1976).

The section on genetic diffusion models is based on material in Ewens (1968) and Moran (1962). A formal description of weak convergence of genetic processes is in Chapter 10 of Ethier and Kurtz (1986). The description of the Fleming–Viot process follows Hochberg and Dawson (1982). The third issue of volume 9 of *Statistical Science* (1994) contains a collection of papers on statistical problems in genetics and molecular biology, indicating some main directions of modern statistical genetics.

6.9. Exercises

Theoretical exercises

1. Derive the Gaussian process continuity criterion (18).

2. Show that $\text{l.i.m.}X_n = X$ iff $\mathbf{E}X_nX_m \to c$ as $n,m \to \infty$.

3. Let X_t be a second order process. Call X **continuous in q.m.** if

$$\underset{t \to t_0}{\text{l.i.m.}} X(t) = X(t_0).$$

(a) Prove that X is continuous in q.m. iff $R(t,u)$ is continuous at $t = u = t_0$.

(b) Show that the Poisson process is continuous in q.m.

(c) Call a process $X'(t)$ the **q.m. derivative of X at t** if

$$\underset{h \to 0}{\text{l.i.m.}} \left| \frac{X(t+h)-X(t)}{h} - X'(t) \right| = 0.$$

Show that X has a q.m. derivative at t iff $\partial^2 R(t,u)/\partial t \partial u$ exists finite at (t,t). Determine the covariance function of X'.

4. For X' the q.m. derivative of X as in Exercise 3(c), find the crosscovariance function $R_{X',X}(s,t) = \text{Cov}(X'(s), X(t))$. If X is a stationary, Gaussian process, show that $X'(t)$ and $X(t)$ are independent.

5. Let $B(t)$ be a standard Brownian motion process, and let $U(t) = \int_0^t \exp(-\beta(t-s))dB(s)$ be an Ornstein–Uhlenbeck process.

(a) Show that $(2\beta)^{-1/2}\exp(-\beta t)B(\exp(2\beta t)-1)$ is an Ornstein–Uhlenbeck process.

(b) Let $X(t)=\int_0^t U(s)ds$ be an integrated Ornstein–Uhlenbeck process. Show that $Z(t)=\beta(X(t)-X(0))+U(t)-U(0)$ is a Brownian motion process.

6. Let $X(t)$ be a stationary, Gaussian Markov process with autocorrelation function $\rho(t)$.

(a) Show that $\mathbf{E}(X(s+t)\,|\,X(s)) = \rho(t)X(s),\ \ t\geq 0$.

(b) For $0\leq s\leq s+t$ show that

$$\mathbf{E}(X(0)X(s+t)) = \rho(t)\mathbf{E}(X(0)X(s)).$$

(c) Deduce that $\rho(t)=\exp(-\beta|t|)$, and that the Ornstein–Uhlenbeck process is the only such process.

7. Let $B(t)$ be a Brownian motion process. Show that the following processes also are Brownian motion processes:

(a) $\sigma W(t/\sigma^2)$, $\sigma>0$;

(b) $W(t+\sigma)-W(\sigma)$, $\sigma>0$;

(c) $tW(1/t)$, $t>0$.

8. There has been much work lately on so-called scaling properties of various phenomena. A stationary process, or a process with stationary increments, is **simple scaling** (or **self-similar**) if its finite-dimensional distributions (or the marginal distribution of its increments) are invariant with respect to rescaling, in the sense that for any $\lambda>0$ there is a constant c_λ such that

$$c_\lambda X(\lambda s) \overset{d}{=} X(s),$$

and similar definitions for the joint distributions at k points.

(a) Show that $c_{\lambda_1\lambda_2} = c_{\lambda_1}c_{\lambda_2}$, and deduce that $c_\lambda=\lambda^{-h}$. The quantity h is called the **scaling exponent**.

(b) Show that the marginal distribution of increments of the Brownian motion process is simple scaling with exponent ½.

(c) Is the Ornstein–Uhlenbeck process simple scaling?

9. (a) Show that the integrated Ornstein–Uhlenbeck process $X(t)=\int_0^t U(s)ds$ is non-Markovian.

(b) Show that the two-dimensional process $(X(t), U(t))$ is a Markov process.

10. Consider the macro description of the Ehrenfest model on page 41. This is a birth and death chain with $\lambda_n=\lambda(N-n)$ and $\mu_n=\mu n$, $n=0,\ldots,N$. Consider the scaled variable $Y=(X-N/2)/N^{1/2}$. Find the infinitesimal mean and variance for the scaled process, and deduce that for large N this is approximately the diffusion process given by the stochastic differential equation

$$dY_t = -2\lambda Y_t + \lambda^{1/2}dB_t.$$

11. Let f be a twice continuously differentiable function. Show that

$$\int_0^t f'(B_s)dB_s = f(B_t) - f(B_0) + \tfrac{1}{2}\int_0^t f''(B_s)ds,$$

expressing $\int f'(B)dB$ as a pathwise integral.

12. Volterra and d'Ancona (1935) suggested a deterministic class of population processes given by

$$\frac{dx(t)}{dt} = f(x_s; s \le t)x(t).$$

Particularly useful choices of the **growth rate** f are the **logistic** model with $f(x_s; s \le t) = \mu - \lambda x_t$, and the **renewable resources** model with $f(x_s; s \le t) = \mu - \lambda \int_0^\tau g(s)x(t-s)ds$. Here μ is the growth rate in the absence of competition, and τ is the time needed to fully renew resources. Kulperger and Guttorp (1981) suggest a stochastic version of these processes, having a random growth rate, so that the process is given by the stochastic differential equation

$$dX_t = (f(X_s, s \le t)dt + \sigma dB_t)X_t,$$

and show that there is a criticality parameter $\phi = \mu - \sigma^2/2$ such that the process becomes extinct with probability one if $\phi \le 0$, while if $\phi > 0$ the process has a stationary distribution.

(a) Derive the mle's for the stochastic version of the two processes given above.

(b) In the logistic case, the mle depends on the stochastic integral $\int X_s dX_s$. Extending Exercise 11, show how to compute this integral as a pathwise integral.

13. Let $dX_i(t) = f_i(t)dt + g_i(t)dB_t$, $i = 1,2$. Show that

$$d(X_1(t)X_2(t)) = X_1(t)dX_2(t) + X_2(t)dX_1(t) + g_1(t)g_2(t)dt,$$

and express $Y_t = X_1(t)X_2(t)$ as a stochastic differential equation with respect to B_t.

14. Suppose a particle performs a Brownian motion in the plane, in the sense that the x- and y-coordinates each follow independent Brownian motion processes with the same variance σ^2, starting from the origin. Find the probability that the particle falls in a circle of radius r around the origin. From independent observations of such particles (after a fixed time t), how would you estimate σ^2?

15. Consider n iid realizations of an Ornstein–Uhlenbeck process with parameters α and σ^2, each observed at times $1, \ldots, p$, yielding an $n \times p$ matrix (x_{ij}).

(a) Show that the likelihood function is

$$L(\alpha, \sigma^2) = (\sigma^{2p}(1-a^2)^{p-1})^{-n/2}$$

$$\times \exp(-\frac{1}{2\sigma^2(1-s^2)}(A_n+(1+\alpha^2)B_n-2\alpha S_n))$$

where

$$A_n = \sum_{i=1}^{n}(x_{i1}^2+x_{ip}^2),$$

$$B_n = \sum_{i=1}^{n}\sum_{j=2}^{p-1}x_{ij}^2$$

and

$$C_n = \sum_{i=1}^{n}\sum_{j=1}^{p-1}x_{ij}x_{i,j+1}.$$

Hint: The covariance matrix of the $X_{i\bullet}$ has a simple structure.

(b) Deduce that the mle of α satisfies a cubic equation in α.

(c) Show that the likelihood ratio test of $\alpha=0$ rejects for large values of $\hat{\alpha}^2$.

Computing exercises

C1. Simulate a Brownian motion process, and the corresponding integrated Ornstein–Uhlenbeck process. Can you tell these processes apart visually?
Hint: You may want to simulate a random walk with very small steps at very close time intervals.

C2. Generate a Brownian motion process with linear drift, except that at the times of a homogeneous Poisson process a random normal variate is added to the process. Look at the marginal distribution of increments of this process, observed at discrete time intervals that are coarse relative to the simulation intervals. Do the increments look normally distributed? This process has been used as a model for stock market prices (e.g., Aase and Guttorp, 1987).

C3. Generate paths of a Brownian motion process with drift, and observe it at discrete time intervals (coarse relative to the simulation intervals). Compare the distribution of a linear regression estimate of the drift with that of the maximum likelihood estimate.

C4. Lee et al. (1991) report results of using nanometer-size colloidal gold probes with video-enhanced microscopy to study the movement of individual molecules in the plasma membranes of living cells. Theory suggests that such movement may be modeled by a Brownian motion process. In order to check this assumption, they estimated the mean square displacement

$$\mathrm{MSD}(t) = \mathbf{E}(X_t^2+Y_t^2)$$

where (X_t, Y_t) is the position of the gold particle on a supported planar membrane at time t, by

$$\widehat{\mathrm{MSD}}(t) = \sum_{1}^{N-n}\frac{(x_{i+n}-x_i)^2+(y_{i+n}-y_i)^2}{N-n}$$

where $t=n\times\Delta t$, Δt is the video frame time (0.033 s), and N is the total number of video frames. Since for the Brownian motion process we have $MSD(t)=4Dt$, where D is the diffusion coefficient, we can estimate D from the slope of a straight line fit of $\widehat{MSD}$ against t, as in the previous exercise. However, for large values of t this relationship was no longer linear.

(a) By simulating paths with $D=4.0\times10^{-13}\,m^2/s$ using a random walk with steps of 0.33 ms, step size being $(4Dt)^{1/2}$, step direction chosen randomly, and 100 steps per observation, with 200 observations per path, assess the linearity of the plot of $\widehat{MSD}(t)$ against t.

(b) In this study, simulations like those in (a) were judged linear for 1–2 s (out of 6.6 s total observation time). Compare the variability of basing the regression estimate of D on 1 second with that of using the full path.

(c) How much better would it be to use the mle of D based on the entire path?

(d) How severe deviations from linearity of the plot of $\widehat{MSD}(t)$ against t are needed to conclude that the Brownian motion process is an inadequate fit?

Data exercises

D1. The larvae of the helminth *Trichostrongylus retortaeformis* are hatched from eggs in the excreta of sheep or rabbits. In nature they wander apparently at random until they climb and remain on blades of grass where they are eaten by another animal in whose intestines the cycle recommences. In an experiment by Dr. H. D. Crofton (Broadbent and Kendall, 1953) the wanderings of 400 larvae on a microscope slide were observed. A few larvae at the time were released at the center of the field of view, and at successive intervals of 5 seconds (up to a total of 30 seconds) the number of larvae in concentric annuli of radii $k\times0.7$ mm, $k=1,\ldots,12$, were counted. Table 6.1 contains the data.

Table 6.1 Crofton's larvae wanderings

	Number of larvae in annulus with outer radius											
Time	1	2	3	4	5	6	7	8	9	10	11	12
5	53	124	118	81	21	3						
10	32	60	142	88	41	20	11	6				
15	20	31	69	124	78	45	31		2			
20	11	40	65	94	96	47	19	6	10	7	5	
25	11	33	54	72	87	69	37	20	8	8		1
30	13	22	43	40	77	90	53	34	17	3	7	

Does a Brownian motion model such as that in Exercise 14 seem reasonable for these data?

D2. Perrin (1913) reported several experiments intended to confirm the Einstein–Smoluchowsky theory of Brownian motion. In an emulsion of gamboge (a vegetable latex dissolved in alcohol and centrifuged to obtain grains of equal size) the displacements at equal time intervals were observed.

(a) What should the distribution of these displacements be?

(b) For the following data set, test the distributional hypothesis in (a).

Table 6.2 Gamboge particle displacements

Displacements	0–1.7	1.7–3.4	3.4–5.1	5.1–6.8
Observed	38	44	33	33
Displacements	6.8–8.5	8.5–10.2	10.2–11.9	11.9–13.6
Observed	35	11	14	6
Displacements	13.6–15.3	15.3–17.0		
Observed	5	2		

D3. Stern (1994) proposed a Brownian motion model for the progress of basketball scores. Based on 493 games between January and April of 1992 in the National Basketball Association of the United States, he found that the point difference at the end of each quarter had a normal distribution with parameters given in Table 6.3. Here $X(t)$ denotes the difference in scores (home team – away team) at time t, where time 1 corresponds to the end of the game.

Table 6.3 Basketball scores by quarter

Quarter	Variable	Mean	Sd
1	$X(0.25)$	1.41	7.58
2	$X(0.50)-X(0.25)$	1.47	7.48
3	$X(0.75)-X(0.50)$	1.51	7.30
4	$X(1)-X(0.75)$	0.22	6.99
Total	$X(1)$	4.63	13.18

(a) Based on the increments $X(i/4)-X((i-1)/4)$, $i=1,...,4$, and assuming a Brownian motion model with drift, estimate the drift parameter and the variance from these data.

(b) Using the estimates from (a), what is the model-based probability of the home team winning the game, given that it is ahead at the end of the ith quarter, $i=1,2,3$?

(c) Compare the estimates in (b) to the observed values 263, 296, and 301, respectively. Does the Brownian motion process model seem adequate?

D4. Data Set 11 consists of average daily temperatures for July at Palmer, Washington, for 1931–1989. The average temperature is calculated as the average of the maximum and the minimum temperature for each day. Hsu and Park (1980) suggest modeling temperature as an Ornstein–Uhlenbeck process, observed at integer times. Thus, the theory of Exercise 15 can be used to test the hypothesis of independence versus Ornstein–Uhlenbeck dependence. Perform such a test.

APPENDIX A

Some statistical theory

A.1. Multinomial likelihood

Let $\mathbf{X}$ be a random r-vector having the multinomial distribution with r categories, n draws and success probabilities $\mathbf{p}=(p_1, \ldots, p_r)$, so that

$$\mathbf{P}(\mathbf{X}=\mathbf{x}) = \frac{n!}{x_1! \cdots x_r!} p_1^{x_1} \cdots p_r^{x_r} \tag{A.1}$$

provided that $\mathbf{x}\mathbf{1}^T=n$. We write this $\mathbf{X} \sim \mathrm{Mult}_r(n;\mathbf{p})$. Having observed the outcome $\mathbf{x}$, we want to estimate $\mathbf{p}$. The **likelihood** $L(\mathbf{p})$ of the parameter $\mathbf{p}$ (given the outcome $\mathbf{x}$) is the probability[1] of the observed value as a function of the unknown parameter. The **maximum likelihood estimator** (mle) $\hat{\mathbf{p}}$ of $\mathbf{p}$ is the value of $\mathbf{p}$ that maximizes $L(\mathbf{p})$. Think of it as the value of $\mathbf{p}$ that best explains the observation $\mathbf{x}$. To perform the maximization, first note that we must have $\mathbf{p}\mathbf{1}^T=1$. It helps to take logarithms, and thus to maximize

$$l(\mathbf{p}) = \log L(\mathbf{p}) = \log \frac{n!}{x_1! \cdots x_r!} + \sum_{i=1}^{r} x_i \log p_i \tag{A.2}$$

The function $l(\mathbf{p})$ is called the **log likelihood function**. Notice that the first term on the right hand side of (A.2) does not depend on $\mathbf{p}$, so that when we maximize $l(\mathbf{p})$ we can ignore it. Generally, the likelihood is only defined up to additive constants (with respect to the unknown parameters).

Since we want to maximize $l(\mathbf{p})$ over all $\mathbf{p}$ that sum to 1, we introduce a Lagrange multiplier λ, and maximize

$$l^*(\mathbf{p},\lambda) = \sum_{i=1}^{r} x_i \log p_i + \lambda(1 - \sum_{i=1}^{r} p_i). \tag{A.3}$$

Taking partial derivatives we get

$$\frac{\partial}{\partial p_i} l^*(\mathbf{p},\lambda) = \frac{x_i}{p_i} - \lambda$$

$$\frac{\partial}{\partial \lambda} l^*(\mathbf{p},\lambda) = 1 - \sum p_i \tag{A.4}$$

[1] If $\mathbf{X}$ has a continuous distribution, the likelihood is defined as the probability density of the observation as a function of the parameter.

and setting them equal to zero yields the equations

$$p_i = \frac{x_i}{\lambda}, \quad \sum p_i = 1 \tag{A.5}$$

whence $\sum x_i/\lambda = 1$, or $\lambda = \sum x_i = n$, so $\hat{p}_i = x_i/n$.

Example (Social mobility) Mosteller (1968) quotes some data on occupational mobility in Denmark. The data are counts of fathers and sons in five different occupational categories. We show the distribution of the 2,391 sons in Table A.1.

Table A.1 Danish social mobility data

Category	1	2	3	4	5
Count	79	263	658	829	562

We estimate the probability p_i that a randomly chosen son from the population studied is in occupational category i, yielding

$$\hat{p}_1 = 0.03 \quad \hat{p}_2 = 0.11 \quad \hat{p}_3 = 0.275 \quad \hat{p}_4 = 0.35 \quad \hat{p}_5 = 0.235 \tag{A.6}$$

□

So far we have derived an estimate of $\mathbf{p}$, i.e., a function $\hat{\mathbf{p}}(\mathbf{x})$ of the data $\mathbf{x}$. When discussing properties of an estimation procedure, it is customary to think of the **estimate** $\hat{\mathbf{p}}(\mathbf{x})$ as an instance of the random variable $\hat{\mathbf{p}}(\mathbf{X})$, called the **estimator**. Note that $X_i \sim \mathrm{Bin}(n, p_i)$, so by the law of large numbers $\hat{p}_i = X_i/n \to p_i$ with probability 1. Furthermore, by the central limit theorem,

$$n^{1/2} \frac{\hat{p}_i - p_i}{(p_i(1-p_i))^{1/2}} \to \mathrm{N}(0,1) \quad \text{in distribution}. \tag{A.7}$$

A.2. The parametric case

Sometimes we are interested in $\mathbf{p}$ that are a function of a lower-dimensional parameter θ. Let θ be one-dimensional for simplicity. We want to maximize $l(\theta) = l(\mathbf{p}(\theta))$. Presumably, $\mathbf{p}(\theta)$ is parametrized to sum to one, so we do not need to constrain the maximization. By the chain rule we must solve

$$\frac{dl(\theta)}{d\theta} = \sum_{i=1}^{r} \frac{\partial l(\mathbf{p}(\theta))}{\partial p_i(\theta)} \frac{dp_i(\theta)}{d\theta} = 0 \tag{A.8}$$

or, equivalently,

$$\sum_{i=1}^{r} \frac{x_i}{p_i(\theta)} \frac{dp_i(\theta)}{d\theta} = 0. \tag{A.9}$$

This equation may or may not have a solution, but it is possible to write down regularity conditions under which it, for n large enough, will have a solution with probability approaching 1 (see Rao, 1973, sec. 5e, for details). Under regularity conditions Rao also proves consistency of $\hat{\theta}$, as well as asymptotic normality in the sense that

$$\left[-l_n''(\hat{\theta})\right]^{1/2}(\hat{\theta}-\theta) \xrightarrow{d} \mathrm{N}(0,1). \tag{A.10}$$

Example (Identical twins) Human twins are either identical or nonidentical. The identical twins arise from the splitting of a fertilized egg, while nonidentical twins occur when two ova are fertilized simultaneously. Assuming that the probability of twins being identical is α, and the probability of male children is $\pi=0.516$, how can we estimate α from data on numbers and genders of twins, but without knowledge of what twins are identical and what are nonidentical?

First note that two twins can both be male either by being identical or by being nonidentical. The former case has probability $\alpha\pi$, while the latter has probability $(1-\alpha)\pi^2$, for a total probability of $\pi(\alpha+\pi(1-\alpha))$. Likewise the probability of female twins is $(1-\pi)(\alpha\pi+1-\pi)$, and that of mixed twins is $2\pi(1-\pi)(1-\alpha)$, since such twins can only be nonidentical. Hence, writing n_1 for the number of male twins, n_2 for the number of female twins, and n_3 for the number of mixed twins, the log likelihood is

$$l(\alpha) \propto n_1\log(\alpha+\pi(1-\alpha)) + n_2\log(\alpha\pi+1-\pi) + n_3\log(1-\alpha) \tag{A.11}$$

Differentiating this and setting the derivative equal to zero yields a quadratic equation in α. □

A.3. Likelihood ratio tests

Suppose we are interested in testing a fully specified hypothesis of the from $\mathbf{p}=\mathbf{p}_0=(p_{10},\ldots,p_{r0})$. General testing theory suggests the usefulness of the **likelihood ratio**

$$\Lambda = \frac{L(\hat{\mathbf{p}})}{L(\mathbf{p}_0} = \prod_{i=1}^{r}\left[\frac{\hat{p}_i}{p_{i0}}\right]^{x_i}. \tag{A.12}$$

As before, it is helpful to take logarithms (and this time multiply by two), to get the **log likelihood ratio statistic**

$$\lambda = 2\log\Lambda = 2\sum_{i=1}^{r} x_i \log\left[\frac{x_i}{np_{i0}}\right]. \tag{A.13}$$

The **likelihood ratio test** rejects for large values of λ. One can show that under the null hypothesis $\mathbf{p}=\mathbf{p}_0$, the statistic λ is approximately distributed as $\chi^2(r-1)$.

For large n, we can write

$$\lambda \approx \sum_{i=1}^{r} \frac{(x_i - np_{i0})^2}{np_{i0}} \equiv \chi^2. \tag{A.14}$$

Since $X_i \sim \text{Bin}(n, p_{i0})$ under the null hypothesis, the quantity np_{i0} is simply the expected value of X_i. The quantity χ^2 is a measure of how far the observations deviate from their expected values, and was introduced by Karl Pearson[1] (1900). To see the validity of the approximation, we do a Taylor expansion of the logarithm,

$$\begin{aligned} \log \hat{p}_i - \log p_{i0} &\approx \frac{\hat{p}_i - p_{i0}}{p_{i0}} - \frac{(\hat{p}_i - p_{i0})^2}{2p_{i0}^2} \\ &= \frac{X_i - np_{i0}}{np_{i0}} - \frac{(X_i - np_{i0})^2}{2n^2 p_{i0}^2}. \end{aligned} \tag{A.15}$$

Thus

$$\begin{aligned} 2\sum X_i \log\left[\frac{\hat{p}_i}{p_{i0}}\right] &\approx 2\sum_{i=1}^{r}\left[\frac{(X_i - np_{i0})^2}{np_{i0}} - \frac{(X_i - np_{i0})^3}{2n^2 p_{i0}^2}\right] \\ &\quad + 2\sum_{1}^{r}\left[(X_i - np_{i0}) - \frac{(X_i - np_{i0})^2}{2np_{i0}}\right]. \end{aligned} \tag{A.16}$$

Some algebra shows that this simplifies to

$$\sum_{i=1}^{r} \frac{(X_i - np_{i0})^2}{np_{i0}} - \sum_{i=1}^{r} \frac{(X_i - np_{i0})^3}{n^2 p_{i0}^2}. \tag{A.17}$$

The last term is bounded by

$$\sum_{i=1}^{r} \frac{(X_i - np_{i0})^2}{np_{i0}} \max_i \frac{|\hat{p}_i - p_{i0}|}{p_{i0}}. \tag{A.18}$$

But $\hat{p}_i \to p_{i0}$ in probability under the null hypothesis, so the last term in (A.17) is negligible compared to the first, and the two statistics λ and χ^2 are approximately the same.

Example (Fairness of dice) One of the most extensive dice experiments, performed by an English mathematician named Weldon and his wife, involved 315,672 throws of dice. These were rolled, twelve at a time, down a fixed slope of cardboard. There were 106,602 instances of the outcome 5 or 6. If the dice were true the probability of 5 or 6 should be 1/3. The expected number

[1] Pearson, Karl (1857–1936), English biometrician. A disciple of Francis Galton. Put biological statistics on a mathematical basis.

of such outcomes would be 315,672/3=105,224. Pearson's χ^2-statistic therefore is

$$\frac{(106{,}602-105{,}224)^2}{105{,}224}+\frac{(209{,}070-210{,}448)^2}{210{,}448}=27.1 \qquad \text{(A.19)}$$

which is highly significant on one degree of freedom. The likelihood ratio statistic is 27.0, virtually identical. □

A.4. Sufficiency

A statistic $T(X_1,\ldots,X_n)$ is called **sufficient** for θ if the conditional density (or probability function) of the data, given the value of $T(X_1,\ldots,X_n)$, does not depend on θ. In other words, if you tell me the value of $T(X_1,\ldots,X_n)$, I don't need to know anything else about the sample in order to infer something about the value of θ.

Example (Binomial case) Let X_1,X_2,X_3 be iid Bin(1, θ), and let $T(\mathbf{X})=(X_1+2X_2+3X_3)/6$. Do we need to know more about the sample than just the value of T in order to make a good guess as to the value of θ? To check that, suppose that the sample is $(X_1,X_2,X_3)=(1,1,0)$, so that the observed value of T is ½. Then

$$\begin{aligned}\mathbf{P}(\mathbf{X}=(1,1,0)\mid T(\mathbf{X})=\tfrac{1}{2}) &= \frac{\mathbf{P}(\mathbf{X}=(1,1,0)\cap T(\mathbf{X})=\tfrac{1}{2})}{\mathbf{P}(T(\mathbf{X})=\tfrac{1}{2})}\\ &= \frac{\mathbf{P}(\mathbf{X}=(1,1,0))}{\mathbf{P}(T(\mathbf{X})=\tfrac{1}{2})}=\frac{\theta^2(1-\theta)}{\mathbf{P}(\mathbf{X}=(1,1,0)\cup\mathbf{X}=(0,0,1))}\\ &= \frac{\theta^2(1-\theta)}{\theta^2(1-\theta)+\theta(1-\theta)^2}=\theta\end{aligned} \qquad \text{(A.20)}$$

Since this probability depends on θ, we need more information about the sample than just the fact that $T(\mathbf{x})=\tfrac{1}{2}$. When $T=\tfrac{1}{2}$ there are two possible explanations, either $\mathbf{X}=(1,1,0)$, which would be likely if θ were large, or $\mathbf{X}=(0,0,1)$, which would be likely if θ were small. Thus, the information that $T=\tfrac{1}{2}$ does not tell is much about the actual value of θ. □

The method used in this example is fine when we want to show that a statistic is not sufficient, but it is not all that helpful in trying to figure out reasonable candidates for sufficient statistics. A criterion for doing this is the following, due to Fisher[1] and Neyman.

Fisher, Ronald Ayles (1890–1962). English statistician and geneticist. Introduced the likelihood approach to statistical inference. Invented the analysis of variance and the randomization approach to experimental design.

Proposition A.1 (Fisher–Neyman factorization criterion) A statistic $T(\mathbf{X})$ is sufficient if and only if the density (or probability function) can be factored as

$$f(\mathbf{x};\theta) = g(T(\mathbf{x});\theta)\,h(\mathbf{x}) \tag{A.21}$$

where g only depends on the x_i's through $T(\mathbf{x})$, while h does not depend on θ.

Example (Binomial case, continued) The density of $\mathbf{X}$ is

$$\mathbf{P}(\mathbf{X}=\mathbf{x}) = \theta^{x_1}(1-\theta)^{1-x_1}\theta^{x_2}(1-\theta)^{1-x_2}\theta^{x_3}(1-\theta)^{1-x_3} \tag{A.22}$$

$$= \left[\frac{\theta}{1-\theta}\right]^{\sum_{i=1}^{3} x_i}(1-\theta)^3.$$

Using for g the entire expression on the right-hand side, and letting $h(\mathbf{x})=1$, we see that g depends on the data $\mathbf{x}$ only through their sum $\sum x_i$, which therefore is a sufficient statistic. To check back with the definition, notice that $\sum_{i=1}^{3} X_i \sim \text{Bin}(3,\theta)$, so that

$$\mathbf{P}(\mathbf{X}=\mathbf{x}\,|\,\textstyle\sum X_i=\sum x_i) = \dfrac{\theta^{\sum x_i}(1-\theta)^{n-\sum x_i}}{\begin{bmatrix}3\\ \sum x_i\end{bmatrix}\theta^{\sum x_i}(1-\theta)^{n-\sum x_i}} = \dfrac{1}{\begin{bmatrix}3\\ \sum x_i\end{bmatrix}}. \tag{A.23}$$

Since the right-hand side is independent of θ, regardless of the value of the x's, we see that $\sum X_i$ is indeed a sufficient statistic. Notice that, in fact, the conditional distribution is uniform over the set of possible outcomes with the given value of the sufficient statistic. □

Example (The normal case) Suppose now that $X_1,\ldots,X_n$ are iid $\text{N}(m,\sigma^2)$. First assume that σ^2 is a known number, so that we are only interested in estimating m. Then

$$f(\mathbf{x};m) = (\sigma\sqrt{2\pi})^{-n}\exp(-\frac{1}{2\sigma^2}\sum_{i=1}^{n}(x_i-m)^2)$$

$$= \exp(\frac{m\sum x_i}{\sigma^2}-\frac{m^2}{2\sigma^2})\exp(-\frac{\sum x_i^2}{2\sigma^2}+n\log(\sigma\sqrt{2\pi})). \tag{A.24}$$

The first exponential function is g, the second is h. We see that $\sum X_i$ is a sufficient statistic.

Now assume instead that m is known, and σ is the parameter of interest. Then we write the density

$$f(\mathbf{x};\sigma^2) = (\sigma\sqrt{2\pi})^{-n}\exp(-\frac{1}{2\sigma^2}\sum_{i=1}^{n}(x_i-m)^2). \tag{A.25}$$

In this case $\sum(X_i-m)^2$ is a sufficient statistic. We have $h(\mathbf{x})=1$.

Finally, if both parameters are unknown, we write the density

$$f(\mathbf{x};m,\sigma^2) = \exp(\frac{m\sum x_i}{\sigma^2}-\frac{\sum x_i^2}{2\sigma^2}-\frac{m^2}{2\sigma^2}+n\log(\sigma\sqrt{2\pi})), \tag{A.26}$$

from which it follows that $(\sum X_i,\sum X_i^2)$ together are sufficient for (m,σ^2). □

APPENDIX B

Linear difference equations with constant coefficients

B.1. The forward shift operator

Many probability computations can be put in terms of recurrence relations that have to be satisfied by successive probabilities. The theory of **difference equations** is the appropriate tool for solving such problems. This theory looks a lot like the theory for linear differential equations with constant coefficients.

In order to simplify notation we introduce the **forward shift operator** E, that takes a term u_n and shifts the index one step forward to u_{n+1}. We write

$$\begin{aligned} Eu_n &= u_{n+1} \\ E^2 u_n &= EEu_n = Eu_{n+1} = Eu_{n+2} \\ &\cdots \\ E^r u_n &= u_{n+r}. \end{aligned} \tag{B.1}$$

The general linear difference equation of order r with constant coefficients can be written

$$\Phi(E)u_n = f(n) \tag{B.2}$$

where $\Phi(E)$ is a polynomial of degree r in E and where we may assume that the coefficient of E^r is 1.

B.2. Homogeneous difference equations

The simplest class of difference equations of the form (B.2) has $f(n)=0$, that is

$$\Phi(E)u_n = 0. \tag{B.3}$$

These are called **homogeneous** equations.

When $\Phi(E)=(E-\lambda_1)(E-\lambda_2)\cdots(E-\lambda_r)$ where the λ_i are constants that are all distinct from each other, the most general solution to the homogeneous equation is

$$u_n = a_1\lambda_1^n + a_2\lambda_2^n + .. + a_r\lambda_r^n \tag{B.4}$$

where $a_1, a_2, \cdots, a_r$ are arbitrary constants.

When $\Phi(E)$ contains a repeated factor $(E-\lambda_\alpha)^h$, the corresponding part of the general solution becomes

$$\lambda_\alpha^n (a_\alpha + a_{\alpha+1} n + a_{\alpha+2} n^{(2)} + \cdots + a_{\alpha+h-1} n^{(h-1)}) \tag{B.5}$$

where $n^{(k)} = n(n-1)(n-2)\cdots(n-k+1) = n!/(n-k)!$

In order to find the nth term of a linear difference equation of order r, one can of course start with r initial values, and then solve recursively for any given n. Thus, if we want our solution to satisfy certain initial conditions we may first determine the general solution, and then (if possible) make it satisfy the initial conditions. There can be no more than r such initial conditions, but they need not (as when we compute the solution recursively) necessarily be conditions on $u_0, \ldots, u_{r-1}$, but can be on any set of r values.

Example (Branching process mean) The equation

$$u_{n+1} - m\, u_n = 0 \tag{B.6}$$

describes the mean of a Bienaymé-Galton-Watson branching process (see Chapter 2). Here

$$\Phi(E) = E - m \tag{B.7}$$

so the general solution is

$$u_n = a\, m^n. \tag{B.8}$$

Using the initial value $u_0 = z_0$, the solution becomes

$$u_n = z_0 m^n. \tag{B.9}$$

□

Example (A harmonic function) Consider the equation

$$u_n - 2u_n + u_{n+2} = 0. \tag{B.10}$$

The equation can be written in the form

$$(E-1)^2 u_n = 0. \tag{B.11}$$

The general solution is therefore

$$u_n = a + bn \tag{B.12}$$

where a and b are constants. □

B.3. Non-homogeneous difference equations

When solving linear differential equations with constant coefficients one first finds the general solution for the homogeneous equation, and then adds any particular solution to the non-homogeneous one. The same recipe works in the case of difference equations, i.e. first find the general solution to

$$\Phi(E)u_n = 0 \tag{B.13}$$

and a particular solution to

$$\Phi(E) = f(n) \tag{B.14}$$

and add the two together for the general solution to the latter equation. Thus to solve these more general equations, the only new problem is to identify some particular solutions. We will only give a few examples here, not attempting to treat this problem in any generality.

B.3.1. Geometric with no factor of $\Phi(E)$

In this case $f(n) = k\mu^n$, $\mu \neq \lambda_i$, $i = 1, 2, \ldots, r$, and

$$u_n = \frac{k\mu^n}{\Phi(\mu)} \tag{B.15}$$

is a particular solution to $\Phi(E)u_n = k\mu^n$. Let $\Phi(E) = \sum a_i E^i$. Then

$$\Phi(E)\frac{k\mu^n}{\Phi(\mu)} = \frac{\sum a_i E^i k\mu^n}{\sum a_i \mu^i} = k\frac{\sum a_i \mu^{n+i}}{\sum a_i \mu^i} = k\mu^n. \tag{B.16}$$

Example (Branching process variance) The difference equation

$$u_n(E-m^2) = \sigma^2 m^n \tag{B.17}$$

describes the variance of a Bienaymé-Galton-Watson process. The solution to the corresponding homogeneous equation is $u_n = am^{2n}$, while a particular solution is given by $\sigma^2 m^n/(m-m^2)$. Using the initial condition $u_0 = 0$, we get $a = \sigma^2/m(m-1)$, whence

$$u_n = \sigma^2 m^n \frac{m^n - 1}{m(m-1)}. \tag{B.18}$$

□

B.3.2. Geometric with a non-repeated factor of $\Phi(E)$

This is the case where $f(n) = k\mu^n$, $\mu = \lambda_i$, and λ_i is a non-repeated factor of $\Phi(E)$. Here a particular solution is given by

$$\frac{kn\mu^{n-1}}{\Phi'(\mu)} \tag{B.19}$$

where $\Phi'(\mu)$ denotes $\left[\frac{d}{dE}\Phi(E)\right]_{E=\mu}$.

B.3.3. Geometric with a repeated factor of $\Phi(E)$

We have $f(n) = k\mu^n$, $\mu=\lambda_i$, where λ_i is a repeated factor of $\Phi(E)$. Suppose that $(E-\lambda_i)$ is repeated h times in $\Phi(E)$. Then

$$\frac{kn^{(h)}\mu^{n-h}}{\Phi^{(h)}(\mu)}, \tag{B.20}$$

where $n^{(k)}=n(n-1)\cdots(n-k+1)$, is a particular solution of the equation $\Phi(E)u_n=k\mu^n$.

Example (A random walk equation) Let

$$(E-1)^2u_n = -2\times 1^n. \tag{B.21}$$

This equation arises in the theory of random walks. The general solution was given above in the harmonic function example, while a particular solution, since $\Phi^{(2)}(1)=2$, is $-2n^{(2)}1^2/2=-n(n-1)$. □

APPENDIX C

Some theory of partial differential equations

C.1. The method of auxiliary equations

This appendix concerns the solution of partial differential equations of the form

$$\frac{\partial G(s,t)}{\partial s} P(s,t,G) + \frac{\partial G(s,t)}{\partial t} Q(s,t,G) = R(s,t,G). \tag{C.1}$$

Define $p = \partial G / \partial s$, $q = \partial G / \partial t$. We can rewrite (C.1) as

$$(p,\ q,\ -1)\,(P,\ Q,\ R)^T = 0. \tag{C.2}$$

In other words, at a particular point $(s_0, t_0, G(s_0, t_0))$ we see that the vector $(p, q, -1)$ is orthogonal to (P, Q, R). Let

$$S = \{(s,t,G(s,t)) : pP + qQ - R = 0\} \tag{C.3}$$

be the solution set of (C.3). The normal to the curve $u = G(s,t)$ has direction $(p, q, -1)$, so relation (C.2) means that for any point $A \in S$, (P,Q,R) lies in the tangent plane to $u = G(s,t)$ at A. Now create a curve C of points in S, such that every point on C is tangential to (P, Q, R).

Theorem C.1

(1) Through every point in S goes a unique C.
(2) These curves do not intersect.

Now consider a point A. An infinitesimal arc along C has $(ds, dt, du) = \alpha(P, Q, R)$ since (P, Q, R) is the tangent (by the definition of C). Thus

$$\frac{ds}{P} = \frac{dt}{Q} = \frac{dG}{R} \tag{C.4}$$

or, for example,

$$\frac{dG}{ds} = \frac{R}{P} \quad \frac{ds}{dt} = \frac{P}{Q}. \tag{C.5}$$

The equations (C.5) are called **auxiliary equations**. Suppose that the first equation in (C.5) has solution

$$g(s,t,u) = \text{constant} \tag{C.6}$$

and the second has solution

$$h(s,t,u) = \text{constant.} \tag{C.7}$$

Then the general solution to (C.1) is the $u = G(s,t)$ solving

$$g(s,t,u) = f(h(s,t,u)) \tag{C.8}$$

for f an arbitrary, differentiable function, which is determined by the initial conditions. To see that this is indeed a solution, compute the partial with respect to s and t of $f(h)-g$, and manipulate to get back the original equation (using, of course, that h and g solve the auxiliary equations (C.5)).

C.2. Some applications

Example (Linear birth process)

$$\frac{\partial G}{\partial t} - \frac{\partial G}{\partial s}\lambda s(s-1) = 0, \tag{C.9}$$

so

$$P(s,t,G) = -\lambda s(s-1) \tag{C.10}$$

$$Q(s,t,G) = 1 \tag{C.11}$$

$$R(s,t,G) = 0. \tag{C.12}$$

The auxiliary equations are $dG/ds = 0$, so G=constant, or $g(s,t,u)=u$, and $ds/dt = -\lambda s(s-1)$, which we rewrite

$$\lambda dt = \frac{ds}{s} - \frac{ds}{s-1} \tag{C.13}$$

so $(s-1)/s \exp(\lambda t)$=constant, so that $h(s,t,u)=(s-1)/s\exp(\lambda t)$. Hence the solution is

$$G(s,t) = f\left(\frac{s-1}{s}e^{\lambda t}\right). \tag{C.14}$$

Suppose that $X_0=1$. Then $G(s,0)=s$, so $f((s-1)/s)=s$, i.e., $f(r)=1/(1-r)$. Hence

$$G(s,t) = \frac{1}{1-\frac{s-1}{s}e^{\lambda t}} = \frac{se^{-\lambda t}}{1-s(1-e^{-\lambda t})}. \tag{C.15}$$

□

Example (The Pólya process) The P\(o'lya process is a birth process with $\lambda_k(t)=\lambda(1+\alpha k)/(1+\alpha\lambda t)$. The pde for the pgf is

$$(1+\alpha\lambda t)\frac{\partial G}{\partial t} - \lambda\alpha s(s-1)\frac{\partial G}{\partial s} = \lambda(s-1)G. \tag{C.16}$$

Hence $P=-\lambda\alpha s\,(s-1)$, $Q=1+\alpha\lambda t$ and $R=\lambda(s-1)G$. We have

$$\frac{dG}{ds} = \frac{\lambda(s-1)G}{-\lambda\alpha s(s-1)}, \tag{C.17}$$

i. e., $dG/ds = G/(-\alpha s)$ whence $\alpha\log G=$ constant $-\log s$, so

$$g(s,t,G(s,t)) = s[G(s,t)]^{\alpha}. \tag{C.18}$$

Also,

$$\frac{ds}{dt} = \frac{-\lambda\alpha s(s-1)}{1+\alpha\lambda t}, \tag{C.19}$$

or $\log(1+\alpha\lambda t) = \log(s/s-1)$+constant, yielding

$$h(s,t,G(s,t)) = \frac{s-1}{s}(1+\alpha t). \tag{C.20}$$

Hence

$$s[G(s,t)]^{\alpha} = f(\frac{s-1}{s}(1+\alpha\lambda t)). \tag{C.21}$$

If $X(0)=0$ we get $G(s,0)=1$, so $s=f((s-1)/s)$, again yielding $f(r)=1/1-r$, and consequently

$$G(s,t) = (1-(s-1)\alpha\lambda t)^{-1/\alpha}. \tag{C.22}$$

□

References

Aalen, O. (1978): Nonparametric inference for a family of counting processes. *Ann. Statist.* **6:** 701–726.

Aase, K. K., and P. Guttorp (1987): Estimation in models for security prices. *Scand. J. Statist.* **4:** 211–224.

Abeles, M. (1982): *Local Cortical Circuits: An Electrophysiological Study.* Berlin: Springer.

Abkowitz, J. L., M. Linenberger, M. A. Newton, G. H. Shelton, R. L. Ott, and P. Guttorp (1990): Evidence for maintenance of hematopoiesis in a large animal by the sequential activation of stem cell clones. *Proc. Nat. Acad. Sci. USA* **87:** 9062–9066.

Abkowitz, J. L., M. L. Linenberger, M. Persik, M. A. Newton, and P. Guttorp (1993): The behavior of feline hematopoietic stem cells years after busulfan exposure. *Blood* **82:** 2096–2103.

Abkowitz, J. L., R. M. Ott, R. D. Holly and J. W. Adamson (1988): Clonal evolution following chemotherapy-induced stem cell depletion in cats heterozygous for glucose-6-phosphate dehydrogenase. *Blood* **71:** 1687–1692.

Abramowitz, M., and I. A. Stegun (1965): *Handbook of Mathematical Functions.* New York: Dover.

Adke, S. R., and S. M. Manjunath (1984): *An Introduction to Finite Markov Processes.* New Dehli: Wiley Eastern.

Andersen, P. K., Ø. Borgan, R. Gill, and N. Keiding (1992): *Statistical models based on counting processes.* Berlin: Springer-Verlag.

Andersen, P. K., and N. Kr. Rasmussen (1982): Admissions to psychiatric hospitals among women giving birth and women having induced abortion. A statistical analysis of a counting process model. Research report **82/6**, Statistical Research Unit, University of Copenhagen.

Andersen, P. K., and N. K. Rasmussen (1986): Psychiatric admissions and choice of abortion. *Statist. in Medic.* **5:** 243–253.

Arley, N. (1943): *On the Theory of Stochastic Processes and Their Application to the Theory of Cosmic Radiation*. New York: Wiley.

Asmussen, S. (1987): *Applied Probability and Queues*. New York: Wiley.

Asmussen, S., and H. Hering (1984): *Branching Processes*. Basel: Birkhäuser.

Atlas, D., and C. Ulbrich (1977): Path- and area-integrated rainfall measurements by microwave attenuation in the 1–3 cm band. *J. Appl. Meteor.* **16:** 1322–1331.

Bachelier, E. (1900): Théorie de la Speculation. *Ann. Sci. Ecole Normale Supérieure* **17:** 21–86.

Baddeley, A., J. Besag, H. Chernoff, P. Clifford, N. A. Cressie, D. J. Geman, B. Gidas, L. S. Gillick, N. Green, P. Guttorp, T. Kadota, A. Lippman, and J. Simpson (1991): *Spatial Statistics and Digital Image Analysis*. Washington: National Academy of Sciences Press.

Baddeley, A., and J. Møller (1989): Nearest-neighbour Markov point processes and random sets. *Int. Statist. Rev.* **57:** 89–121.

Bailey, N. T. J. (1975): *The Mathematical Theory of Infectious Diseases and Its Applications* (2nd ed.). London: Griffin.

Barlow, R. E., D. J. Bartholomew, J. M. Bremner, and H. D. Brunk (1972): *Statistical Inference Under Order Restrictions*. London: Wiley.

Bartholomew, D. J. (1973): *Stochastic Models for Social Processes* (2nd ed.). New York: Wiley.

Bartlett, M. S. (1963): The spectral analysis of point processes. *J. Roy. Statist. Soc. B* **25:** 264–296.

Bartlett, M. S. (1967): Inference and stochastic processes. *J. Roy. Statist. Soc. A* **130:** 457–477.

Bartlett, M. S. (1975): *The Statistical Analysis of Spatial Pattern*. London: Chapman & Hall.

Batista, J. L. F. (1994): *Spatial Dynamics of Trees in a Brazilian Atlantic Tropical Forest under Natural and Managed Conditions*. PhD dissertation, College of Forest Resources, University of Washington.

Becker, N. (1976): Estimation for an epidemic model. *Biometrics* **32:** 769–777.

Becker, N. (1977): Estimation for discrete-time branching processes with applications to epidemics. *Biometrics* **33:** 515–522.

Besag, J. E. (1972): Nearest-neighbor systems and the auto-logistic model for binary data. *J. Roy. Statist. Soc. B* **34:** 75–83.

Besag, J. E. (1974): Spatial interaction and the statistical analysis of lattice systems. *J. Royl Statist. Soc. Ser. B* **36:** 192–225.

Besag, J. E. (1986): Statistical analysis of dirty pictures. *J. Roy. Statist. Soc. Ser. B* **48:** 259–302.

Besag, J. E. (1989): Towards Bayesian image analysis. *J. Appl. Statist.* **16:** 395–407.

Besag, J., and P. Clifford (1989): Generalized Monte Carlo significance tests. *Biometrika* **76:** 633–642.

Besag, J., P. Green, D. Higdon, and K. Mengersen (1995): Bayesian computation and stochastic systems (with discussion). *Statistical Science* **:10**: 3-66.

Bhattacharya, R. N., and E. C. Waymire (1990): *Stochastic Processes with Applications*. New York: Wiley.

Bickel, P. J., and K. A. Doksum (1977): *Mathematical Statistics: Basic Ideas and Selected Topics*. San Francisco: Holden-Day.

Bienaymé, I. J. (1845): De la loi de multiplication et de la durée des familles. *Soc. Philomath. Paris* **5:** 37–39.

Billingsley, P. (1961): *Statistical Inference for Markov Processes*. Chicago: University Press.

Billingsley, P. (1979): *Probability and Measure*. New York: Wiley.

Blackwell, D. (1958): Another countable Markov process with only instantaneous states. *Ann. Math. Statist.* **29:** 313–316.

Blanchard, O. S., and M. Watson (1982): Bubbles, rational expectations and fincancial markets. In P. Wachtel (ed.): *Crises in the Economic and Financial Structure*. Lexington: Lexington Books.

Böker, F. (1987): A stochastic first purchase diffusion model: a counting process approach. *J. Marketing Res.* **24:** 64–73.

Brecher, G., S. L. Beal, and M. Schneiderman (1986): Renewal and release of hemopoietic stem cells: Does clonal succession exist? *Blood Cells* 12: 103–112.

Breiman, L. (1968): *Probability*. Reading: Addison-Wesley.

Brillinger, D.R. (1975): Statistical inference for stationary point processes. In M.L. Puri (ed.): *Stochastic Processes and Related Topics* 55–99. New York: Academic Press.

Brillinger, D. R. (1978): Comparative aspects of the study of ordinary time series and of point processes. In P. Krishnayah (ed.): *Developments In Statistics,* vol. 1: 33–133. New York: Academic Press.

Brillinger, D. R. (1988): Some statistical methods for random process data from seismology and neurophysiology. *Ann. Statist.* **16:** 1–54.

Brillinger, D. R. (1992): Nerve cell spike train data analysis: A progression of technique. *J. Amer. Statist. Assoc.* **87:** 260–271.

Brillinger, D.R., J. Guckenheimer, P. Guttorp, and G. Oster (1980): Empirical modelling of population time series: The case of age and density dependent rates. In G. Oster (ed.): *Some Mathematical Questions in Biology.* Lectures on Mathematics in the Life Sciences **13:** 65–90. Providence: American Mathemathical Society.

Broadbent, S. R., and D. G. Kendall (1953): The random walk of *trichostongylus Retortaeformis. Biometrics* **9:** 460–466.

Brook, D. (1964): On the distinction between conditional probability and joint probability approaches in the specification of nearest neighbor systems. *Biometrika* **51:** 481–483.

Brown, M. (1972): Statistical analysis of nonhomogeneous Poisson processes. In P. A. W. Lewis (ed.): *Stochastic point processes: statistical analysis, theory, and applications*, 67–89. New York: Wiley-Interscience.

Cannings, C., E. A. Thompson, and M. H. Skolnick (1978): Probability functions on complex pedigrees. *Adv. in Appl. Probab.* **10:** 26–61.

Chaisson, E. J. (1994): *The Hubble Wars.* New York: Harper Collins.

Chandrasekhar, S. (1943): Stochastic problems in physics and astronomy. *Rev. Mod. Phys.* **15:** 1–89. Reprinted in N. Wax (ed.): *Noise and Stochastic Processes*: 3–91. New York: Dover, 1954.

Cochran, W. G. (1936): The statistical analysis of field counts of diseased plants. *J. Roy. Statist. Soc.*, Suppl. Vol. **III:** 49–67.

Cohen, J. E. (1969): Natural primate troops and a stochastic population model. *Amer. Naturalist* **103:** 455–477.

Cox, D. R. (1955): Some statistical methods connected with series of events. *J. Roy. Statist. Soc. B* **17:** 129–164.

Cox, D. R. (1965): On the estimation of the intensity function of a stationary point process. *J. Roy. Statist. Soc. B* **27:** 332–337.

Cox, D. R. (1972a): Regression models and life tables (with Discussion). *J. Roy. Statist. Soc. B* **34:** 187–220.

Cox, D. R. (1972b): The statistical analysis of dependencies in point processes. In P. A. W. Lewis (ed.): *Stochastic Point Processes*, 55–66. New York: Wiley.

Cox, D. R., and V. Isham (1980): *Point processes.* London: Chapman & Hall.

Cox, D. R., and P. A. P. Lewis (1966): *The Statistical Analysis of Series of Events.* London: Methuen.

Cox, D. R., and H. D. Miller (1965): *The Theory of Stochastic Processes*. London: Chapman & Hall.

Cox, D. R., and D. Oakes (1984): *Analysis of Survival Data*. London: Chapman & Hall.

Cramér, H., and M. R. Leadbetter (1967): *Stationary and Related Stochastic Processes*. New York: Wiley.

Cressie, N. A. C. (1991): *Statistics for Spatial Data*. New York: Wiley.

Daley, D. J. (1974): Various concepts of orderliness for point processes. In E. J. Harding and D. G. Kendall (eds): *Stochastic Geometry*, 148–161. Chichester: Wiley.

Daley, D. J., and D. Vere-Jones (1988): *An Introduction to the Theory of Point Processes*. Berlin: Springer-Verlag.

De Gruttola, V., N. Lange, and U. Dafni (1991): Modelling the progression of HIV infection. *J. Amer. Statist. Assoc.* **86:** 569–577.

Denzel, G. E. and G. L. O'Brien (1975): Limit theorems for extreme values of chain-dependent processes. *Ann. Probab.* **3:** 773–779.

Di Gésu, V. (1992): *Data Analysis in Astronomy IV.* New York: Plenum Press.

Diggle, P. J. (1983): *Statisical Analysis of Spatial Point Patterns*. London: Academic Press.

Dobrushin, R. L. (1960): Properties of sample functions of a stationary Gaussian process. *Theory Prob. Appl.* **5:** 120–122.

Dovland, H., and E. Mohn (1975): An analysis of precipitation data from Birkenes (in Norwegian). Tech. Rep. TN17/75, SNSF project, Ås, Norway.

Doyle, P. G., and J. L. Snell (1984): *Random Walks and Electric Networks*. Math. Assoc. of America.

Edwards, A. W. F. (1985): *Likelihood.* Cambridge University Press.

Efron, B. (1978): *The Jackknife, the Bootstrap, and Other Resampling Plans.* CBMS-NSF Regional Conference Series in Applied Mathematics, vol. 38. Philadelphia: SIAM.

Eggenberger, F., and G. Pólya (1923): Über die Statistik verketterer Vorgänge. *Z. Angew. Math. Mech.* **3:** 279–289.

Ehrenfest, P., and T. Ehrenfest (1906): Über eine Aufgabe aus der Wahrscheinlichkeitsrechnung, die mit der kinetischen Deuterung der Entropievermerung zusammenhängt. *Math.-naturw. Bl.* Nos. 11, 12.

Einstein, A. (1905): Über die von der molekular-kinetischen Theorie der Wärme geforderte Bewegung von in ruhenden Flüssigkeiten suspendierten Teilchen. *Ann. der Physik* **17:** 549–60.

Einstein, A. (1926): *Investigations on the Theory of the Brownian Movement.* London: Methuen. Reprinted in 1956 by Dover, New York.

Elston, R. C., and J. Stewart (1971): A general model for the genetic analysis of pedigree data. *Human Heredity* **21:** 523–542.

Engel, E. M. R. A. (1992): *A Road to Randomness in Physical Systems.* Lecture Notes in Statistics **71**. Berlin: Springer-Verlag.

Erdös, P., W. Feller, and H. Pollard (1949): A theorem on power series. *Bull. Amer. Math. Soc.* **55:** 201–204.

Ethier, S. N., and T. G. Kurtz (1986): *Markov Processes—Characterization and Convergence.* New York: Wiley.

Ewens, W. J. (1968): *Population Genetics.* London: Methuen.

Fama, E. (1970): Efficient captial markets: A review of theory and empirical work. *J. Finance* **25:** 383–416.

Feigelson, E. D., and G. J. Babu (1992): *Statistical Challenges in Modern Astronomy.* New York: Springer.

Feller, W. (1939): Die Grundlagen der Volterraschen Theorie des Kampfes ums Dasein in wahrscheinlichkeitstheoretischer Behandlung. *Acta Biotheoretica* **5:** 11–40.

Feller, W. (1968): *An Introduction to Probability Theory and its Applications,* vol. 1 (3rd ed.). New York: Wiley.

Feller, W. (1971): *An Introduction to Probability Theory and its Applications,* vol. 2 (2nd ed.). New York: Wiley.

Fisher, R. A. (1930): *The Genetical Theory of Natural Selection.* Oxford: Clarendon.

Fleming, T. R., and D. P. Harrington (1991): *Counting Processes and Survival Analysis.* New York: Wiley.

Fleming, W. H., and M. Viot (1978): Some measure-valued population processes. In *Proc. International Conf. Stoch. Anal., Northwestern University*: 97–108. New York: Academic.

Fredkin, D. R., M. Montal, and J. A. Rice (1985): Identification of aggregated Markovian models: applications to the nicotinic acetylcholine receptor. In L. M. LeCam and R. A. Olshen (eds): *Proc. Berkeley Conf. in Honor of Jerzy Neyman and Jack Kiefer.* Monterey: Wadsworth.

Fredkin, D. R., and J. A. Rice (1992): Bayesian restoration of single channel patch clamp recordings. *Biometrics* **48:** 427–448.

Freedman, D. (1983): *Markov Chains.* New York: Springer-Verlag.

Furry, W. H. (1937): On fluctuation phenomena in the passage of high-energy electrons through lead. *Phys. Rev.* **52:** 569.

Fürth, R. (1918): Statistik und Wahrscheinlichkeitsnachwirkung. *Phys. Z.* **19:** 421–426.

Gabriel, K. R., and J. Neumann (1962): A Markov chain model for daily rainfall in Tel Aviv. *Quart. J. Roy. Met. Soc.* **88:** 90–95.

Geman, S., and D. Geman (1984): Stochastic relaxation, Gibbs distributions and the Bayesian restoration of images. *IEEE Trans. Pattern Analysis and Machine Intelligence* **6:** 721–741.

Geyer, C. J. (1991): Markov chain Monte Carlo maximum likelihood. *Proc. 23rd Symp. Interface*: 156–163.

Geyer, C. J. (1992): Practical Markov chain Monte Carlo. *Statist. Science* **7:** 473–483.

Geyer, C. J., and E. A. Thompson (1992): Constrained Monte Carlo maximum likelihood for dependent data. *J. Roy. Statist. Soc. B* **54:** 657–699.

Ghent, A. W. (1963): Studies of regeneration of forest stands devastated by the Spruce Budworm. *For. Sci.* **9:** 295–310.

Gibbs, W. S. (1902): *Elementary Principles in Statistical Mechanics, Developed with Especial Reference to the Rational Foundation of Thermodynamics.* New York: Charles Scribner's Sons.

Gilberg, A., L. Gilberg, R. Gilberg, and M. Holm (1978): Polar Eskimo genealogy. *Meddelelser om Grønland* **203**, no. 4.

Golde, D. W. (1991): The stem cell. *Sci. Amer.* **265 (6):** 86–93.

Grandell, J. (1976): *Doubly Stochastic Poisson Processes.* Springer Lecture Notes in Mathematics **529**. Berlin: Springer-Verlag.

Grandell, J. (1985): *Stochastic Models of Air Pollutant Concentrations.* Springer Lecture Notes in Statistics **30**. Berlin: Springer-Verlag.

Griffiths, R. B. (1964): Peierl's proof of spontaneous magnetization in two dimensional Ising ferromagnet. *Phys. Rev.* **A 136:** 437–438.

Grimmett, G. R., and D. R. Stirzaker (1982): *Probability and Random Processes.* Oxford: University Press.

Guckenheimer, J., G. Oster, and A. Ipaktchi (1976): Density dependent population models. *J. Math. Biol.* **4:** 101–114.

Gudmundsson, G., and K. Saemundsson (1980): Statistical analysis of damaging earthquakes and volcanic eruptions in Iceland from 1550–1978. *J. Geophys.* **47:** 99–109.

Guttorp, P. (1980): *Statistical Modelling of Population Processes.* PhD dissertation, Department of Statistics, University of California at Berkeley.

Guttorp, P. (1986): Models for transportation and deposition of atmospheric pollutants. SIMS Technical Report **95.**

Guttorp, P. (1988): Analysis of event based precipitation data with a view towards modeling. *Water Resources Research* **24:** 35–44.

Guttorp, P. (1991): *Statistical Inference for Branching Processes.* New York: Wiley.

Guttorp, P., and M. L. Thompson (1991): Estimating second order parameters of volcanicity from historical data. *Journal of the American Statistical Association* **86:** 578–583.

Guttorp, P., and A. Walden (1987): On the evaluation of geophysical models. *Geophys. J. Roy. Astronom. Soc.* **91:** 201–210.

Hammersley, J. M., and P. Clifford (1971): Markov fields on finite graphs and lattices. Unpublished manuscript, Oxford University.

Hanisch, K.-H., and D. Stoyan (1979): Formulas for the second-order analysis of marked point processes. *Math. Operationsforsch. Statist.* **10:** 555–560.

Hardy, G. H. (1908): Mendelian proportions in a mixed population. *Science* **28:** 49–50.

Harris, T. E. (1963): *Branching Processes.* New York: Springer-Verlag.

Haskey, H. W. (1954): A general expression for the mean in a simple stochastic epidemic. *Biometrika* **41:** 272–275.

Hastings, W. K. (1970): Monte Carlo sampling methods using Markov chains and their applications. *Biometrika* **57:** 97–109.

Hatzikostandis, G., and S. Howe (1967): A study of a multi-channel queueing process in a communications terminal. In R. Cruon (ed.): *Queueing Theory: Recent Developments and Applications*: 167–175. New York: American Elsevier.

Hawkes, A. G. (1971): Spectra of some self-exciting and mutually exciting point processes. *Biometrika* **58:** 83–90.

Heyde, C. C., and E. Seneta (1977): *I. J. Bienaymé—Statistical Theory Anticipated.* Studies in the History of Mathematics and Physical Sciences, vol. 3. New York: Springer-Verlag.

Higdon, D. (1994): *Spatial Applications of Markov Chain Monte Carlo for Bayesian Inference.* PhD dissertation, Department of Statistics, University of Washington.

Hiscott, R. N. (1981): Chi-square tests for Markov chain analysis. *Math. Geol.* **13:** 69–80.

Hochberg, K. J., and D. A. Dawson (1982): Wandering random measures in the Fleming-Viot model. *Ann. Probab.* **10:** 554–580.

Holgate, P. (1967): The size of elephant herds. *Math. Gaz.* **51:** 302–304.

Hooper, J. E., and M. Scharff (1958): *The Cosmic Radiation*. London: Methuen.

Hsu, Y. S., and W. J. Park (1980): Some statistical inferences on O-U processes. *Comm. Statist., Part A—Theory and Methods* **9:** 529-540.

Huber, P. J. (1981): *Robust statistics*. New York: Wiley.

Hughes, J. P. (1993): *A Class of Stochastic Models for Relating Synoptic Atmospheric Patterns to Local Hydrologic Phenomena.* PhD dissertation, Department of Statistics, University of Washington.

Ising, E. (1925): Beitrag zur Theorie des Ferromagnetismus. *Z. Phys.* **31:** 253–258.

Jacobsen, M. (1982): *Statistical Analysis of Counting Processes*. Springer Lecture Notes in Statistics **12**. New York: Springer-Verlag.

Jacobsen, M., and N. Keiding (1990): Markovkæder (3rd ed., 3rd printing). Institute of Mathematical Statistics, University of Copenhagen.

Jain, S. (1986): Markov chain model and its application. *Computers and Biomedical Research* **19:** 374–378.

Jeffreys, H. (1939): *Theory of Probability*. International Series of Monographs on Physics. Oxford: University Press.

Jeffreys, H. (1961): *Theory of Probability* (3rd ed.). Oxford: Clarendon Press.

Johnson, N. L., and S. Kotz (1970): *Continuous Univariate Distributions*, vol. 2. Boston: Houghton Mifflin.

Joseph, L., C. Wolfson, and D. B. Wolfson (1990): Is multiple sclerosis an infectious disease? Inference about an input process based on the output. *Biometrics* **46:** 337–349.

Juang B. H., and L. R. Rabiner (1991): Hidden Markov models for speech recognition. *Technometrics* **33:** 251–270.

Kalman, R.E. (1960): A new approach to linear filtering and prediction problems. *J. Basic Eng. (Trans. ASME ser. D)* **82:** 34–45.

Kaluzny, S. (1987): *Estimation of trends in spatial data* . PhD dissertation, Department of Biostatistics, University of Washington.

Karr, A. F. (1991): *Point Processes and their Statistical Inference* (2nd ed.). New York: Dekker.

Katz, R. W. (1977): An application of chain-dependent processes to meteorology. *J. Appl. Probab.* **14:** 598–603.

Katz, R. W. (1981): On some criteria for estimating the order of a Markov chain. *Technometrics* **23:** 243–249.

Kay, H. G. M. (1965): How many cell generations? *Lancet* **2:** 418.

Keiding, N. (1977): Statistical comments on Cohen's application of a simple stochastic population model to natural primate troops. *Amer. Naturalist* **111:** 1211–1219.

Keiding, N. and S. Lauritzen (1978): Marginal maximum likelihood estimates and estimation of the offspring mean in a branching process. *Scand. J. Statist.* **5:** 106–110.

Keller, J. (1986): The probability of heads. *American Math. Monthly* **93:** 191–196.

Kendall, D. G. (1948): The generalized "birth and death" process. *Ann. Math. Statist.* **19:** 1–15.

Kendall, D. G. (1953): Stochastic processes occurring in the theory of queues and their analysis by means of the imbedded Markov chain. *Ann. Math. Statist.* **24:** 338–354.

Kindermann, R., and J. L. Snell (1980): *Markov Random Fields and their Applications*. Contemporary Mathematics, vol. 1. Providence: American Math. Soc.

King, I. R., S. A. Stanford, P. Seitzer, M. A. Bershady, W. C. Keel, D. C. Koo, N. Weir, S. Djorgovski, and R. A. Windhorst (1991): The current ability of *HST* to reveal morphological structure in medium-redshift galaxies. *Astron. J.* **102:** 1553–1568.

Klug, W. S., and Cummings, M. R. (1991): *Concepts of Genetics* (3rd ed.). New York: Macmillan.

Kolmogorov, A. N. (1933): *Grundbegriffe der Wahrscheinlichkeitsrechnung*. Ergebnisse der Mathematik, vol. 2, no. 3. Berlin: Springer-Verlag.

Kulperger, R., and P. Guttorp (1981): Criticality conditions for some random environment population processes. *Stoch. Proc. Appl* **11:** 207–212.

Kurtzke, J. K., and K. Hyllested (1986): Multiple sclerosis in the Faroe Islands II. Clinical update, transmission and the nature of MS. *Neurology* **36:** 307–328.

Labarca, P., J. Rice, D. Fredkin, and M. Montal (1984): Kinetic analysis of channel gating: application to the cholinergic receptor channel and the chloride channel from *Torpedo californica*. *Biophys. J.* **47:** 469–478.

Landau, E. (1930): *Grundlagen der Analysis*. Leipzig: Akad. Verl.

Langevin, P. (1908): Théorie du mouvement Brownien. *C. R. Acad. Sci. Paris* **146:** 530–533.

Le, N. D. (1990): *Modeling and Bootstrapping for Non-Gaussian Time Series*. PhD dissertation, Department of Statistics, University of Washington.

Lee, G. M., A. Ishihara, and K. A. Jacobson (1991): Direct observation of Brownian motion of lipids in a membrane. *Proc. Nat. Acad. Sci. USA* **88:** 6274–6278.

Lee, W. H. K., and D. R. Brillinger (1979): On Chinese earthquake history—an attempt to model an incomplete data set by point process analysis. *Pure and Applied Geophysics* **117:** 1229–1257.

Leroux, B. G., and M. L. Puterman (1992): Maximum-penalized-likelihood estimation for independent and Markov-dependent mixture models. *Biometrics* **48:** 545–558.

Lewis, P. A. W. (1964): A branching Poisson process model for the analysis of computer failure patterns. *J. Roy. Statist. Soc. B* **26:** 398–456.

Lewis, P. A. W., and G. S. Shedler (1979): Simulation of nonhomogeneous Poisson processes by thinning. *Naval Res. Logistics Quarterly* **26:** 403–413.

Lin, S. (1993): *Markov Chain Monte Carlo Estimates of Probabilities on Complex Structures*. PhD dissertation, Department of Statistics, University of Washington.

Lin, S., E. Thompson, and E. Wijsman (1994): Finding noncommunicating sets for Markov chain Monte Carlo estimations on pedigrees. *Am. J. Hum. Genetics* **54:** 695–704.

Loschmidt, J. (1876): Zustand der Wärmegleichgewichts einer System von Körpern mit Rücksicht auf die Schwerkraft. *Wien. Ber.* **73:** 139.

Luria, S., and M. Delbrück (1943): Mutations of bacteria from virus sensitivity to virus resistance. *Genetics* **28:** 491–511.

Maistrov, L. E. (1974): *Probability Theory—A Historical Sketch*. New York: Academic Press.

Mardia, K. V., and G. K. Kanji (1993): *Advances in Applied Statistics: Statistics and Images 1*. Abingdon: Carfax.

Matérn, B. (1960): *Spatial Variation*. Medd. Statens Skogsforskningsinstitut **49**: no. 5.

Matérn, B. (1986): *Spatial Variation* (2nd ed.). Springer Lec. Notes Statist. **36**. Berlin: Springer.

McCullagh, P., and J. A. Nelder (1989): *Generalized Linear Models* (2nd ed.). London: Chapman & Hall.

McKendrick, A. G. (1914): Studies on the theory of continuous probabilities, with special reference to its bearing on natural phenomena of a progressive nature. *Proc. London Math. Soc.* **13:** 401–416.

McKendrick, A. G. (1925): Applications of mathematics to medical problems. *Proc. Edin. Math. Soc.* **44:** 1–34.

McQueen, G., and S. Thorley (1991): Are stock returns predictable? A test using Markov chains. *J. Finance* **66:** 239–263.

Metropolis, N., A. W. Rosenbluth, M. N. Rosenbluth, A. H. Teller, and E. Teller (1953): Equations of state calculations by fast computing machines. *J. Chemichal Physics* **21:** 1087–1091.

Miller, R. S., D. B. Botkin, and R. Mendelssohn (1974): The Whooping Crane (*Grus Americana*) population of North America. *Biol. Cons.* **6:** 106–111

Mills, T. M., and E. Seneta (1989): Goodness-of-fit for a branching process with immigration using sample partial autocorrelation functions. *Stoch. Proc. Appl.* **33:** 151–161.

Molina, R., A. Del Olmo, J. Perea, and B. D. Ripley (1991): Bayesian deconvolution in optical astronomy. *Astron. J.* **103:** 666–675.

Molina, R., and B. D. Ripley (1989): Using spatial models as priors in astronomical image analysis. *J. of Appl. Stat.* **16:** 193–206.

Mooley, D. A. (1981): Applicability of the Poisson probability model to the severe cyclonic storms striking the coast around the Bay of Bengal. *Sankhyā* **43B:** 187–197.

Moran, P. A. P. (1962): *The Statistical Processes of Evolutionary Theory.* Oxford: Clarendon.

Morgan, B. J. T., and S. A. Watts (1980): On modelling microbial infections. *Biometrics* **36:** 317–321.

Mosteller, F. (1968): Association and estimation in contingency tables. *J. Amer. Statist. Assoc.* **63:** 1-28.

Neyman, J. (1939): On a new class of 'contagious' distributions, applicable in entomology and bacteriology. *Ann. Math. Statist.* **10:** 35–57.

Neyman, J., and E. L. Scott (1959): Large scale organization of the distribution of galaxies. *Handbuch der Physik* **53:** 416–444.

O'Brien, G. L. (1974): Limit theorems for sums of chain-dependent processes. *J. Appl. Probab.* **11:** 582–587.

Ohta, T., and M. Kimura (1973): A model of mutation appropriate to estimate the number of electrophoretically detectable alleles in a finite population. *Genet. Res. Camb.* **22:** 201–204.

Onsager, L. (1944): Crystal lattices I. A two-dimensional model with an order-disorder transition. *Phys. Rev.* **65:** 117–149.

Pearson, K. (1900): On the criterion that a given system of deviations from the probable in the case of a correlated system of variables is such that it can be reasonably supposed to have arisen from random sampling. *Phil. Mag., 5th Series,* **50:** 157–175.

Peierls, R. E. (1936): On Ising's ferromagnet model. *Proc. Camb. Phil. Soc.* **32:** 477–481.

Perrin, J. (1913): *Les Atomes*. Paris: Libraire Félix Alcan. Reprinted in English translation as *Atoms* by Ox Bow Press, Woodbridge, 1990.

Phelan, M. J. (1992): Aging functions and their nonparametric estimation in point process models of rainfall. In A. Walden and P. Guttorp (eds), *Statistics in the Environmental and Earth Sciences*: 90–116. London: Edward Arnold.

Pincus, M. (1968): A closed form solution of certain programming problems. *Oper. Res.* **16:** 690–694.

Poincaré, H. (1896): *Calcul des Probabilités*. Paris: Gauthier-Villars.

Preisler, H. K. (1993): Modelling spatial patterns of trees attacked by bark-beetles *Appl. Statist.* **42:** 501–514.

Preston, C. (1976): *Random Fields.* Springer Lec. Notes. Math. **24**. Berlin: Springer-Verlag.

Quetelet, A. (1852): Sur quelques propriétès curieuses que présentent les résultats d'une série d'observations, faites dans la vue de déterminer une constant, lorsque les chances de rencontrer des écarts en plus et en moins sont égales et independantes les unes des autres. *Bull. Acad. Royale Belgique* **19** Parte 2: 303–317.

Raftery, A. E. (1985a): A model for high-order Markov chains. *J. Roy. Statist. Soc. B* **47:** 528–539.

Raftery, A. E. (1985b): A new model for discrete-valued time series: autocorrelations and extensions. *Rassegna di Metodi Statistci ed Applicazoioni* **3–4:** 149–162.

Raftery, A. E., and S. Tavaré (1994): Estimation and modelling repeated patterns in high order Markov chains with the mixture transition distribution model. *Appl. Statist.* **43:** 179–199.

Rasch, G. (1960): *Probabilistic Models for some Intelligence and Attainment Tests*. Copenhagen: Danish Educational Research Institute.

Rathbun, S. L., and N. Cressie (1994): A space-time survival point process for a longleaf pine forest in Southern Georgia. *J. Amer. Statist. Assoc.* **89:** 1164–1174.

Reed, W. J., and H. R. Clarke (1990): Harvest decisions and asset valuation for biological resources exhibiting size-dependent stochastic growth. *Int. Econ. Rev.* **31:** 147–169.

Reid, A. T., and H. G. Landau (1951): A suggested chain process for ratiation damage. *Bull. Math. Biophys.* **13:** 153–163.

Reid, N. (1988): Saddlepoint methods and statistical inference (with discussion). *Statist. Sci.* **3:** 213–238

Renwick, J. A. (1989): *Short-Range Prediction of Fog Occurrence at Christchurch Aerodrome.* N. Z. Met. Service.

Resnick, S. (1987): *Extreme Values, Regular Variation, and Point Processes.* New York: Springer.

Rida, W. N. (1991): Asymptotic properties of some estimators for the infection rate in the general stochastic epidemic model. *J. Roy. Statist. Soc. Ser. B* **53:**a 269–283.

Ripley, B. D. (1976): The second-order analysis of stationary point processes. *J. Appl. Probab.* **13:** 255–266.

Ripley, B. D. (1981): *Spatial Statistics.* New York: Wiley.

Ripley, B. D. (1988): *Statistical Inference for Spatial Processes.* Cambridge University Press.

Ripley, B. D. (1991): The use of spatial models as image priors. In A. Possolo (ed.): *Spatial Statistics and Imaging*, 309–340. Hayward: Institute of Mathematical Statistics.

Rodriguez-Iturbe, I., V. K. Gupta, and E. Waymire (1984): Scale considerations in the modeling of temporal rainfall. *Water Resour. Res.* **20:** 1611–1619.

Rosenblatt, M. R. (1971): Curve estimation. *Ann. Math. Statist.* **42:** 1815–1842.

Royal Statistical Society (1992): Meeting on Chaos. *J. Roy. Statist. Soc. B* **54:** 301–474.

Ruelle, D. (1969): *Statistical Mechanics: Rigorous Results.* New York: Benjamin.

Ruelle, D. (1991): *Chance and Chaos*. Princeton University Press.

Rutherford, E. and Geiger, H. (1910): The probability variations in the distribution of α-particles. *Phil. Mag. Ser. 6,* **20:** 698–707.

Sartwell, P. E. (1950): The distribution of incubation periods of infectious disease. *Amer. J. Hygiene* **51:** 310–318.

Schimert, J. (1992): *A High Order Hidden Markov Model.* PhD dissertation, Department of Statistics, University of Washington.

Sheehan, N. (1990): *Genetic Restoration of Complex Pedigrees.* PhD dissertation, Department of Statistics, University of Washington.

Sheehan, N. (1992): Sampling genotypes on complex pedigrees with phenotypic constraints: The origin of the B allele among the Polar Eskimos. *IMA J. Math. Applied in Med. & Biol.* **9:** 1–18.

Sheehan, N., and A. Thomas (1993): On the irreducibility of a Markov chain defined on a space of genotypic configurations by a sampling scheme. *Biometrics* **49:** 163–176.

Silverman, B. W. (1986): *Density Estimation for Statistics and Data Analysis.* London: Chapman & Hall.

Smith, J. A. (1993): A marked point process model of raindrop size distributions. *J. Appl. Meteor.* **32:** 284–296.

Smith, J. A., and A. F. Karr (1983): A point process model for summer rainfall occurrences. *Water Resour. Res.* **19:** 95–103.

Smith, J. A., and A. F. Karr (1985): Statistical inference for point process models of rainfall. *Water Resour. Res.* **21:** 73–79.

Spitzer, F. (1975): Markov random fields on an infinite tree. *Ann. Probab.* **3:** 387–398.

Steinijans, V. W. (1976): A stochastic point-process model for the occurrence of major freezes in Lake Constance. *Appl. Statist.* **25:** 58–61.

Stern, H. S. (1994): A Brownian motion model for the progress of sports scores. *J. Amer. Statist. Assoc.* **89:** 1128–1134.

Stern, R. D. (1982): Computing a probability distribution for the start of the rains from a Markov chain model for precipitation. *J. Appl. Meteor.* **21:** 420–423.

Takacs, L. (1960): *Stochastic Processes: Problems and Solutions.* London: Methuen.

Taylor, J. M. G., W. G. Cumberland, and J. P. Sy (1994): A stochastic model for analysis of longitudinal AIDS data. *J. Amer. Statist. Assoc.* **89:** 727–736.

Thompson, E. A. (1976): Estimation of age and rate of increase of rare variants. *Am. J. Hum. Genet.* **28:** 442–452.

Tierney, L. (1991): Exploring posterior distributions using Markov chains. *Proc. 23rd Symp. Interface*: 563–570.

Udias, A., and J. Rice (1975): Statistical analysis of microearthquake activity near San Andreas Geophysical Observatory, Hollister, California. *Bull. Seism. Soc. Amer.* **65:** 809–827.

Uhlenbeck, G. E., and Ornstein, L. S. (1930): On the theory of the Brownian motion. *Phys. Rev.* **36:** 823–841.

Van Kampen, N. G. (1981): *Stochastic Processes in Physics and Chemistry.* Amsterdam: North-Holland.

Venkataraman, K. N. (1982): A time series approach to the study of the simple subcritical Galton-Watson process with immigration. *Adv. Appl. Probab.* **14:** 1–20.

Vere-Jones, D., and T. Ozaki (1982): Some examples of statistical estimation applied to earthquake data I. Cyclic Poisson and self-exciting models *Ann. Inst. Statist. Math.* **34:** 189–207 (Corr: **39:** 243).

Viterbi, J. (1967): Error bounds for convolutional codes and an asymptotically optimal decoding algorithm. *IEEE Trans. Inf. Theory* **13:** 260–269.

Volterra, V., and U. D'Ancona (1935): *Les Associations Biologiques, Au Point De Vue Mathematiques.* Paris: Hermann.

Waymire, E. and V.K. Gupta (1981): The mathematical structure of rainfall representations 1–3, *Wat. Resour. Res.* **17:** 1261–1294.

Weidlich, W. (1971): The statistical description of polarization phenomena in society. *Br. J. Math. Statist. Psychol.* **24:** 251–266.

Weinberg, W (1908): Über den Nachweis der Vererbung beim Menschen. *Jh. Ver. vaterl. Naturk. Württemb.* **64:** 368–382.

Westgren, A. (1916): Die Veränderungsgeschwindigkeit der lokalen Teilchenkonzentration in kolloiden Systemen (Erste Mitteilung). *Ark. Mat. Astron. Fys.* **11:** 1–24

Whittle, P. (1954): On stationary processes in the plane. *Biometrika* **41:** 434–449.

Whittle, P. (1986): *Systems in Stochastic Equilibrium.* Chichester: Wiley.

Whittle, P. (1992): *Probability via Expectation* (3rd ed.). New York: Springer-Verlag.

Wiener, N. (1923): Differential space. *J. Math. Phys.* **2:** 131–174.

Winkel, P., P. Gæde, and J. Lyngbye (1976): Method for monitoring plasma progesterone concentration in pregnancy. *Clin. Chem.* **22:** 422–428.

Wittman, B. K., D. W. Rurak, and S. Taylor (1984): Real-time ultrasound observation of breathing and body movements in fetal lambs from 55 days gestation to term. Abstract presented at the XI Annual Conference, Society for the Study of Fetal Physiology, Oxford.

Woolhiser, D. A. (1992): Modeling daily precipitation—progress and problems. In A. T. Walden and P. Guttorp (eds): *Statistics in the Environmental and Earth Sciences*, 71–89. London: Edward Arnold.

Wright, S. (1931): Evolution in Mendelian populations. *Genetics* **16:** 97–159.

Yule, G. U. (1924): A mathematical theory of evolution, based on the conclusions of Dr. J. C. Willis, F. R. S. *Phil. Trans. Roy. Soc. London, Ser. B* **213:** 21–87.

Zawadzki, I. I. (1973): Statistical properties of precipitating patterns. *J. Appl. Meteorol.* **12:** 459–472.

Zehna, P. W. (1966): Invariance of maximum likelihood estimators. *Ann. Math. Statist.* **37:** 744.

Zucchini, W. and P. Guttorp (1991): A hidden Markov model for space-time precipitation. *Water Resourc. Res.* **27:** 1917–1923.

Zwanzig, R., A. Szabo, and B. Bagchi (1992): Levinthal's paradox. *Proc. Nat. Acad. Sci. USA* **89:** 20–22.

Index of results

Applications and examples

Index of notation

Index of terms

Data sets

The following data sets are needed for some of the exercises. They are available from the Statistics Department of the University of Washington using anonymous ftp from ftp.stat.washington.edu. There is a README file containing the latest information about the data sets. The procedure is as follows:

```
unix> ftp ftp.stat.washington.edu (or ftp 128.95.17.34)
Name: anonymous
Password: <your userid>
ftp> cd pub
ftp> get stoch.mod.data
ftp> quit
```

Data set 1: Snoqualmie Falls precipitation occurrence data. Each line contains data for one year. The data are precipitation amounts in 1/100th of an inch (0.254 mm). Data obtained from Dennis Lettenmaier, University of Washington..

Data set 2: Bi-daily counts of a blowfly population in a laboratory cage. There are three columns: number of emerging flies, number of deaths, population size. Each row corresponds to counts every other day. The cage started with 1000 pupae, not all of which emerged (Brillinger et al., 1980).

Data set 3: Hourly average wind directions at Koeberg weather station in South Africa. May 1, 1985 through April 30, 1989. Data are integers between 1 and 16 where 1 is N and 16 NNW. Data obtained from Iain MacDonald, University of Cape Town.

Data set 4: Successive counts of gold particles observed in a fixed small volume of water every 1.39 seconds from Westgren (1916). The missing data at counts number 123, 238, 423, 479, 559, 580, 601, 657, 991, 1174, and 1410, have been estimated using the maximum of the conditional distribution given the previous observation (Guttorp, 1991, Appendix).

Data set 5: Spotted wilt on tomato plants in Australia (Cochran, 1936). There are 24 rows with 60 plants in each. The data, showing for each row the location of diseased plants, correspond to the first count on December 18.

Data set 6: Additional instances of wilt at the second count on December 31. Coding and layout as in data set 5.

Data set 7: Mountain pine beetle infections of an unmanaged stand of lodgepole pines in Oregon. There are eight columns, each corresponding to a tree. The data has an index in the first column, x- and y-coordinates in the second and third, while column 4 is a measure of vigor (in g stemwood per m^2 of crown leaf area), column 5 is the diameter at breast height (in inches), column 6 is a measure of leaf area (in m^2), column 8 is the age (in years at breast height), and column 9 the response variable (0 for not attacked, 1 for attacked). Data obtained from Haiganoush Preisler, United States Forest Service.

Data set 7: Counts of wombat warrens in a grassland reserve in South Australia. Each count corresponds to a 450×450 m^2 quadrat. Data obtained from Stephen Kaluzny, StatSci/Mathsoft.

Data set 8: Arrival times of severe cyclonic storms striking the Bay of Bengal coast (Mooley, 1981). Dates are given.

Data set 10: Volcanic eruptions in Iceland (Gudmundsson and Saemundsson, 1980). Dates (as accurately as known) are given.

Data set 11: Daily average July temperatures at Palmer, WA, 1931–1989. Each line is one year. Data obtained from Jim Hughes, University of Washington.

Milton Keynes UK
Ingram Content Group UK Ltd.
UKHW021118111024
449327UK00051B/1368